Electrochemistry in Industry

NEW DIRECTIONS

Electrochemistry in Industry

NEW DIRECTIONS

Edited by

Uziel Landau
Ernest Yeager

and

Diane Kortan

Case Center for Electrochemical Sciences
Case Institute of Technology
Case Western Reserve University
Cleveland, Ohio

SPRINGER SCIENCE+BUSINESS MEDIA, LLC

Library of Congress Cataloging in Publication Data

International Symposium on Electrochemistry in Industry: New Directions (1980: Case Institute of Technology)
Electrochemistry in industry, new directions.

"Proceedings of an International Sympsoium on Electrochemistry in Industry: New Directions, held October 20–22, 1980, at Case Institute of Technology, Case Western University, Cleveland, Ohio, in celebration of the centennial of Case Institute of Technology"—T. p. verso.
Includes bibliographical references and index.
1. Electrochemistry, Industrial—Congresses. I. Landau, Uziel. II. Yeager, Ernest B., 1924– .III. Kortan, Diane. IV. Case Institute of Technology. V. Title.
TP250.I57   1980   660.2'97   82-5207
ISBN 978-1-4684-4240-3   ISBN 978-1-4684-4238-0 (eBook)   AACR2
DOI 10.1007/978-1-4684-4238-0

Case Centennial Celebration
1880 - 1980

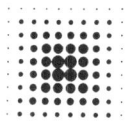

Proceedings of an International Symposium on Electrochemistry in Industry: New Directions, held October 20 – 22, 1980, at Case Institute of Technology, Case Western University, Cleveland, Ohio, in celebration of the centennial of Case Institute of Technology

© 1982 Springer Science+Business Media New York
Originally published by Plenum Press, New York in 1982
Softcover reprint of the hardcover 1st edition 1982

SPONSORING COMPANIES

Funds to help cover the travel expenses of speakers were provided by the following sponsoring companies:

Celanese Research Company
Diamond Shamrock Corporation
The Dow Chemical Company
General Motors Corporation
Hooker Chemicals & Plastics Corporation
International Business Machines Corporation
The Lubrizol Corporation
PPG Industries
Xerox Corporation

PREFACE

This volume represents the proceedings of the International
Symposium on Electrochemistry in Industry - New Directions, held
at Case Institute of Technology of Case Western Reserve University
on October 20-22, 1980. This symposium was one of a number held
at Case Institute during the 1980 calendar year as part of its
centennial celebration.

The following faculty members from Case Institute of Technology constituted the organizing committee for the symposium:

> Uziel Landau, Chairman
>   Associate Professor of Chemical Engineering
>
> Robert Hehemann
>   Professor of Metallurgy
>
> C. C. Liu
>   Professor of Chemical Engineering
>
> Ernest Yeager
>   Director of CLES and Professor of Chemistry

All lectures at this symposium were by invitation. The manuscripts as received for all but two of the lectures are herein
published in the order of presentation. Discussion submitted by
participants in written form appears at the end of each paper.
Part of the panel discussion on Future Trends in Major Electrochemical Industries has also been included in this volume.

CONTENTS

INTRODUCTION

THE CASE INSTITUTE OF TECHNOLOGY CENTENNIAL CELEBRATION

On February 24, 1877, Leonard Case, Jr. signed a trust deed
willing his estate to endow "The Case School of Applied Science,"
in the city of Cleveland in order to provide the community with a
technical institution. Since its founding in 1880, Case Institute
of Technology has developed its educational curricula to include
comprehensive and contemporary undergraduate and graduate programs
in science, engineering, mathematics and management; and internationally recognized interdisciplinary research centers. With the
federation of Case Institute of Technology and Western Reserve
University in 1967, the science and mathematics departments of
Western Reserve became part of Case Institute and Case Institute
became part of a larger university, particularly distinguished by
its professional schools including the Medical School with its
extensive research as well as educational program. Case Institute
maintains a prestigious physical plant within the University as a
whole and in the unique University Circle cultural setting in Cleveland. The campus contains some of the most advanced technological
and scientific instrumentation, facilities and laboratories in
current use.

Case Institute has a distinguished faculty, dedicated to excel-
lence in education for talented men and women and at the same time
committed to basic and applied research. The faculty's reputation
in research dates back to Dr. Albert A. Michelson, the first Case
professor of physics, who later became the first American Nobel
Laureate in the sciences. One of the cornerstone experiments of
relativistic physics, the Michelson-Morley ether drift experiment
was carried out on the campus by Prof. Michelson in collaboration
with Prof. Edward Morley, the renowned chemist of Western Reserve
University, in the 1890's. Today Case Institute of Case Western
Reserve University numbers 1700 undergraduates and 600 graduate
students plus many postdoctoral students and visiting scientists
involved in research. Case students rank among the highest in the
country on placement test scores and related scholastic criteria.

Case has a long trandition of strong interaction with industry. Particularly noteworthy for this conference concerned with electro-chemistry is the key role played by Case faculty together with Herbert C. Dow, a Case alumnus, in setting up the Dow Chemical Company, one of the principal companies involved in electrochemistry internationally. It is indeed appropriate to have a symposium on the theme "Electrochemistry in Industry--New Directions" as part of the Case Institute of Technology Centennial Celebration.

CASE LABORATORIES FOR ELECTROCHEMICAL STUDIES

Case Laboratories for Electrochemical Studies (CLES) was formed within Case Institute of Technology in 1976 with the objective of pulling together the resources within six science and engineering departments to focus on research and education in electrochemistry. More recently, CLES has also had input from the Engineering Design Center as a collaborating organization.

Through CLES, Case is able to undertake research projects in electrochemistry which are beyond the resources of any one research group or department. CLES research funding during the current year totals approximately $2,000,000 and is derived from a number of government agencies, private foundations and industry. CLES with its 19 faculty and senior research staff, 16 postdoctoral research associates and 32 graduate students is one of the largest electro-chemistry groups in an American university.

CLES and its participating departments also offer one of the most comprehensive graduate programs in electrochemistry in the United States with five graduate courses in electrochemistry, an extensive seminar series, special short courses and international conferences.

CLES has not reached its peak. During the coming year we expect to strengthen our relations with industry through more joint research projects and cooperative efforts. The distance between the funda-mental discoveries and applications is shorter in electrochemistry than in other technical areas and it is to the great advantage of academic and industrial scientists to interact with each other.

THEME AND OBJECTIVES OF THE CONFERENCE

Ernest Yeager

Case Laboratories for Electrochemical Studies
Case Western Reserve University
Cleveland, Ohio 44106

Electrochemistry is undergoing a renaissance which involves both fundamental and applied electrochemistry. In the applied area this renaissance has been stimulated by the energy problem, the environmental problem and the materials problems faced by industry and individual countries in a competitive world. The surge of activity in applied electrochemistry has spilled over into the fundamental research area. An additional factor is a new level of vigor that has entered into the science of electrochemistry over the past decade. This is in turn attracting outstanding graduate students and younger scientists into electrochemical research.

A number of exciting developments have occurred in applied electrochemistry over the past decade. Lithium non-aqueous batteries have come of age in response to the needs of an exploding electronics industry and also military-space needs. Fuel cells finally found an important, highly successful application in providing the non-propulsive power for the Apollo moon missions and continue to be used in the shuttle. The dimensionally stable anode with its continued high catalytic activity for the chlorine generation reaction and stability has virtually replaced the carbon anode in the chlorine industry in the United States and in good part also internationally. Ion selective membranes are now finding application in large scale industrial electrolytic cells - particularly chlor-alkali cells. Electrochemical machinery is now used routinely in the manufacture of such items as automobile engines, turbine blades and rifle barrels. Advanced electrodeposition and other electrolytic processes are now used in the electronics industry in the manufacture of integrated circuit chips, memory devices and contacts. A wide range of electrochemical sensors are now available for monitoring ionic

3

and non-ionic species in medical as well as industrial applications.

The end for new applications of electrochemistry is by no means in sight. There is a general atmosphere of excitement in electrochemistry. Many companies, including some who have not traditionally worked in the field of electrochemistry, are asking the question "what next?" This question has prompted Case Laboratories for Electrochemical Studies to arrange this symposium.

The symposium, which is international in scope, is designed to focus on new and developing trends in electrochemistry, particularly those which have industrial implications. Symposium participants come from the industrial, academic and government research communities. Invited plenary lectures are being presented by world-renowned experts of both the academic and industrial communities, who are asked to present not only an overview of a particular area, but to consider the critical future issues and directions in their field. Time has been set aside for questions and discussions which are expected to follow the presentations. The program also features a panel discussion exploring the future of electrochemistry in major industries throughout the world.

SPONSORING COMPANIES

Case expresses appreciation to the following sponsoring companies whose support has made it possible to bring together an outstanding group of speakers for this symposium.

CELANESE RESEARCH COMPANY

DIAMOND SHAMROCK CORPORATION

THE DOW CHEMICAL COMPANY

GENERAL MOTORS CORPORATION

HOOKER CHEMICALS & PLASTICS CORPORATION

INTERNATIONAL BUSINESS MACHINES CORPORATION

THE LUBRIZOL CORPORATION

PPG INDUSTRIES

XEROX CORPORATION

THE OUTLOOK FOR THE ELECTROCHEMICAL INDUSTRY

OPENING ADDRESS

Vittorio de Nora

Diamond Shamrock Electrosearch S.A.
Geneva
Switzerland

Mr. Dean, Members of the Faculties, Ladies and Gentlemen :

Good morning.   What an impressive assembly of scientists
for the celebration of the Centennial of Case Institute of
Technology, now part of this famous center of knowledge which
is the Case Western Reserve University. This reunion gives us
the occasion to have fruitful meetings and learn directly from
experts the present and the future of different fields of the
electrochemical industry.

Ernest Yeager's idea of celebrating this anniversary by
holding an international symposium on "ELECTROCHEMISTRY IN
INDUSTRY" is full of hope for the future of this important
branch of applied science.

The Case Institute of Technology has the same age as the
electrochemical industry.

One hundred years ago, this center of studies was established
to prepare technical people required for the industrial development of this country.

About one hundred years ago, the electrochemical industry
began to produce chlorine and caustic through brine electrolysis,
hydrogen and oxygen through water electrolysis and aluminum
through molten salt electrolysis.

Then other metals were obtained or purified by electrolytic
processes. Electro-organic reactions began to be developed and
utilized. Different types of batteries started to be manufactured and used.

Case Western Reserve University has since become an
important center of studies and research in the field of
electrochemistry.

The contribution of Professor Yeager and his team at Case
Laboratories for Electrochemical Studies has been outstanding
not only to the knowledge and teaching of electrochemistry
but also to the understanding of some of the basic electrochemical principles relating to charge transfer at electrode-
electrolyte interfaces, to the mechanism of electrochemical
reactions, electrosorption and electrocatalysis, and to the
development of the oxygen electrode and oxygen electrocatalysts.

The culture and inspiring enthusiasm of Ernest Yeager is an
invitation to students and graduates to be involved in electrochemical research, which is carried out here so successfully.

I am proud to be here today and I feel honored to have been
chosen to speak to you, distinguished guests, on the future of
electrochemistry.

It is exciting to celebrate a Centennial and forecast what
might happen in the next few years or until the next Centennial.

However, I am not a fortune teller but, together with you I
could take some guesses about the progress of electrochemistry
and wish with you that the contribution of electrochemistry to
the well-being of mankind be great.

Progress, usually, must be reported on logarithmic scale to
look linear. Hopefully, it may only take ten years now to
make the same step forward which took one hundred years in the
past, and sometimes I like to think that, even on earth, time
may have a different standard unit, like in the neutron stars.

Fobert Forward has imagined, with the astronomer Frank Brake,
life forms having features of neutron stars beings. Such
creatures would see progress much faster than we see it.

The description of possible life on such rapidly spinning
balls of ultra dense liquid neutrons, with a density of more than

700 million tons/cc, and a gravity field at its surface
70 billion times as strong as on earth, is fascinating and to
imagine the behavior of a neutron star being is thrilling.

For instance, since the molecules in a neutron star being
would react a million times faster than ours, these beings would
live, think, reproduce and die a million times faster than human
beings.

Therefore, a human year would be equal to a million of theirs
and one human day would equal almost 3000 of their years.

Fortunately this morning, I am not speaking to a neutron
star group of beings because, if you belonged to that race, you
would be terribly bored by my talk and I would be exposed probably
to radiation and certainly to knowledge that I could not possibly
handle either physically or emotionally. In addition, the length
of my talk would cover the life's span of a neutron star being
or the 100 years of the Centennial we are celebrating.

But let us return on earth at Case Western Reserve University
in Cleveland.

In preparing this conversation with you, there came to my
mind many subjects of interest : the molten salt electrolysis,
dimensionally stable electrodes, electro-organic synthesis, solar
hydrogen production, primary and secondary batteries, solid electrolytes, ion selective membranes, oxygen electrodes and brine
electrolysis. Then I received the list of the topics for the
invited speakers and found that there were very few subjects to
be added to what we are going to hear from these outstanding
experts, so I decided to talk about the past and the future,
leaving the present to be told by others. Let me start from what
I like to call "the beginning of the electrochemical science on
earth".

That was 200 years ago, when the first electrochemical experiments of Luigi Galvani and Alessandro Volta gave to science the
first battery "the copper zinc pile" and provided scientists with
the first continuous source of direct current.

When Volta invented the pile, one of the great discoveries
of all times, a new field of research was opened to scientists
utilizing the wonderful tool which is the electrolytic cell.
Many human and industrial processes are based on electrolytic
cells.

Our brain is an assembly of minute fuel cells in which information bits are stored in and retrieved from. Conscious thought and transmission of messages along nerves in our bodies are electrolytic phenomena. Heart beats are the results of impulses also of electrochemical nature.

A century of progress in industrial electrochemistry has brought the cells to a high degree of efficiency and to numerous variations in components, shapes, sizes and operating conditions. Electrolytic cells have been going through a big development and their capacity has reached many 100 thousands amperes.

Electrochemistry will play a major role in the world of tomorrow particularly because in the future electricity will be the indispensable intermediate form of energy, whether derived from fossil fuels or the sun, the seas, the deep earth, or from nuclear reactions.

Worldwide electrochemical processes already use almost 10 % of the total electric energy produced and most of it is used to manufacture chlorine and aluminum. However in the United States the electrochemical industry is involved in more than 20 % of the gross national product and in the future could have a bigger share with new processes and more efficient cells.

THE CELL

To increase the efficiency of the cell, its three components

"Electrodes",   "Separators",   and "Electrolytes"

must undergo substantial improvements and new cell designs will have to be introduced.

New permanent ion selective electrodes and high active area electrodes together with new ion selective separators and solid polymer electrolytes as well as the utilization of high temperatures and pressures in the operating conditions of cells of new geometry will be the major contributors to more efficient processes.

ELECTRODES

The development of dimensionally stable electrodes (DSE®) has permitted great improvements in electrochemical processes, so far particularly in brine electrolysis to produce chlorine, caustic, hypochlorite and chlorate by reducing the energy consumption and by increasing the purity of the products.

DSE[®] have permitted the design of new types of cells, like the high capacity monopolar or bipolar cells and membrane cells for the production of chlorine and the cells for the direct electrolysis of sea water. The design of such cells, particularly the bipolar types, would have been impossible without the use of dimensionally stable electrodes.

There is no reason why in the future, the principle of bipolar cells with monoface or biface electrodes could not be applied successfully also to the design of batteries by utilizing DSE[®]. The contribution that these new electrodes brings to the manufacture of nickel and cobalt is already considerable, but the future processes for electrowinning of metals will be based on the utilization of DSA[®].

It is the development of dimensionally stable electrodes with high electrocatalytic activity which can be made selective for each application that has given a versatile tool to the electro-chemist.

DSE[®] can operate at high current densities with low overvoltages which permit high current efficiencies, and they can be made in configurations that have low bubble effect, which often permits bigger voltage-savings than that due to high electrocatalytic activity.

It might be interesting to mention that while the invention of DSA[®] was due to a brilliant idea of Henry Beer, the improvement of these electrodes has not been empirical, but has taken into account thermodynamical concepts and kinetical mechanisms.

High area activated electrodes will be developed and used in many of the future cell designs. Promising are the electrode supports made of cellular conductive vitreous carbon, obtained by heat treating polyurethanes under controlled atmosphere, or of porous cellular metals or metal alloys.

High porosity electrodes permitting high current densities will also be manufactured utilizing sintered metallic powders, having low electrical resistivity and high mechanical and chemical stability.

Particulate electrodes having a few tenths of a millimeter diameter will also be used in fluidized bed cells especially for the recovery of metals, treatment of effluents and electroorganic reactions.

Coating and doping of electrodes will be used to a greater extent to improve their electrical conductivity and electro-catalytic activity and to increase the energetic gap between the unwanted and wanted reaction.

Finally new non metallic materials such as conductive polymers will be utilized for the structure or the surface of the electrodes and will find wide applications.

In addition to metallic oxides or oxyhalides, in particular the unstoichiometric oxides, spinels and cermets will also be used.

SEPARATORS

The separators being used today such as asbestos diaphragms have many limitations due to pollution, short life, poor chemical and mechanical stability, plugging, high ohmic drop, low ionic selectivity and empirical preparation techniques. That is why, modified and ion active diaphragms and membranes are now beginning to be used and will be used to a greater extent in the future. The modified and ion active diaphragms which have been suggested have a porous structure of synthetic materials, ceramics, carbon cloth or polyurethane resin strands, treated with stabilising and ion active agents containing poly-valve metal hydroxides, doped poly-valve metal hydroxides, or sulphon-ated or carbosilated organic resins, as well as phosphate or polyphosphates of valve metals.

However, it will be the permionic membranes, having ionic conductivity and ionic selectivity but no porosity or electronic conductivity, that will find large applications as soon as they can be manufactured at lower cost and with longer life.

Permionic membranes include organic stable polymers such as polyacrylate, polyfluoroethylene, polystyrene, polyvinyl-benzene, and ion active groups such as sulphonates, phosphates, polyphosphonates, carbosilates, sulphocarbosilates, and amine derivatives.

ELECTROLYTES

Electrolytes of high purity and high conductivity will be required for high cell efficiency.   It is the solid polymer electrolyte (S.P.E.) cell which has an attractive configuration by having active electrode materials applied directly on both sides of the solid polymer which acts at the same time as separator, electrolyte and electrodes.

The solid polymer systems can be organic or inorganic systems having an ionic conductivity while the electronic conductivity is practically absent.

The electrochemical reactions which can be carried out in S.P.E. cells are : Na Cl electrolysis, H Cl electrolysis, water electrolysis, organic electrosyntheses, molten salts electrolysis and of course those for the production and storage of electricity in batteries and fuel cells.

To improve electrochemical processes new electrolytes have to be developed and their quality better controlled.

INDUSTRIAL PROCESSES

In addition to the largely utilized existing processes of brine electrolysis, water electrolysis, molten salts electrolysis and the aqueous electrowinning or refining of metals, other processes will be developed and utilized in the future.

THE ELECTROWINNING OF METALS

Not only aluminum, magnesium and sodium, but other metals will be manufactured by molten salt electrolysis. Should an economical cell be developed there can be expected a big increase in the electrowinning of metals which are now produced in small amounts such as berylium, indium, tantalum, zirconium, hafnium, potassium and a greater utilization of such metals as magnesium and sodium.

New cells could be monopolar or bipolar and have an horizontal or vertical geometry. In addition, they might operate at a temperature low enough to permit the electrodeposition of metals of high quality in solid form.

Sintered electrodes have to be developed being chemically stable to and possibly wettable by molten salts, liquid metals

and other electrolysis products. Of course, they must have a high
electronic conductivity and negligible ionic conductivity and
show low overpotentials for the gases when produced.

The metallurgy of metalwinning will one day be based on
electrowinning methods. Ores, concentrated or dilute, will be
treated to give sulphate or chloride solutions which after
treatment by solvent extraction or ion exchange will give
electrolytes containing sulphates and possibly mainly chlorides.

Deep sea manganese nodules as a potential source of nickel,
copper and cobalt, when harvested, will utilize probably hydro-
metallurgy and electrochemistry for the recovery of their metal
contents.

The electrochemical recovery of particular metals and/or
metal oxides like uranium, rhenium, manganese, molybdenum,
yttrium, vanadium, silver and other trace metals from ocean
waters, rivers and lakes might also be possible. This is very
attractive even though the proposed techniques, including biological extraction, may seem unrealistic.

The development of DSA$^{®}$ suitable to operate at very high
current density or having a high active area, and without any
appreciable electrochemical passivation will one day allow the
electrowinning of these metals to become economical.

If a suitable electrochemical technology were to be developed,
permitting the economical recovery of the most useful elements
from the low concentration ores of the world, which are however
enormous in volume, the availability of such elements to mankind
could be assured well beyond the present estimates and certain
elements could be made available in quantities larger than the
presently known reserves now permit.

THE ELECTROCHEMICAL MANUFACTURE AND STORAGE OF HYDROGEN

The electrochemical manufacture and storage of hydrogen will
play an important role in the energy problems of tomorrow.
However, the electrolysis of water is today still inefficient from
the energetic point of view. Therefore, new cells have to be
developed such as those utilizing solid polymer electrolytes
and operating at high temperatures and pressures.

New porous electrodes obtained by plasma jet technique or by
other methods can decrease the hydrogen and oxygen potentials.

In addition, on the cathode side the electrocatalytic activity can be improved by adding to the electrolyte particular organic or inorganic compounds which act as carriers of protons from the bulk of the solution towards the cathode film or as electron donors.

To decrease the voltage on the anode side, it has been suggested to use cobalt, cobalt oxide, sulphides, selenides, and tellurides which seem to improve the electrocatalytic activity of the anode.

What is important is to have a new electrode material able to operate at low overpotentials without having ageing or passivation problems.

The storage of hydrogen at low pressure in the form of hydrides which can be cheaper and safer than the storage as a gas under pressure may improve the overall economical picture.

As fuel cells utilizing hydrogen constitute the ideal electricity generator since chemical energy is directly transformed in electrical energy, more attention should be paid to the development of fuel cells.

ENERGY PRODUCTION AND STORAGE

Energy production and storage will take a big share of tomorrow's electrochemical industry with the manufacture of power cells, fuel cells, dry cells, and other types of batteries. There is an increasing demand for high energy density, high power density, highly reliable and long life primary and secondary batteries and cells.

It is difficult to say which type of battery will be the most economical for load levelling or for vehicles.

Each one of the batteries under improvement or development today : the nickel-iron, the nickel-zinc, the zinc-chlorine, or those operating at high temperature, has advantages and disadvantages, but none is good enough.

It is probable, however, that the metal-air battery in which the recharging would consist of replacing the metal electrodes, which are consumed during electricity production, and the spent electrolyte, will be those having greater chances of success. It has taken one hundred years from the first commercial

batteries to such beautiful products of technology as the fuel
cells upon which the Gemini and Apollo space flights have relied
for their electrical power.

I am sure it will take only a few years more to arrive at
the efficient economic batteries needed.

ELECTROCHEMICAL ORGANIC SYNTHESES

Many possibilities would exist in the field of electrochemical
organic syntheses if suitable electrode materials were developed.
Such electrodes should (a) contain an active coating with high
valency dopants which catalise in situ the organic reactions
(heterogeneous catalysis) (b) be able to regenerate the catalyst
in solution or in suspension (homogeneous catalysis) (c) be able
to dissolve themselves (sacrificial anodes) to generate active
metal complexes and (d) be able to allow oxidation or reduction
by having a high overpotential for the unwanted reactions, i.e.
oxygen or hydrogen evolution.

The selected electrodes should also have very high active
surfaces, very high reaction selectivity, high chemical and
electrochemical stability, slow catalytic ageing rate, high
adsorption rate of the reactants and high desorption rate of
the products.

It should also be mentioned that the utilization of fluorine
and bromine in the manufacture of chemicals and the use of inter-
halogens produced electrochemically (the most important being
Br Cl) can be interesting for several processes such as selective
halogenation, oxydation and hydroxylation.

OTHER INDUSTRIAL PROCESSES

The manufacture of perchlorates, persulphates, perborates,
hydrogen peroxide and manganese dioxide are also important fields
of industrial electrochemistry.

Of course, the electrochemical industry is also electroplating,
electroforming, electrocleaning, electroflocculation, electro-
flotation, electroosmosis, electrodialysis, electrophoresis, as
well as anodic or cathodic protection.

All these processes will find wider applications in the future.

SOLAR ENERGY

The utilization of solar energy for the generation of electricity is one of the higher priority topics in electrochemical research and technology.

While it is expected that heating will be the first major application of solar energy, photochemical use of the sun will probably reach industrial importance and the photochemical production of hydrogen through solar electrolysis may become feasible before the year 2000.

The production of hydrogen through solar energy presents still many problems and low conversion efficiencies are obtained with currently used semiconductors. But many improvements are already in sight and new ideas such as the utilization of solar radiation to produce hydrogen by water photolysis in artificial asymmetric membranes may one day become a reality.

To conclude, let me make some guesses about the future of electrochemical industry.

1. The chlorine requirements will continue to increase but rather than shipping liquid chlorine from larger plants, on-site production in smaller plants will be preferred. For these plants membrane cells will be the chosen cells.

However, chlorate and hypochlorite obtained from direct electrolysis of brine or sea water will probably substitute for chlorine in some applications.

2. For the production of aluminum and magnesium from molten salts, cells will be designed with lower energy consumption by utilizing non-carbonaceous cathodes and permanent anodes which will permit reduction of the electrode gap.

Other metals whether already manufactured commercially or not will be obtained from molten salts.

3. The electrowinning of metals from aqueous solutions will find more and more applications particularly for ores having a low metal content or for the recovery of metals from industrial waste and surface waters.

4. Hydrogen production, storage, distribution, and utilization may become of paramount importance to solve energy problems.

5. The electrical utility industry will require means of sorting energy for load levelling for better utilization of the consumption diagrams. Batteries that efficiently store and return energy will be developed.

6. The automotive industry will require batteries with high power and energy density. We may expect to have soon batteries that will last longer and will cost less than the existing ones - possibly the fuel for these batteries will also be manufactured by electrochemical processes whether it be hydrogen or aluminum or other metals.

7. Electrochemical routes have fewer process steps for organic reactions, higher efficiencies, greater selectivity and offer higher purity products and we may expect many applications of electroorganic syntheses.

8. Finally, we will utilize more and more solar energy and there the contribution of electrochemistry will be considerable.

One hundred years ago the invention of the dynamo gave an impetus to research in the electrochemical field which, until that time, had been quite academic and permitted the development of the electrochemical industry.

However, it was only the discovery of the static rectifiers that provided electrochemists with high currents and, therefore, permitted the design of modern cells of large capacity.

What a great challenge -- what big incentives have electro-chemists today !

Development of efficient electrochemical processes including those producing energy is essential to the future economy of the world.

To have electrochemists capable of detecting fields of major interest and inspiring young scientists, is of paramount importance.

However, one should not look at spectacular successes, but at concrete results, and these can be reached, as in any techno-logical search, by the teamwork of men of different minds and backgrounds. The pure scientists, the engineers, the observers, the inventors, the discoverers, and also the managers with experience and vision.

Failure to recognize the problems, establish the goals, define objectives and strategies and have the grasp of the realities based on economical situations and feasible solutions would make progress slow.

The energy crisis really exists and we need a coherent vision of an industrial world in which electrochemistry plays a big role.

How fast, how soon, and how much will electrochemistry contribute to the industrial development of the world by utilizing electric energy, or by producing and storing energy, I cannot forecast.

However, I know that electrochemistry will have a bright future. What we will need to make it very bright are not only :

- high area, highly electrocatalytic active ion selective electrodes

- mechanically and chemically stable ion active separators having low ohmic resistance

- highly conductive, ion selective, solid electrolytes, and

- efficient narrow gap cells of new design, but also

- many highly intelligent, highly trained, highly skilled "aim selective" electrochemists.

All of them, including you, having long life.

DISCUSSION

Prof. Ernest Yeager, Case Western Reserve University: A rather negative attitude is developing in the United States concerning chlorine and organic chlorides such as vinyl chloride. This may intensify the search for alternatives to chlorine and could result in a decrease in the demand for chlorine. Do you foresee such a possibility?

Prof. V. de Nora: There might be some limitations imposed in the use and transportation of chlorine. However, if the manufacture of some chlorinated products will be reduced or eliminated, others will be developed. I am of the opinion that the chlorine consumption will continue to increase.

Prof. W. Vielstich, Bonn University: You did suggest metal-air systems like Al-air for a large use. Due to the recycling of the metal via electricity we have to consider an overall energy efficiency factor. Could you comment on this problem in the light of the energy situation?

Prof. V. de Nora: Primary battery systems where the dissolving metal is produced in the right conditions seem to be more efficient than those in which the recharging is done under non-ideal conditions with respect to current density, temperature, electrolyte concentration and cost of power.

Prof. W. Vielstich: What should be the primary energy for H_2-production around the year 2000, realizing that (a) the net energy factor of today's photovoltaic cells is only about 1-2, and (b) electrochemical solar cells are still in the academic state? In Germany so far we see hydrogen production only via water electrolysis using nuclear energy.

Prof. V. de Nora: I agree with you that efficient water electrolysis will be for many years the only practical source of hydrogen. However, progress sometimes is made more rapidly than expected and that is why I have hope in solar cells for hydrogen production.

Prof. Karl Kordesch, Technical University, Graz, Austria: What is your reason for emphasizing polymer-membrane cells for fuel cells and electrolysers?

Prof. V. de Nora: The combination of the separator, the electrolyte and the electrodes seems to be the most efficient way to reduce energy consumption and increase overall efficiency.

Dr. Adam Heller, Bell Laboratories, Murray Hill, New Jersey: For innovations in the electrochemical industries very substantial capital outlays are and will be required. How effectively does and will the U.S. compete for this capital with countries such as Japan, Germany, and other countries in which the rates of savings, as well as the rate of increase in productivity, have outpaced the U.S. in the past several years?

Prof. V. de Nora: In spite of temporary decreased productivity, savings, and research efforts in the United States, I have no doubt that the resources and incentives of this country will permit the development of the electrochemical industry faster than that of any other country.

DIMENSIONALLY STABLE ANODES

Henri B. Beer

Scientific Research Society N.V.
Heikantvenstraat 41
2190 Essen, Belgium

I consider it a great honour to have been requested to read a paper in this famous university and I have accepted this opportunity with great pleasure. It is now more than twenty years ago that I came to the States for the first time to visit the Patent Office for a discussion about one of my patent applications on metal anodes. Little could I have thought at that time that I would ever receive such an honourable invitation from your Institute, and I am extremely happy to find myself in the company of so many well-known and illustrious scientists at this commemoration of the centennial of the Case Institute of Technology.

My paper today shall be a general review of how my inventions of the dimensionally stable metal electrodes were made somewhat accidentally and the further developments in practice.

When I say practice, I mean not only technically, but also commercially. For this last part a strong patent position over the world was needed and this has not been achieved without great difficulties. This took me to many patent hearings in many different countries personally. I will come back to this later.

As usual in my papers I should like to say a few things about the noble art of inventing itself and the innumerable difficulties inherent in it, especially for the independent not too rich inventor.

Like every invention destined to cause a revolution in the existing technology, the metal anodes have also taken a considerable amount of time, about fifteen years, to finally acquire their definite form and success.

However, the savings of energy which are the immediate result
of the replacement of graphite anodes by dimensionally stable anodes
in the chlorine-alkali cells alone, have been very considerable.
From 1970 up to the end of 1979 the savings of electric energy in
North America which can be attributed directly to the use of DSA^R
in 70 different chlorine plants totals considerably over 19 billion
kilowatt-hours.

One of the reasons why it took such a long time before a real
interest was roused for this innovation is, for instance, that the
chlorine industry, one of the main possible future customers, was
at that time quite content with the existing, and even more with
the constantly improving, graphite anodes and naturally was therefore reluctant to invest much money in a product of which so little
was known and with which hardly any experience had been gained.
This was of course understandable as there was not only the risk
connected with the use of an entirely new type of electrode, but,
as is the case in most industries, the production of chlorine is
only one of the links in the chain of a total manufacturing process.
The chlorine, therefore, is in the first place a basic raw material,
so that a possible interruption or even a slow-down of its production
would affect the whole plant and so lead to unforeseeable and grave
consequences.

Another great obstacle for the development of many inventions
is often the mysteriousness and secretiveness which is practiced by
some manufacturers who think that they possess a very special know-
how on account of which their plant works better or more efficiently
than that of their competitors. This often proves to be an illusion.
The "not invented here" syndrome was another difficulty.

The metal anodes were faced with numerous general problems due
to the fact that anodes are not only an important part of the equipment of an electrolysis plant, but they also have to be adaptable
and perform well in various types of cells. In order to achieve
all this, the operation of the different parts of the cells such as
the electrical conductors and circuits, the regulating device for
the adjustment of the gap between anode and cathode, the circulating
speed of the brine, etc., all had to be taken into account in order
to obtain the best possible and most economical results.

Much spadework had to be done with regard to the substructure
of the electrodes from titanium metal, since at that time (1956)
very little information was available about the special tools and
methods needed to work with titanium, be it with a saw or a file,
or on a lathe, and above all, how it should be welded. I had to
find out too which was the most suitable design of the substrates,
the size of the perforation or the space between the rods, to let
the chlorine gas pass freely. The titanium tube containing the

current-conducting copper rod which should convey the current to
the substrate with the least possible resistance and distribute it
evenly to all parts of the anode presented quite a few problems to
be solved.

Then a solution had to be found for the pretreatment of the
titanium substrate such as etching, ultrasonic cleaning and finally
the application of the active coating itself. In the beginning,
when relatively small quantities of electrodes were produced, the
coating was applied by electro-plating. It goes without saying
that the absolute flatness of the anode surface is imperative, a
requirement which was difficult to meet and needed expertise
and professional skill.

Although through the years I had acquired a fair amount of
experience with electrodes, I lacked most of the knowledge of the
technique and practice to make a complete chlorine cell run efficiently, and that is another reason why it took me many years of
collaboration with various manufacturers to gradually obtain a
better understanding of this intricate matter and to arrive at a
satisfactory performance of the metal anodes (or dimensionally
stable anodes, as they are called today) in every respect. To
become conversant with the complicated knowledge of any industrial
process is usually extremely difficult for an outsider and was for
me perhaps even more so, as all my life I had been working as a
free-lance inventor not connected with any particular industry.
On the contrary, my activities in research have always been rather
widespread, as I have worked on a great variety of subjects, for
instance, air-conditioning, electric heating, dry batteries, wine-
making, ozone production, atomizers for spraying devices used in
agriculture, metal oxides, cathodic protection, ferrites, the extraction of air from seawater for submarines, the extraction of carbonic
acid from the air in submarines and cold-stores, the biochemical
generation of electricity, and much more. My first invention was
just an improvement for a better working type of scarecrow for
farmers and fruit-growers. On all these subjects and ideas I have
been working with various degrees of success, but I managed to live
on it.

The greatest problem for me, and probably for most independent
inventors, has always been to find the financial means for the
necessary research work, not to mention how much money is needed
after the research has been successful. It might even be said that
in many cases it is less difficult to make an invention than to find
the money to work it out. I have a feeling that in the future this
will become an even greater difficulty for the free-lance inventor,
not least because inventions tend to become more and more technical
and sophisticated, which implies that longer and costlier research
work will have to be done. For many inventors this will be a

serious additional obstacle. Most of the cheap inventions have
already been made. Independent inventors were always paid as the
Chinese doctors in the past: no cure, no pay.

Still, I am of the opinion that for many years to come this
type of solitary inventor will continue to contribute to the advance
of technology and to the expansion of knowledge of all kinds, and
to cite the words of Albert Einstein:

"All the history of Science shows that it is not through
organization and planning that the great advances are
achieved. The view of some individual who must get the
spark, and finally the freedom of the individual scholar
is the main condition for scientific progress."

To come back to electrodes now, records show that in the course
of the last decade the following materials have been used:

1) solid platinum metal - too expensive, not resistant to
chlorine and too energy-consuming

2) retort carbon - not resistant to most electrochemical
reactions

3) graphite - Although the use of graphite was a great improvement compared to the old retort-carbon anodes, many of the disadvantages of the retort carbon still adhere to graphite anodes, as for
instance, the consumption of some pounds of graphite per ton of
chlorine produced, the possible generation of chlorinated organic
compounds which will clog pumps and pipes used for the transportation
of chlorine, etc. Further, they are difficult to mold in the
desired form and have high overvoltages at interesting current densities. Contamination of the electrolyte with graphite particles
creates stray-currents resulting in loss of energy. One of the
greatest disadvantages of the graphite electrode, however, is that
on account of the erosion of the graphite anodes, it is impossible
to maintain a constant gap of a few mm between anode and cathode
and to keep diaphragms and membranes free from graphite particles
that will clog them up.

To obtain the highest efficiency of supplied energy, it is
necessary to narrow this gap (space) between anode and cathode as
much as possible in order to get the lowest possible resistance
through the electrolyte between the electrodes.

4) magnetite anodes, anodes of lead-silver, manganese dioxide
or lead peroxide on lead or graphite substrate anodes - All these
have found some applications, for instance in chlorate production,
cathodic protection, chromium plating, in galvanizing technology,

etc., but none of these anodes is entirely resistant against chemical corrosion. Further they have a high overvoltage, high ohmic
resistance and are not easily given intricate forms in large quantities.

   5) anodes having tantalum, niobium, zirconium, etc. as a base
metal and coated with layers of platinum metals - They are used
for applications such as cathodic protection, in the plating industry, etc., but have never found use in large industrial applications
such as the chlor-alkali industry, one reason being the high cost
of these anodes and their less than completely satisfactory performance.

   In 1955 I was doing research work in the field of very pure
single and multiple oxides, which I intended to use for the manufacture of ferrites, the mixed oxides of which magnets are made.
The basic principle of my idea was to produce those oxides by electrolysis, by means of metal anodes at different current densities.
The salts thereof were then coprecipitated as mixed oxides.

   Apart from experimenting with this process for the production
of oxides of metals like iron, zinc, nickel, cobalt, etc., I also
did experiments in order to see whether I could obtain such oxides
starting from titanium metal with other oxides, to be used as piezo
crystals. During these experiments I gradually became aware of the
strong endurance of titanium if used anodically in electrolytes,
especially in chlorine-containing electrolytes. This brought me
to the idea of using titanium as a substrate for anodes and this
effectively produced a decisive break towards progress in the
development of the dimensionally stable metal anodes. Coated with
alloys of platinum metals or their oxides, they have many advantages.

   Wherever the active coating might become porous or damaged,
titanium will form a protective oxide layer. One might say that
the titanium substrate is anodically self-healing.

   These coated titanium substrates preserve their shape and low
overvoltage characteristics even under the most severe anodic conditions, especially those prevailing when electrolysis is applied
to hot aqueous solutions of alkali metal chlorides. Therefore,
they are called dimensionally stable anodes in contrast to most
other anodes which have a significant wear rate and thereby inevit-
ably change their characteristics.

   The metallic platinum metal coating on the first titanium anodes
also had a great wear rate and the tendency to passivate in chlorine-
containing electrolytes. This could be drastically reduced by the
addition of 10% to 30% iridium metal to the platinum coating. Still

the cost of these anodes based on the present and prior price of
platinum and iridium metals makes their use on a large scale not
so attractive.

Further, due to the substantially higher current densities
permitted on activated titanium anodes and essential to make them
economically attractive compared to graphite anodes, their use is
of particular interest in mercury cathode cells. However, coatings
consisting of platinum metals or their alloys in the metallic state
have additional drawbacks in that they are much too sensitive to
current reversal, which occurs whenever the cells are short-circuited
for maintenance. Further, any accidental contact with the mercury
or amalgam cathode will bring about a sizable loss of noble metals
and a depassivation of the coating. As this contact with the mercury or amalgam cathode is unavoidable and every kind of cell has
to be short-circuited for maintenance, I had to look for different
kinds of coatings which would not have these drawbacks and would
be cheaper if possible.

The philosophy I followed to contrive these new sorts of exceptionally stable electrocatalytic coatings proceeded along two main
and converging routes. For one thing, the greater thermodynamic
stability that a state of aggregation of two or more elements, be
it an alloy or an oxide, may acquire in principle over its single
components was persistently in my mind at the time of my experiments
with platinum alloys. Secondly, I was well aware that certain mineralogical varieties of mixed oxides, such as for instance, magnetite or ilmenite and similar compounds pertaining to the spinel
family, or those displaying the polycrystalline structure similar
to ilmenite and tungsten bronzes, are not only endowed with outstanding chemical inertness, but also acquire conductive properties
owing to the complete lack of stoichiometry of the compound. In
fact, these tertiary compounds have since long been classified as
"mixed oxide semiconductors" or "polycrystalline semiconductors"
or "ceramic semiconductors". The ohmic resistance of a thin layer
made therefrom, as would be required for an electrocatalytic coating
on titanium, is negligible.

I came to ask myself whether some oxides of this sort could
be applied by chemideposition, or some other method, on a metal
having structural and barrier-film forming properties such as,
for instance, titanium. Going even further, I wondered if it were
possible to formulate some "tailor made" mixed oxides displaying
an even more satisfactory stability together with great electrocatalytic properties. Then going through the mineralogical tables,
I was struck by the fact that some noble metals like iridium and
ruthenium form oxides belonging to the rutile type which have lattice parameters very close to titanium dioxide, so that they also
would seem to be ideal to build up mixed oxide compounds with

titanium. While oxides of other platinum group metals and other
valve metals may be combined for different purposes, the mixed
crystals of titanium/ruthenium oxides were found to be the best
for the production of chlorine. Titanium or tantalum-iridium
mixed oxide crystals are highly electrocatalytic for all electrochemical processes, as for instance, oxygen evolution. Following
up these ideas and doing a few thousand more experiments, I found
the ultimate coatings, now known as DSA^R and described in my patents
of 1965 and 1967. These coatings are applied to the titanium base
by thermal decomposition from aqueous or organic solutions.

The advantages of these coatings can be summarized as: low
overvoltage at low and high current densities, resistance against
short periods of contact with amalgams, resistance against the reactant products, no wear losses, resistance against current reversal,
great mechanical resistance, lower investment costs and also easy
manufacture.

The invention was now ready to be commercialized and for tech-
nicians to set up manufacturing facilities to produce the anodes to
fit in the different kinds of cells. A result is that these anodes
have revolutionized the design of chlorine cells in many ways, particularly in the field of diaphragm cells and membrane cells. This
last type of cell could only be made to work with metal anodes.

Mercury cells, the performance of which has been described
above, are now about as close to perfection as the industry can
reach since they produce a caustic lye of great concentration at
high current efficiencies, completely free of chloride or other
impurities. However, due to the harmful effect of mercury in waste
effluents, the use of mercury cells may be forbidden in the future.
This is already the case in Japan where the government has decreed
that mercury cells must be replaced by 1983.

Industry has turned to membrane cells as a suitable alternative
to the mercury cells, and much design work and new constructions
have come forward due to the successful adoption of dimensionally
stable metal anodes.

A very interesting metallic anode construction is used in the
Diamond Shamrock diaphragm cell, in which an expandable metal anode
is used. This anode is maintained in a retracted position while
the cell is being assembled; once the cathode is placed on the cell
base, the retainers are removed so that the anodes will expand to
contact spacers situated on the cathode tubes. This means that the
anode/cathode gap is controlled by the diameter of the spacer and
the cell can be easily assembled with a minimum gap. Recently,
dimensionally stable diaphragms have been developed which consist
of a mixture of asbestos and plastic fibres. They have the advan-

tage of not swelling when in operation, which means that the electrode gap can be reduced and that the lifetime of such diaphragms
is much longer and they are therefore much more economical in many
ways. All these innovations have been possible only with metallic
anodes.

As said before, the preferred electrocatalytic coating is the
ruthenium/titanium oxide combination for chlorine production. However, this coating can be modified for other particular uses by
the addition of other metal oxides such as those of tin, antimony,
etc. Moreover, the application fields of these anodes are constantly
growing. Dimensionally stable anodes have met with great success
in the electrowinning of metals from aqueous solutions of metal
chlorides or metal sulfates. Other coating modifications are being
used in chlorate production and in bromine evolution.

The chlorine industry in the western world produces at the moment
yearly more than 12 million tons of chlorine with metal anodes.
Metal anodes have further spurred much research which resulted in
many new patents in order to improve the form of anodes, anode
fabrication, electrocatalytic coating materials, cell designs, cell
construction materials, cathodes, diaphragms and membrane materials
and means to prevent short circuits, particularly in mercury cells.
The importance of electrochemistry has certainly been advanced by
the dimensionally stable anodes. The result is that over the past
years the number of scientific reports relating to electrochemistry
has been greatly increased.

With respect to my two basic patents for metal anodes, one is
called the "single oxides patent," and was filed in 1965. It is
directed to electrodes with a valve metal substrate, with an electrocatalytic coating of one or more oxides of the platinum group metals
such as platinum, rhodium, ruthenium or iridium alone or a mixture
of these oxides with the oxides of non-noble metals as iron, copper,
silicium, etc.

The second patent application, which describes the most widely
used electrode in the chlorine industry, deals with what are called
"the mixed crystal coatings," which claim electrodes with valve
metal substrates with an electrocatalytic coating consisting of
mixed crystals of platinum group metal oxides and film-forming metal
oxides such as titanium, tantalum or zirconium, etc. This patent
was filed in 1967.

The first application was filed in 26 countries, and the second
was filed in 44 countries. Practically all the patents have been
granted, so that a generally strong patent position has been built
up nearly all over the world. Hopefully, in the few places where
the procedure is still going on we will get to a conclusion in the
near future, although I understand that it is not unheard of for

a patent to be finally granted after the period of protection has
expired.

The credit for the rapid development of these electrodes from
the laboratory stage to commercial use goes to Vittorio de Nora.
He is now Chairman of the Board of a newly created subsidiary of
Diamond Shamrock Corporation of Cleveland, called Diamond Shamrock
Technologies S.A., Geneva, and of many other companies. Vittorio
de Nora has been successful in convincing the large chlorine producers of the enormous advantages which these new anodes would
have for their industries and he persuaded them to try out these
anodes in their plants. This was by no means an easy job, as industrial firms are often rather sceptical when it comes to the introduction of new technologies in their plants, especially when it
would require an investment of millions of dollars and as a successful outcome is never certain. Especially, the idea not to sell but
to lease the anodes to customers and to guarantee their performance
has to a great extent contributed to this successful enterprise,
with the result that these anodes could be commercialized on a
large scale and did not remain a laboratory curiosity.

I should like to emphasize once more the great importance in
energy savings and the contribution to pollution control which are
being obtained by the application of dimensionally stable anodes,
especially in the chlorine-alkali electrolysis. It has recently
been calculated that all energy savings in this field in the United
States of America, taken together from the year 1973 until now,
would make it possible to supply 2 million houses of middle-sized
families with electricity for a whole year. Taking into account
that in the States almost half of all the western-world production
of chlorine is being manufactured, it would mean more than twice
as much for the total western hemisphere and three times for the
entire world. This is the contribution of the dimensionally stable
anode to the energy problem.

As far as pollution is concerned, after the installation of
DSA^R anodes, the cell rooms could be improved and much less pollution was the result.

To conclude how metal anodes were developed, I would like to say
that I found the unique properties of titanium as a substrate in
electrolytes, especially those containing chlorides, by accidental
observation. The successful coatings thereon, however, were only
created after years of research in close cooperation with the industry.

At the end of this paper I should like to express my gratitude
to all the people of this University who made it possible for me
to read this paper and to all those who have assisted me in the past
morally, technically, commercially and financially to bring this
venture to a good end.

DISCUSSION

Prof. Karl Kordesch, Technical University, Graz, Austria: The beneficial effect of depositing noble metals like Pt, Pd, Ru, etc., on
porous carbon or spinel structures is a well-known method for making
fuel cell anodes. Do you ascribe this beneficial effect to an en-
largement of the active surface or to the stabilization of the
noble metal catalyst? Is there any theory known?

Dr. Henri Beer: The problems and conditions encountered in chlor-
alkali electrolysis are very different from those in fuel cells,
so I cannot really offer an answer to the question. A vast amount
of knowledge has been accumulated on dimensionally stable anodes
for chlor-alkali electrolysis, but so far this has not had any
impact on fuel cell technology just as, in the past, the teachings
of the fuel cell art were of no help in solving the problems faced
in using metal anodes for brine electrolysis.

OXYGEN ELECTRODES FOR INDUSTRIAL ELECTROLYSIS

AND ELECTROCHEMICAL POWER GENERATION

Ernest Yeager
Case Laboratories for Electrochemical Studies
and The Chemistry Department
Case Western Reserve University
Cleveland, Ohio 44106

I. INTRODUCTION

Electrochemistry is one of the technologies which can make important contributions toward solving the energy problem. A significant portion ($\sim 8\%$) of this nation's electrical power is consumed for industrial electrolytic processes, most of which are carried out with relatively low energy efficiency. Improvements in the efficiencies of these processes will result in substantial power savings. Further, electrochemical energy conversion and storage systems using various fuel cells, batteries, and electrolysis cells offer substantial promise for stationary and vehicle applications.

A key component in the optimization of these energy important electrochemical systems is the oxygen electrode, both as an O_2 consuming cathode and O_2 generating anode. The overall electrode reaction in aqueous electrolytes (expressed in the cathode direction) and the corresponding standard thermodynamic potentials at $25^{\circ}C$ are as follows:

acid electrolytes $E^{\circ}(V)$ vs. NHE

$$O_2 + 4\ H^+ + 4\ e^- \longrightarrow 2\ H_2O \qquad\qquad 1.23\ V \qquad (1)$$

alkaline electrolytes

$$O_2 + 2\ H_2O + 4\ e^- \longrightarrow 4\ OH^- \qquad\qquad 0.401\ V \qquad (2)$$

In practice, these O_2 reduction reactions and the corresponding reverse anodic reactions are very irreversible at low and moderate temperatures and the operating potentials deviate very substantially from the thermodynamic values.

29

Various cells using O_2 electrodes are listed in Table 1 together with typical cell voltages and the voltage losses associated with the O_2 electrode. These voltage losses result in very substantial energy losses and seriously compromise the applicability of many of the cells listed in Table 1. For example, if presently available H_2-O_2 alkaline fuel cells and water electrolyzers were to be used for energy storage with H_2 as the storage medium, the overall energy out/in efficiency would be only $\sim$ 50% principally because of the irreversibility of the O_2 electrodes. Today, water electrolysis is of relatively minor importance. This situation would change, however, if and when nuclear and/or solar-electric become the predominant energy sources and the H_2 economy concept is adopted in whole or part. With rechargeable metal-air cells such as the iron-air cell under consideration for vehicle propulsion, the energy out/in efficiency is also only $\sim$ 50% principally because of the polarization of the O_2 positive electrode on discharge and charge.

Further problems associated with the air-consuming O_2 cathode in fuel cell and metal-air battery-applications are limited life (ranging from a few weeks to 5 years) and relatively high cost, particularly if platinum is used as the catalyst at substantial loadings.

In addition to these applications for O_2 electrochemistry the anodic generation of O_2 plays an important role in most battery systems as a competing reaction at the positive electrode. If it were not for the irreversibility of the O_2 generation reaction and the large O_2 overpotential, most aqueous batteries would be impractical. The positive electrode materials generally have potentials more positive than that of the reversible O_2 electrode and hence would otherwise spontaneously self-discharge with the oxidation of water to O_2.

O_2 electrochemistry also plays an important role in corrosion processes. In the presence of air, O_2 reduction is often the cathodic process which drives the potential of iron and ferrous alloys into the passivation range where the corrosion is inhibited.

The principal concern of this lecture is with the factors controlling the performance of air cathodes in aqueous solutions. Before proceeding with such, however, the promising applications for the air cathode in the chlor-alkali industry and a few other electrolysis applications will be discussed. Over the past several years, Case Laboratories for Electrochemical Studies and the Diamond Shamorck Corporation have been involved in a cooperative program concerned with air cathodes with one important near-term application expected to be their use as cathodes in membrane-type chlor-alkali cells.

Table 1. Applications for O_2 electrodes in batteries and industrial electrolyzers

System	Temperature	Typical cell voltage[a]	Voltage loss at O_2 cathode	anode
1. Fuel Cells				
H_2|H_3PO_4|air	190°C	+0.6V	0.45V	
H_2|KOH|air	100	+0.8	0.3	
H_2|carbonate|air	650	+0.9	0.1	
2. Metal-air cells: primary				
Zn|KOH|air	25	+1.3	0.3	
Fe|KOH|air	25	+0.85	0.3	
Al|NaOH|air	40	+1.6	0.35	
Al|NaCl|air	25	+1.2	0.4	
Na(Hg)| NaOH |air	90	+1.5	0.3	
3. Metal-air: secondary (charging mode)				
Fe|KOH|air(O_2↑)	25	−1.6	0.4	
4. Water electrolyzer				
↑H_2|KOH|O_2↑	80	−1.8		0.6
5. Chlor-alkali cell: membrane				
Cl_2|NaCl⋮NaOH|O_2	80	−3.5	0.4	
6. Amalgam denuder				
Na(Hg)|NaOH|air	90	+1.5	0.3	
7. HCl electrolyzer: membrane-solid polymer electrolyte				
↑Cl_2|SPE|air	60	−0.6	0.3	
8. Peroxide generating cells				
↑O_2|KOH|HO_2^-,KOH|air	40	−1.5	0.1[b]	0.6
H_2|KOH|HO_2^-,KOH|air	40	+0.6	0.1[b]	
9. O_2 concentration				
↑O_2|SPE|air	30	−1.0	0.3	0.4

[a] Power generating:+ ; power consuming: −.
[b] Relative to the reversible O_2\|HO_2^-,OH^- electrode.

II. APPLICATIONS FOR O_2 CATHODES IN THE CHLOR-ALKALI INDUSTRY

The H_2 generated in present diaphragm and membrane chlor-alkali cells is usually burned for its heat value rather than used for chemical purposes. In such instance the replacement of the H_2 generating cathodes with air consuming O_2 cathodes offers the opportunity for substantial energy savings. A cell voltage saving of ~ 1.0 V is anticipated, taking into account the thermodynamic potentials as well as polarization characteristics of the O_2 and H_2 cathodes and the voltage losses within the bubble field in front of the H_2 cathodes. The reversible potentials and expected cell terminal potentials are listed in Table 2 for membrane cells with H_2 generating and air consuming cathodes.

Membrane-type cells are particularly appropriate for air cathodes since the catholyte is of high purity and much less likely to present poisoning problems with respect to air cathode catalysts. Further, the caustic produced in membrane cells may be sufficiently concentrated and free of NaCl to avoid the need for further evaporative concentrating, thus eliminating the large thermal energy input required for the evaporators used with present diaphragm cells. With presently available ion selective membranes, caustic concentrations in the range 9 to 14 $\underline{M}$ are anticipated for the catholyte.

The estimated cell potential saving of 1.0 V with the air cathode is based on an operating potential of 0.80 V vs. the reversible hydrogen electrode in the same electrolyte (RHE) for the air cathode, excluding ohmic losses in the electrolyte external to the porous cathode (see Table 3). The overpotential of the H_2 generating cathode which it replaces has been taken as -0.3 V with conventional H_2 cathodes and -0.1 V with highly catalyzed H_2 cathodes (1-3). Since the bubble field adjacent to the H_2 cathode can be avoided with the O_2 cathode, a thinner membrane-cathode gap can be used and hence some further voltage saving can be realized because of the lower ohmic loss. Even in comparison with highly catalyzed H_2 cathodes, the air cathode is still expected to afford a voltage saving of 1.0 V.

While the cell voltage saving with the in-situ air cathode is expected to be $\sim 28\%$, the total energy saving will be less since the heat value of the H_2 will no longer be available. Taking this into account together with the present efficiencies of coal fired electric generating plants and the ancillary power requirements for air processing, the total energy saving is projected to be $\sim 16\%$. In addition to the total energy saving, an important consideration is the reduction of 28% in the needed electric generating plant capacity and the corresponding saving in capital investment.

Table. 2 Chlor-alkali applications for O_2 cathodes in conjunction with membrane cells

1. H$_2$ generating membrane cell

   (DSA) Cl$_2$|NaCl (15%) |NaOH (30%)|H$_2$(Fe,cat.) (85°C)

   Reaction: 2 NaCl + 2 H$_2$O $\longrightarrow$ 2 NaOH + Cl$_2$ + H$_2$

   E(Rev) = 2.2 V
   E (Actual) = 3.6 V

2. O$_2$ (air) consuming membrane cell

   (DSA) Cl$_2$|NaCl(15%)|NaOH(30%)|air (cat.) (85°C)

   Reaction: 2 NaCl + 1/2 O$_2$ + H$_2$O $\longrightarrow$ 2 NaOH + Cl$_2$

   E(Rev) = 1.0 V
   E(Actual) = 2.5 V

3. Recovery of electrical energy from H$_2$ with H$_2$-air fuel cell

   (DSA) Cl$_2$|NaCl(15%)|NaOH (30%)|H$_2$(Fe,cat.) (85°C)
   H$_2$(cat.)|H$_3$PO$_4$|air(cat.)   (190°C)

   Overall reaction: 2 NaCl + 1/2 O$_2$ + H$_2$O $\longrightarrow$ 2 NaOH + Cl$_2$

   E(fuel cell-actual)* = $\sim$ 0.7 V
   E(net) = E (chlor-alkali)-E (fuel cell) = $\sim$ 2.9 V

---

*Assuming phosphoric acid fuel cell operating at $\sim$ 190°C with cogeneration of heat for evaporation to concentrated caustic.

Table 3. Cell voltages estimated for membrane type chlor-alkali
          cells at 300 mA/cm^2 at 85°C with 15% NaCl anolyte and
          30% NaOH catholyte with H_2 and O_2 (air cathodes)

	$\underline{H}_2$ Cathode	Air Cathode[d]
Thermodynamic potential	2.25 V	1.08 V
Voltage losses[a]		
anode polarization	0.1	0.1
anolyte ohmic loss	0.2	0.2
membrane	0.5	0.5
catholyte ohmic loss	0.4	0.2[b]
cathode polarization	0.3(0.1)[c]	0.4[d]
	1.5 V(1.3 V)[c]	1.4 V
Terminal Voltage	3.8 V(3.6 V)[c]	2.5 V

[a]Includes all losses within cells.

[b]Assumes no bubble field in catholyte.

[c]Parenthesis numbers based on use of low H_2 overpotential
catalyst such as Raney nickel.

[d]Assumes air at 1 atm.

The relatively high cost of ion selective membranes such as du
Pont's Nafion leads to high current density requirements, because
of capital investment considerations; for example 300 mA/cm^2, operating at 0.8 V vs. RHE in 10 $\underline{M}$ NaOH as 85°C. This is a rather
high current density for O_2 electrodes, particularly operating on
air rather than pure O_2 in concentrated NaOH. Furthermore, the
operational life should be a minimum of one year with very low
failure statistics at shorter times. The electrode cost is also a

critical factor and should probably not exceed $\sim$ \$50 per kW-year
of electric energy saved or $\sim$ \$15 per sq. ft. for an operating life
of one year.

These requirements generally exceed those of presently available
low cost O_2 cathodes designed to operate on air in concentrated
caustic electrolytes, even though very substantial research and
development work have been directed to O_2 electrodes over the past
25 years in conjunction with fuel cells. Despite the attractive-
ness of in situ air cathodes for chlor-alkali cells, there is still
some uncertainty whether the operational and cost requirements can
be met within the coming decade. For a significant penetration
of air cathodes into the chlor-alkali industry during the present
century, electrodes filling these requirements must become available within the next five to ten years so that they may be incorporated into new membrane-cell installations. To retrofit installed
membrane cells with air cathodes is likely to be unattractive for
economic reasons. Further, if the air cathode option is not available within the next decade, the chlor-alkali industry will probably
relocate its chlorine plants and/or modify its chemical installations
as a whole so as to find chemical uses for by-product H_2 from chlor-
alkali cells, particular as H_2 from hydrocarbons becomes more expensive.

Operation of O_2 cathodes at high current densities is considerably more difficult on air then pure O_2 because of the higher polarization and usually shorter life. Preliminary considerations
of capital investment and operating costs, however, indicate that
the economics will probably not be favorable for the use of pure
O_2 rather than CO_2-free air with the possible exception of very
large chlor-alkali installations ($\sim$ 1000 tons Cl_2 per day) or installations combined with other large scale users of pure O_2.

In principle, the air cathode can also be used to recover
electrical energy in conjunction with mercury cells through the
use of an electrochemical amalgam denuder with a net electrical
energy saving of $\sim$ 35% (see Table 4). A high performance cell of
the form Na(Hg)|NaOH|O_2 was built in the early 1960's in the United
States for military application and operated on pure O_2 at 1.5 V
with a vertical amalgam electrode of the flowing thin Hg film type
and hence a very small mercury inventory (4). To make such an
electrical energy recovery system practical for the chlor-alkali
application requires the use of very high current densities (>1A/
cm^2) to keep the mercury inventory low. Even so, the capital expenditure for additional cell hardware would be high. Furthermore,
new mercury cell construction is unlikely in most countries because
of real and imagined environmental problems. The better approach
is not to generate amalgam in the first place; i.e., the use of
the in situ air cathode.

Table 4.   Recovery of electricity with an amalgam denuder
           using an air cathode

1. <u>Sodium amalgam – Cl_2 generating cell</u>

   (DSA) Cl_2 | NaCl(30%) | Na(Hg)-liq.

   <u>Reaction:</u> 2 NaCl + (Hg) $\longrightarrow$ 2 NaCl(Hg) + Cl_2

      E(rev) = 3.3 V
      E(actual) = 3.8 – 4.5 V

2. <u>O_2 (air) cathode-sodium amalgam denuder cell</u>

   O_2 (air) | NaOH(50%) | Na(Hg)

   <u>Reaction:</u> 4 Na(Hg) + O_2 + 2 H_2O $\longrightarrow$ 4 NaOH + (Hg)

      E(rev) = 2.4 V
      E(actual) = $\sim$ 1.5 V

3. <u>Overall reaction for cells 1 and 2</u>

   4 NaCl + O_2 = 2 H_2O $\longrightarrow$ 4 NaOH + 2 Cl_2

      E(net,rev) = 0.9 V
      E(net,actual) = 2.3 – 3.0 V
      Electricity saving: 33 to 40%

An alternative to the <u>in situ</u> air cathode is to use the H_2 in H_2-air fuel cells such as the United Technologies phosphoric acid system. These cells operate at temperatures up to 200°C and the cogenerated heat would be available for concentrating the caustic particularly with diaphragm chlor-alkali cells. Further, unprocessed air can be used because of the CO_2 tolerance of the acid fuel cell systems. A major advantage is that these cells could be retro-fitted to existing diaphragm chlor-alkali cells.

Such fuel cells are presently being installed in a 4.8MW system for peak shaving-load levelling in New York City by the Consolidated Edison Electric Company. This system will operate on H_2 containing a small amount of CO, derived from naphtha with a catalytic reformer-shift reactor. With this type of fuel cell using alloy platinum catalysts for the air cathode it should be possible to achieve $\sim$ 5 years life. Thus the H_2 air fuel cell approach, utilizing by-product H_2 from chlor-alkali cells, has the advantage that the hardware is in good part available. The operating voltage, however,

is presently $\sim$ 0.6 V and hence the recovered electric power is only
$\sim$ 60% of that with the in situ air cathodes. Furthermore, a
second set of cells together with additional electrical switch-con-
trol equipment and additional plumbing are required. The estimated
cost for the fuel cell system (excluding the H_2 reformer-shift
reactor) is $350 per kW in 1979 dollars.

The alkaline type fuel cell using concentrated KOH or NaOH as
the electrolyte affords considerably higher operating potentials
(0.85 - 0.90 V) and hence higher equivalent thermal efficiency
($\sim$ 60%). This higher efficiency would probably more than offset
the disadvantage that CO_2 must be removed from the air feed to the
cathodes. In addition, it is possible to use such cells directly
to concentrate the caustic while simultaneously recovering
electrical energy from the by-product H_2 produced in the chlorine
cells. A cation permeable membrane such as Nafion would be used to
separate the anolyte and catholyte compartments of the H_2-air cells.
The caustic from the catholyte side of the chlor-alkali diaphragm
cells ($\sim$ 12% NaOH) would be fed to the anolyte side of the H_2-air
cell. The Na^+ would migrate through the membrane of the H_2-air
cell and together with the O_2-cathode half cell reaction would result in the build-up of the concentration of caustic in the catholyte. The demands on the performance of the air cathode would be
similar to those for the in-situ air cathode in chlor-alkali membrane cells. The energy savings, however, will be considerably
less than with the in-situ air cathode membrane cell arrangement
because the voltage losses in two different cells are involved including the voltage losses in the bubble-filled catholyte of the
H_2 producing chlorine-caustic cell (e.g., 0.75 V vs. 1.0 to 1.1 V).

An alternative to the fuel cell is to consume the by-product
hydrogen in a convention steam powered electric plant designed to
provide cogeneration of heat for caustic evaporation. With acid
fuel cells the equivalent thermal efficiencies are comparable
while the life of the steam powered electric plant is far longer.

The air cathode also can be used in place of the H_2 generating
cathode in HCl electrolysis cells producing Cl_2. One possibility
(Fig. 1) would involve a solid polymer membrane with a platinum
catalyzed air cathode pressed into one side of the membrane and
the other side facing into an HCl anolyte with a conventional Cl_2
generating dimensionally stable anode (DSA). The membrane would
be of the Nafion type with the transport number of the proton essentially unity. The thermodynamic and estimated operating potentials
are as follows:

$$Cl_2 | HCl | SPE | H_2 \qquad\qquad \begin{aligned} E(rev) &= 1.2 \text{ V} \\ E(act) &= 1.8 \end{aligned}$$

$$Cl_2 | HCl | SPE | air \qquad\qquad \begin{aligned} E(rev) &= 0.1 \text{ V} \\ E(act) &= 0.6 \end{aligned}$$

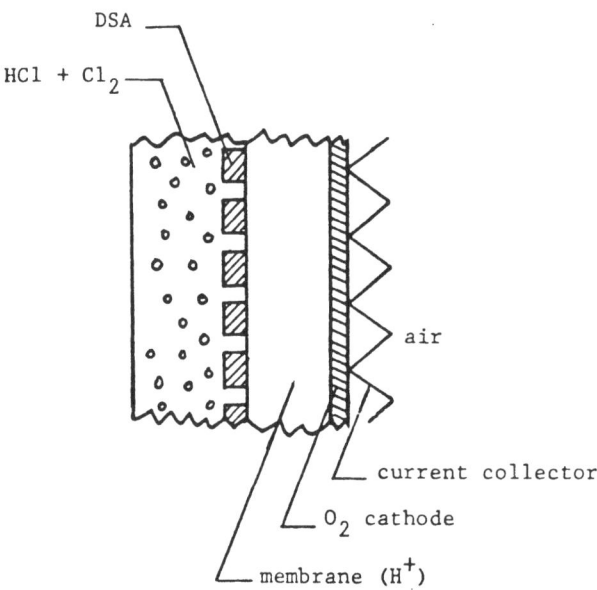

Figure 1. Possible structure for membrane HCl electrolysis cell
with air cathode (not to scale).

A voltage saving of $\geq$ 1.0 V and energy saving of > 60% appears
feasible with the air cathode.

III. OTHER ELECTROLYSIS APPLICATIONS FOR O_2 ELECTRODES

Another promising application for O_2 cathodes is the generation
of hydrogen peroxide. In alkaline electrolytes on graphite and
carbon surfaces of low ash content, O_2 is reduced to the peroxide
according to the reaction

$$O_2 + H_2O + 2\ e^- \longrightarrow HO_2^- + OH^- \tag{3}$$

where O_2H^- is the ionized form of H_2O_2. In the absence of catalysts
promoting the reduction or decomposition of HO_2^-, substantial yields
are obtained of HO_2^- and the concentration can build up to quite
high values (up to 1 $\underline{M}$ HO_2^-). The peroxide producing O_2 consuming
cathode can be incorporated in either of the two types of cells
indicated in Table 5. The first involves O_2 generation as the
complementary anodic process with this O_2 together with air fed to
the cathode. In the second type the anodic reaction involves the
oxidation of H_2 assuming that pure O_2 is available from, for example, a conventional chlor-alkali cell. This cell is capable of
producing some electric output but at relatively modest potentials
($\leq$ 0.5 V). KOH is the preferred electrolyte over NaOH if either
of these cells are to be operated at low temperatures because the
solubility of sodium peroxide in concentrated NaOH is quite limited

Table 5.   Peroxide producing electrochemical cells

A.   Power consuming peroxide producing cell

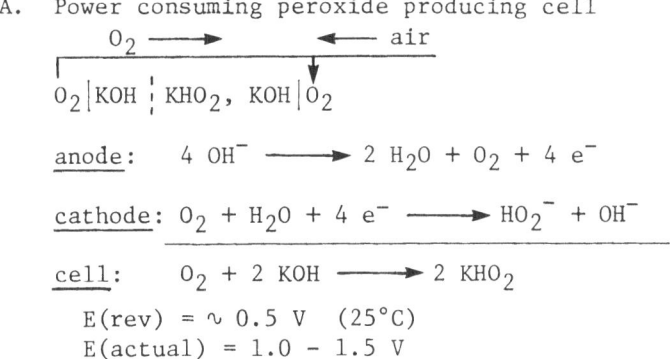

anode: $4\ OH^- \longrightarrow 2\ H_2O + O_2 + 4\ e^-$

cathode: $O_2 + H_2O + 4\ e^- \longrightarrow HO_2^- + OH^-$

cell: $O_2 + 2\ KOH \longrightarrow 2\ KHO_2$

   $E(rev) = \sim 0.5\ V\quad (25°C)$
   $E(actual) = 1.0 - 1.5\ V$

B.   Power producing peroxide producing cell

   $H_2 | KOH \vdots KHO_2,\ KOH | air$

anode: $H_2 + 2\ OH^- \longrightarrow 2\ H_2O + 2\ e^-$

cathode: $O_2 + H_2O + 2\ e^- \longrightarrow HO_2^- + OH^-$

cell: $H_2 + O_2 + KOH \longrightarrow KHO_2 + H_2O$

   $E(rev) = \sim 0.76\ V\quad (25°C)$
   $E(actual) = \sim 0.4\ V$

(e.g., 0.03 $\underline{M}$ in 10 $\underline{M}$ NaOH at 25°C). The solubility of the peroxide in NaOH, however, increases rapidly with temperature and is
no longer a problem at 85°C. Oloman (32) has reported producing
up to 0.78 $\underline{M}$ NaHO$_2$ in 5 $\underline{M}$ NaOH with a cell of type A in Table 5
using a trickle-bed reactor. This type of cell may prove promising
for on-site generation of peroxide for the paper industry.

   The O_2 producing anode and air consuming cathode can also be
used as an O_2 concentrator. Such a cell can use either a phosphoric
acid or alkaline electrolyte. The solid polymer electrolyte of the
type used in the General Electric Gemini fuel cell would be quite
attractive although the cost for the presently used membrane and
catalyst (platinum) may be a deterrent to some applications. Such
an oxygen concentrator can be used to provide O_2 for example for
hospital patients.

III. GENERAL FEATURES OF AIR CATHODES FOR OPERATION IN CONCENTRATED
      CAUSTIC

Oxygen has very low solubility in concentrated caustic; for
example, 2×10^{-5} $\underline{M}$ in 9 $\underline{M}$ NaOH ($\sim$30% by weight) at 85°C. Consequently,
to achieve practical current densities, it has been necessary to use
porous electrodes with the O_2 supplied through the porous structures
to the electrode-electrolyte interfaces within the porous structures.
Two types of approaches have been used: 1) completely hydrophilic
porous electrode in which the contact angles promote wetting and
2) semi-hydrophobic porous structures with part of the electrode
pore structure having non-wetting internal surfaces (See Fig. 2a,b).
The hydrophilic electrodes have usually been constructed with sintered metal particles such as silver. Such electrodes have a fine
pore layer on the side of the electrode towards the bulk electrolyte and a coarse pore layer on the rear gas side. A sufficient
excess O_2 or air pressure is maintained on the rear side to maintain the electrolyte miniscus in the transition region between the
fine and coarse pore layers and yet not to blow air into the electrolyte. This requires very close control of the pore size distribution in the manufacture of the electrodes. Much of the oxygen
reduction reaction appears to take place in the liquid film covering the pores to the rear of the meniscus. High electrode-electrolyte surface areas are required to support the O_2 reduction reaction
at reasonable overall current densities. To achieve such, the surfaces of the pores are usually made microporous, for example through
the use of Raney metal technology.

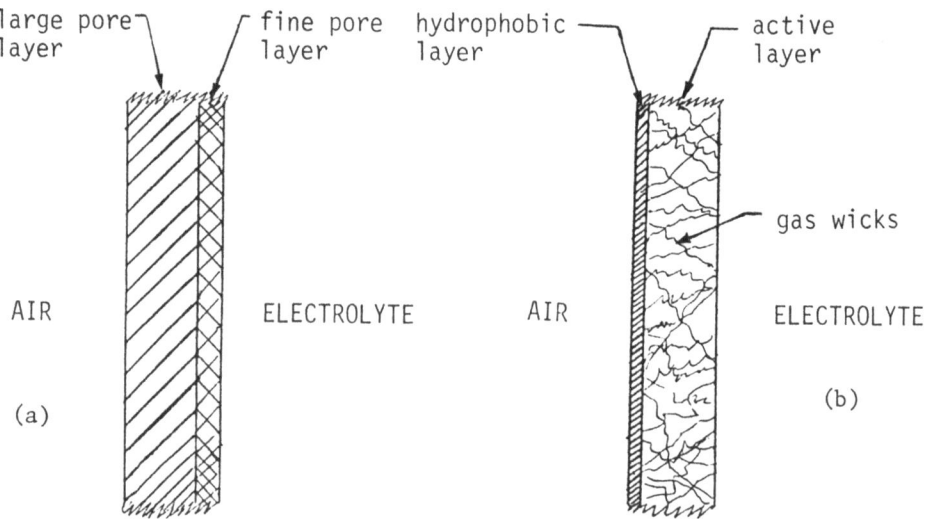

Figure 2a. O_2 cathode structure: hydrophilic type (not scaled).
      2b. Semi-hydrophobic type (not scaled).

The semihydrophobic type electrodes rely on contact angles greater than 90° rather than excess gas pressure to maintain the meniscus in a stable configuration within the porous electrodes. A thin very hydrophobic layer prepared from Teflon or a Teflon-carbon mixture is often used as an air permeable porous backing layer through which electrolyte will normally not penetrate. On the electrolyte side is a wetted very high area layer containing the catalysts consisting of usually carbon plus an additional catalyst such as platinum or silver. Gas-filled capillaries or wicks made up of Teflon particles in chain like clusters or fibrils extend into and through this active layer and expedite transport of the O_2 into the whole active layer through principally Knudsen diffusion. In most instances the hydrophobic backing layer is made with the pore diameters comparable to or smaller than the mean free path in the gas phase so that Knudsen diffusion is also predominant in the layer.

With concentrated NaOH, the solubility of Na_2CO_3 is very low ($<10^{-3}$M in 9 M NaOH at 85°C). It is necessary to remove almost all of the CO_2 from the air supply (even with a Knudsen diffusion backing layer) to prevent carbonate precipitation within the active layer.

Figure 3 indicates the type of semihydrophobic electrode structures used at CWRU for evaluation of high area O_2 catalysts in Teflon bonded porous carbon electrodes. Structural support is provided by the nickel grid structure.

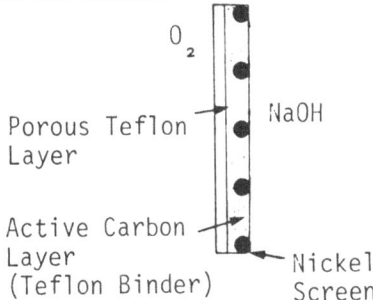

Figure 3. Semihydrophobic electrode structure used at CWRU.

For the in situ chlor-alkali and primary metal-air battery applications the hydrophilic type O_2 cathodes appear less attractive than the semihydrophobic type. Sintered metal electrodes tend to be considerably more expensive to fabricate particularly when the pore size distribution must be carefully controlled and it is difficult to make them very thin without structural support-cracking problems. The gas filled rear layer of the hydrophilic type electrodes is usually considerably thicker than the corresponding backing layer of the semihydrophobic type electrodes. This presents some additional difficulty with the hydrophilic type electrodes in operation on air because of the counter stagnant diffusion layer problem (4,5). The pore radii in this layer are generally greater than 10^{-4} cm and Knudsen diffusion is not involved. As O_2 is consumed from within the rear porous layer, N_2 accumulates and must back diffuse out of the layer, thus impeding the transport of O_2 to the active region of the electrode. Oxygen transport in the gas filled rear layer is then controlled by the diffusion of O_2 through a stagnant inert layer rather than viscous flow. This problem becomes more severe as the rear gas filled layer becomes thicker and leads to excessive polarization in operation on air and a limiting current density.

Most O_2 semihydrophobic carbon type cathodes now available for alkaline electrolytes have been optimized for concentrated KOH with the electrolyte, at the most, only occasionally replaced. The chlor/alkali application involves principally NaOH with this electrolyte continuously generated and removed from the cathode compartment. Concentrated NaOH differs from KOH in the solubility of carbonate and also the solubility of peroxide as discussed earlier. In concentrated NaOH, the addition of hydrogen peroxide can result in the formation of a precipitate or gel. For example with 9 $\underline{M}$ NaOH at 25°C, precipitation occurs first with the addition of sufficient HO_2^- to reach 0.03 $\underline{M}$ and then gel formation with further peroxide addition, probably because of the formation of a hydrated sodium peroxide. This has not been observed with concentrated KOH. Peroxide is formed during O_2 reduction in porous carbon cathodes and can reach concentrations well in excess of 0.03 $\underline{M}$ unless the electrode contains an effective peroxide elimination catalyst or the carbon itself is a good peroxide eliminator. As the peroxide undergoes decomposition, bubbles form within the gel and cause mechanical stresses which disintegrate the electrode. Fortunately the gel formation and peroxide solubility have a large temperature dependence and have not been found to be a problem at 85°C. Care must be exercised, however, in cell start up not to pass large current until higher temperatures are reached.

IV. OXYGEN ELECTROREDUCTION CATALYSTS

Typically O_2 cathodes deviate from the thermodynamic potential by 0.3 to 0.4 V in alkaline electrolytes at operating temperatures of 60° to 80°C, exclusive of obscure losses. The large majority of this polarization is associated with the irreversibility of the overall O_2 reduction reaction:

$$O_2 + 2 H_2O + 4 e^- \longrightarrow 4 OH^- \tag{4}$$

Much effort has been directed to finding effective catalysts for this reaction, principally in conjunction with fuel cells and metal-air batteries. So far, the most effective catalyst has been highly dispersed platinum in carbon supports. Even so, the activation overpotential is still very high by battery standards. The cost is of concern even with platinum loadings below 1 mg/cm^2. Furthermore, it is difficult to maintain the high surface area of the platinum on carbon supports under the operating conditions of the O_2 cathode and the activity declines with time. Under some circumstances platinum may catalyze the oxidation of the carbon support, thus causing detachment of the platinum particles from the support and leading to electrode failure. Thus, the motivation to find alternatives to platinum is great.

A. Pathways for O_2 reduction

The search for more effective O_2 reduction catalysts has been generally guided by the following mechanistic considerations. The O_2 reduction is usually considered to proceed by two types of pathways:

1. The peroxide pathway (the series process):

$$
\begin{array}{c}
HOH + O_2 \xrightarrow{\ 2e^-\ } HO_2^-\text{(ads)} + OH^- \\
\quad 2e^- \searrow \quad \updownarrow \\
\qquad\qquad HO_2^- + OH^-
\end{array}
\tag{5}
$$

The peroxide is then either electroreduced further to OH^- or catalytically decomposed; i.e.,

peroxide reduction

$$
\begin{array}{c}
HOH + HO_2^- \\
\downarrow \uparrow \qquad \searrow\ 2e^- \\
HOH + HO_2^-\text{(ads)} \xrightarrow[\ 2e^-\ ]{} 3\ OH^-
\end{array}
\tag{6}
$$

peroxide catalytic decomposition

$$
\begin{array}{c}
2\ HO_2^- \\
\updownarrow \qquad \searrow \\
2\ HO_2^-\text{(ads)} \longrightarrow 2\ OH^- + O_2
\end{array}
\tag{7}
$$

The overall reaction is the 4-electron reduction

$$
O_2 + 2\ H_2O + 4\ e^- \longrightarrow 4\ OH^-
\tag{8}
$$

regardless of whether the peroxide elimination occurs via reaction 6 or 7 since the O_2 resulting from reaction 7 is recycled through reaction 5.

2. The direct 4-electron pathway:

This pathway involves a series of steps in which O_2 is reduced to OH^- or water without hydrogen peroxide being produced in the solution phase. This does not mean that the reduction process does not involve an adsorbed peroxide intermediate but rather that the reduction does not involve an adsorbed intermediate which leads to peroxide in the solution phase. When both pathways are operating on a given electrode surface, the reduction is referred to as involving parallel mechanisms (see ref. 6). The distinction between these two pathways mechanistically can be quite diffuse since the extent to which an adsorbed peroxide intermediate desorbs or not depends on various impurities in the electrolyte as well as electrode potential and temperature.

Illustrations of catalysts on which the peroxide pathway 1 is clearly predominant include carbon (9,10), graphite (9,10), and gold (42) in alkaline electrolytes, while pathway 2 is predominant on clean platinum (6,28,33) and also certain transition metal macrocyclics such as adsorbed iron tetrasulfonated phthalocyanine (7, 35) and the face-to-face dicobalt porphyrin complexes on graphite (34). In the presence of impurities, pathway 1 can become predominant even on platinum.

B. O_2 reduction in porous carbon electrodes

Most O_2 cathodes in air cell batteries and fuel cells operating at low and moderate temperatures use high area carbon in the active layer together with various high surface area catalysts supported on the carbon; e.g., platinum, silver, various oxides, and partially pyrolyzed transition metal macrocyclic complexes. In alkaline solutions the carbon itself is a very effective catalyst for the reduction of O_2 to peroxide (9-11). The exchange current density for reaction 5 is typically 10^{-4} to 10^{-3} A/cm^2 (true area) for graphite and carbon in 1 $\underline{M}$ KOH + 10^{-3} $\underline{M}$ HO$_2^-$ (9,10). When account is taken of the very high area carbons used in most O_2 cathodes for alkaline electrolytes (e.g., 10^4-10^5 cm^2 of true area per cm^2 of superficial area), this means that there is very little activation polarization with respect to reaction 5 and that the predominant process is the peroxide pathway 1.

Under these circumstances the electrode potential within the electrode responds to the local O_2, OH$^-$, HO$_2^-$ and H$_2$O concentrations through the Nernst equation: i.e.,

$$E = E_o - \frac{RT}{2F} \ln \frac{[HO_2^-][OH^-]}{[O_2][H_2O]} \tag{8}$$

where the [ ] quantities correspond to the activities of the indicated species and the other symbols have their usual meaning. The standard electrode potential E_o for the O_2-HO$_2^-$ couple corresponds to E_o = -0.065 V vs SHE at 25°C (32).

The further reduction or decomposition of peroxide on carbon and graphite is very slow compared to reaction 5. While some carbons and graphites, particularly with a high ash content, may exhibit moderately high rates for the peroxide elimination step initially, their catalytic activity for peroxide elimination may drop off even after just a few days to relatively low values. This can result in the build up of substantial peroxide in the electrolyte within the porous electrode; for example, to concentrations in the 10^{-1} $\underline{M}$ range when operating at high current densities (>100 mA/cm^2). This has an adverse effect on the electrode potential as can be seen from eq. 8. Furthermore, the high peroxide concentration can lead to high $O_2^{\cdot-}$ radical concentration within the electrode because of the reaction

$$HO_2^- + O_2 + OH^- \rightleftharpoons 2 \ O_2^{\cdot-} + H_2O \qquad (9)$$

The equilibrium constant for this reaction is estimated to be $\sim 10^{-8}$ (33) at 25°C. In air saturated 10 $\underline{M}$ NaOH containing 10^{-1} $\underline{M}$ HO_2^-, the equilibrium concentration of $O_2^{\cdot-}$ would be $\sim 10^{-4}$ $\underline{M}$ or almost an order of magnitude higher than the concentration of O_2. The $O_2^{\cdot-}$ radical ion and also such species as OH· radicals produced as intermediates during the homogeneous peroxide decomposition may attack the carbon surfaces and accelerate oxidation of the carbon, modifying the hydrophobic properties and producing premature electrode failure. As indicated earlier with NaOH, a sodium peroxide gel can form in the pores of the carbon electrode at high current densities, particularly at lower temperatures.

Suppression of the peroxide concentration in porous carbon electrodes in alkaline electrolytes is usually accomplished through the incorporation of a separate catalyst for this purpose. In alkaline H_2-air fuel cells and metal-air cells catalysts which have been used include silver, manganese dioxide and the nickel-cobalt spinels on high area carbon supports. While these catalysts may be capable of catalyzing the O_2 reduction to peroxide and in some instances, the overall 4-electron reduction, this is usually not their principal function in the porous carbon electrode. The O_2 reduction to a peroxide is sufficiently fast on the high area carbon in the electrode structure that this process tends to be predominant in alkaline electrolytes.

With very active catalysts for the 4-electron reduction such as high area platinum and some transition metal macrocyclics supported on carbon, the 4-electron and peroxide producing processes occur in parallel on the catalyst and carbon surface, respectively, in gas fed carbon type electrodes. Platinum is also reasonably effective as a peroxide elimination catalyst but it may still be desirable in platinum catalyzed O_2 cathodes to augment the peroxide elimination by using a second specific catalyst for peroxide elimination (e.g., a spinel).

In principle, if the peroxide elimination catalyst is sufficiently active and well dispersed within the porous electrode structure, it would be possible to depress the HO_2^- activity to the equilibrium value for reaction 7 and the electrode potential would approach
the thermodynamic value for the overall 4-electron reduction (reaction 4). In practice this is not feasible if the peroxide must be
transported through the electrolyte to the peroxide elimination cata-
lyst. The equilibrium peroxide concentration (10^{-15} $\underline{M}$ in 10 $\underline{M}$ KOH
at 25°C) is too low for effective transport over even a few Angstroms
at any reasonably overall current density. The only possible means
by which such low values might be achieved in high area carbon electrodes is to have some type of spillover effect where the surface
precursor to the solution-phase peroxide is transferred by surface
migration to adjacent or near adjacent surface catalyst sites. This
requires an atomic level or near atomic level of dispersion of the
peroxide elimination catalyst on the carbon support. Such levels of
dispersion are approached with platinum clusters on carbon but so
far there is little evidence for such a spillover mechanism, perhaps
because platinum is not sufficiently active as a peroxide elimination
catalyst.

Figure 4 compares the short-term polarization curve for an O_2
fed cathode of the type shown in Fig. 3 with and without the spinel
$NiCo_2O_4$ supported on a high area carbon (RB carbon, Pittsburgh Carbon Co.) in 9.2 $\underline{M}$ NaOH at 85°C. The polarization is decreased substantially by the inclusion of this peroxide elimination catalyst
but not by an unusually large amount. The deviation from Tafel
linearity at higher current densities is probably caused by O_2 mass
transport limitations in the active layer. The hysteresis shown
for the electrode without the spinel is typical of that encountered

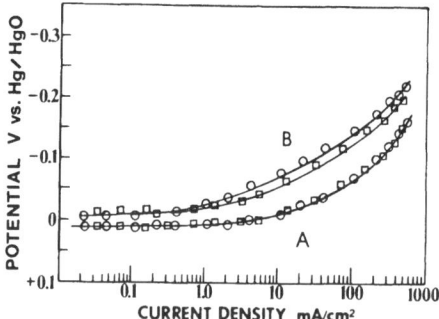

Figure 4. Polarization curves for O_2 reduction with CWRU test gas
            fed O_2 cathode with 20% $NiCo_2O_4$/RB carbon catalysis +
            RB carbon + 30% Teflon. $NiCo_2O_4$/RB catalyst prepared by
            freeze-dry method. (15% $NiCo_2O_4$ + 85% RB). Electrolyte:
            9.2 $\underline{M}$ NaOH; Temp: 85°C; Pure O_2, 1 atm. Curve A with
            catalyst; curve B without. O increasing current
            □ decreasing current.

with such electrodes in the early stages of their use.

Figure 5 indicates the performance of an electrode with a Pt-Mo alloy in highly dispersed form as the catalyst on a lower area carbon, Vulcan XC-72 (200 m^2/g). This electrode on pure O_2 shows quite low polarization at high current densities. The deviation from Tafel linearity at high current densities on air is caused by O_2 transport limitations. Even so, the polarization characteristics on air for this electrode meet the short term performance requirements for the chlor-alkali membrane cell application (i.e., 300 mA/cm^2 at 0.80 V vs RHE or -0.13 V vs Hg/HgO).

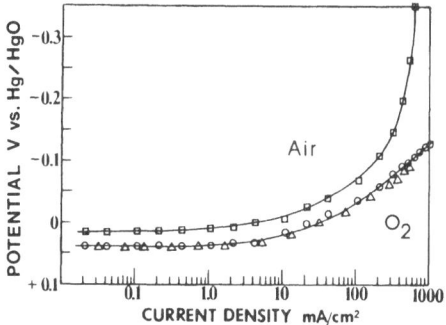

Figure 5. Polarization curves for nickel screen gas-fed test O_2 cathodes using CWRU Pt-Mo catalyzed XC-72 carbon plus 25% Teflon. Loading: Pt - 0.58 mg/cm^2; Mo - 0.28 mg/cm^2. Electrolyte: 9 $\underline{M}$ NaOH. Temp: 85°C. O forward; Δ back.

Some of the catalysts which have been examined in alkaline electrolytes in our laboratory are listed in Table 6. So far, the highest catalytic activities at reasonable loadings in gas fed carbon type electrodes have been obtained with platinum, platinum alloys and several of the transition metal macrocyclics. Several of the Pt alloy catalysts including most of those listed in Table 6 have activity for O_2 reduction comparable or even slightly higher than pure Pt per unit catalyst surface area and hence reduce the weight of Pt required. The principal interest in such alloy catalysts, however, is the possibility for longer catalyst life because of greater stability, particularly with some of the intermetallics.

Gold also has quite high activity for O_2 reduction to peroxide and in high area form has been used successfully in high performance O_2 cathodes involving gold or gold alloy-Teflon sinters. Its performance at low loadings on carbon supports, however, is not as effective as platinum.

Table 6. O_2 reduction electrocatalysts investigated at Case Western
           Reserve University in alkaline electrolytes

Metals

    Pure: e.g., Pt, Pd, Au, Ag

    Alloys: e.g., Pt with Zr, V, Mo, Pd, Co, Ag, Cu

    Underpotential deposited layers: metal element species
        adsorbed on Pt, Au, Ag, including Pb, Tl, Pd, Cu, Cd, Bi

Oxides

    Spinels: nickel-cobalt

    Perovskites: lanthanum-cobalt, strontium

    Others (mixed oxides of manganese, nickel, silver;
        passivation layers; pyrochlores)

Intercalation compounds: graphite

Transition metal complexes

    Phthalocyanines: monomeric, polymeric

    Porphyrins

    Tetraazannulenes

    Naphthalocyanines

    Others, including dimetal complexes

Carbons and graphites

    Pure

    Doped

    Surface treated

    Silver has also been used successfully on carbon supports in
alkaline electrolytes but much higher loadings (weight per superficial area of electrode) are required to obtain reasonable performance. Life at high current densities is usually rather limited
(less than a few thousand hours).

    The polarization curves obtained with some transition metal
macrocyclic complex catalysts are quite exciting in short term tests
and will be discussed in the next section of this lecture.

    C. <u>Prospects for major improvements in the operating potentials</u>
<u>of O_2 electrodes</u>

    The achievement of a quantum jump in the potential performance
of O_2 cathodes in aqueous media requires the development of more

effective catalysts for the overall 4-electron reduction. While
this process is catalyzed by Pt and some other metal surfaces under
very clean conditions, it does not appear likely that any pure
metals will be found that have sufficient activity even in high area
forms to approach the reversible value (1.2 V vs RHE at 25°C) under
practical operating conditions at temperatures below 100°C. Transition metal oxides and macrocyclic complexes appear more likely
candidates, provided adequate stability can be achieved. Underpotential deposited (UPD) foreign metal layers on noble metal substrates
may also be interesting possibilities (12-15,36). In highly dispersed forms, such UPD/substrate systems resemble the binary dispersed catalysts which have been developed for the hydrocarbon
industry, particularly by the Exxon Corporation.

The search for effective 4-electron catalysts is guided by the
models in Figure 6 for the interaction of O_2 and related oxygen species with adsorption sites. O_2 reduction in aqueous solutions requires
a strong interaction with the electrode surface for the reaction to
proceed at a reasonable rate. Three types of interactions have been
proposed. The Griffiths model (31) involves a lateral interaction
of the π-orbitals of the O_2 interacting with empty d_z^2 orbitals of
a transition element, ion or metal atom with back bonding from at
least partially filled d_{xz} or d_{yz} orbitals of the transition element
to the π^*-orbitals of the O_2. A strong metal-to-oxygen interaction
results in a weakening of the O-O bond and an increment in the length
of this bond (16). Sufficiently strong interaction of this type
may lead to the dissociative adsorption of O_2 with probably simultaneous proton addition and valency change of the transition element
in the manner represented by Pathway I in Fig. 6, followed by reduction of the $M(OH)_2$ to regenerate the catalyst site. Sandstede et al.
(17) have attempted to explain oxygen reduction with square pyramidal
Co(II), Fe(II) and Fe(III) complexes as well as on the thio-spinels
on the basis of such bonding. Tseung, Hobbs, and Tantram (18) have
proposed that O_2 reduction on Li-doped NiO changes from a non-disso-
ciative to dissociative mechanism above the Neel point (200°C for
their ~10-atom % Li doped NiO) in order to explain the increment in
catalytic activity in KOH hydrate melts above the Neel temperature.

With most transition metal catalysts, the most probable structure of O_2 adsorption is the Pauling model (19) in which π^*-orbitals
of O_2 interact with d_z^2 orbitals of the transition metal. The square
pyramidal complexes of Fe(II) and Co(II), which have good activity
for O_2 reduction in acid and alkaline solutions, appear to involve
such end-on interaction on the basis of esr and other evidence (20).
This adsorption of O_2 is expected to be accompanied by at least a
partial charge transfer to yield a superoxide and then peroxide
state, as represented by Pathway II in Fig. 6. The adsorption of
the O_2 on the square pyramidal complexes of Fe(II) and Co(II) may
lead directly to the superoxide state. The change in valency state
of the transition metal coupled with the change in O_2 oxidation

state during formation of the O_2 adduct corresponds in principal to the redox electrocatalyst concept proposed by Beck et al. (22,23).

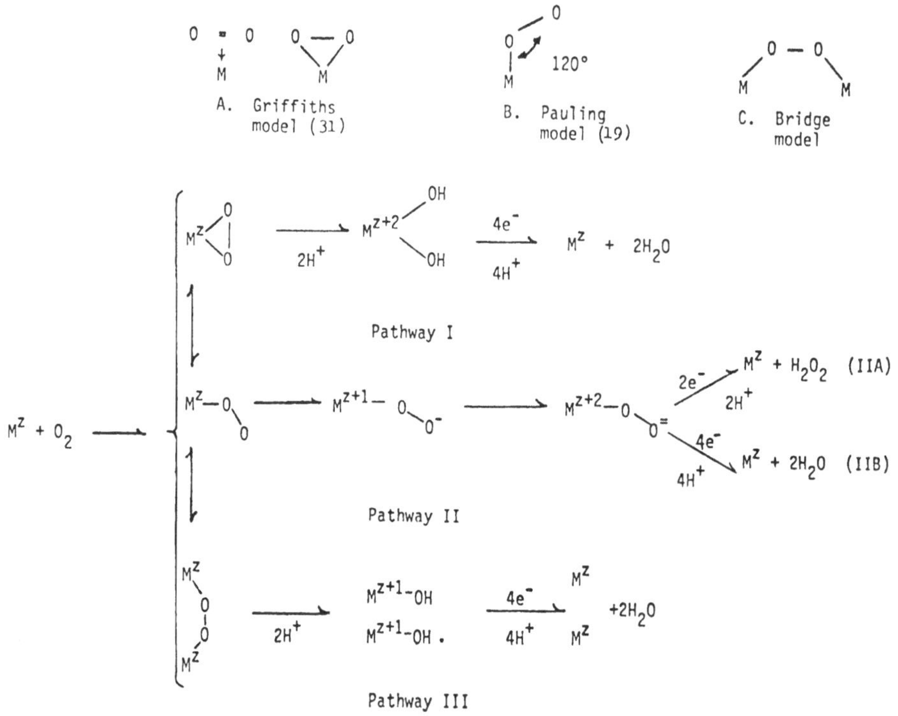

Figure 6. Reaction pathways for O_2 electroreduction in acid electrolytes [Yeager (21)].

The further reduction of the O_2 beyond the peroxide state requires rupture of the O-O bond. Such can occur in Pathway IIB
through the formation of O^- or HO free radicals in solution or the
simultaneous reduction-bond cleavage (electrochemical desorption)
to yield H_2O or OH^-. Neither of these processes is likely to be
sufficiently fast at practical operating potentials for O_2 cathodes
but the electrochemical desorption is a better candidate.

Pathway III in Fig. 6 provides an alternate means for bringing
about rupture of the O-O bond through the formation of an O-O bridge.
Such a mechanism may come into play with the proper surface spacing
of transition metal atoms or ions in a metal, oxide or thiospinel
or in a bimetal complex such as a macrocycle. The formation of the
bridge species also requires that the two metal species have partially filled d orbitals to participate in bonding with the π*-
orbitals of the O_2. Bimetal macrocyclic complexes with the proper
M-M distance have been synthesized [e.g., see ref.(24-27)] and
appear to occur naturally in hemeythrin. For O_2 to bond to M will

generally require the replacement of a water molecule or anion of the electrolyte--a situation which would normally be expected to be unfavorable to O_2 unless the O_2 adduct has a pronounced dipolar character ($M^{z+1}O-O^-$) (29,30).

Of the various possibilities for catalysts which promote the 4-electron reduction, the transition metal macrocyclics appear particularly promising. In our laboratory these are adsorptively attached to high area conducting substrates as monolayers. The water soluble iron tetrasulfonated phthalocyanine (Fe-TSPc)(Fig. 7) strongly adsorbs on graphite and at monolayer levels has high activity for the overall 4-electron reduction of O_2 in neutral and dilute alkaline electrolytes. The catalytic activity per unit surface area is higher than for Pt. Rotating disk-ring electrode measurements indicate no detectable peroxide over a substantial potential range (35). The principal problem is the stability of the iron macrocyclic in concentrated caustic and acid solutions. In contrast to the Fe-TSPc, the adsorbed Co-TSPc catalyzes the reduction of O_2 to the peroxide in alkaline solutions (35,37).

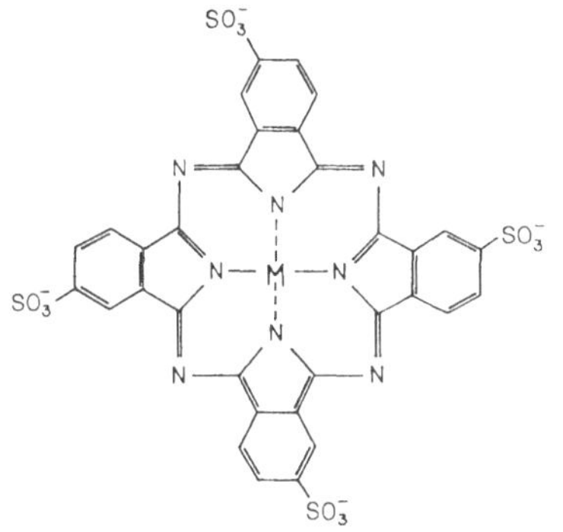

Figure 7. Transition metal tetrasulfonated phthalocyanine (M-TSPc)

The interaction of adsorbed Co-TSPc and Fe-TSPc with the substrate electrode surface (graphite, Pt, and Au) has been studied using visible reflectance spectroscopy (38). The reflectance spectra of the adsorbed Fe-TSPc and Co-TSPc monolayers undergo substantial changes at constant potential with the introduction of O_2 into the electrolyte. This is believed to be caused by the formation of an O_2 adduct with the adsorbed species. The Raman spectrum of the Co-TSPc has also been obtained with this complex adsorbed on silver, which provides a large surface enhancement of the Raman signal (39). Further work is needed to interpret the Raman spectra and their potential dependence. On the basis of the strong adsorption and the Raman data, we believe that the Co-TSPc is adsorbed on the silver predominantly as shown in Fig. 8 with one of the Co-TSPc units in the O_2 bridge dimer interacting directly with the metal substrate. We do not yet have Raman data for the Fe-TSPc but suspect that the

adsorption may be in a similar configuration. Dimeric -O-O- complexes
have been proposed in aqueous solutions for both the Fe-TSPc and
Co-TSPc, principally on the basis of the UV-visible absorption spectra. The -O-O- bridged complex would be a likely candidate in the
case of the adsorbed Fe-TSPc to explain the 4-electron reduction.
The Co-TSPc apparently also forms such a bridge but because of its
redox properties or some other factor, does not undergo reduction
via a 4-electron pathway (37).

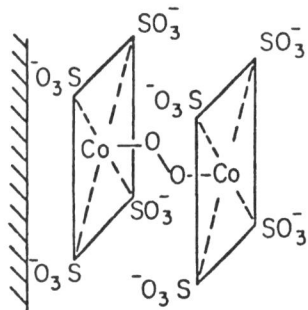

Figure 8.   Possible configuration for CoTSPc adsorbed on an elec-
            trode surface [Kötz and Yeager (39)].

Collman, Anson, and their co-workers (34) have recently synthesized covalently linked face-to-face di-cobalt porphyrin dimers
with the proper spacing to form a Co-O-O-Co bridge. With relatively
thick layers of this complex on graphite, rotating disk-ring electrode measurements also indicate a 4-electron reduction in acid
electrolytes, verifying that the bridged complex can promote the
overall 4-electron reduction.

On the other hand, recent studies by J. Molla (40) at CWRU
using the rotating ring-disk technique have shown O_2 reduction on
cobalt tetramethoxy phenyl porphyrin (the monomer) adsorbed on carbon also to proceed by the 4-electron pathway in 1 $\underline{M}$ NaOH. This
implies that the formation of bridge species such as shown in Fig. 8
are not specific to the CoTSPc and FeTSPc complexes or to the face-
to-face porphyrins, but can also occur spontaneously with the mono-
meric porphyrins.

A large number of macrocyclic complexes of the first, second,
and third row transition metals have been examined in gas fed electrodes of the type shown in Fig. 3 in the author's laboratory. Of
these, electrodes using the cobalt tetramethoxy phenyl porphyrin
exhibit the lowest polarization at high current densities - somewhat
lower than even with highly dispersed platinum at reasonable loadings
at high current densities. A typical polarization curve with this
catalyst in 9 $\underline{M}$ NaOH at 85°C is shown in Fig. 9.

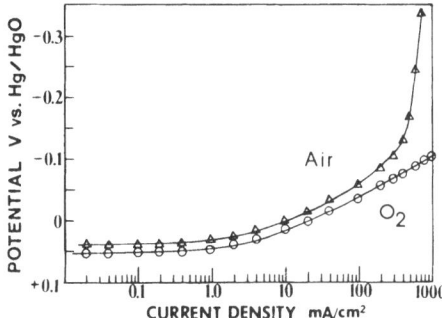

Figure 9.   Polarization curves obtained with CWRU nickel screen
            type gas-fed O_2 test cathode using cobalt tetramethoxy
            phenyl porphyrin catalyzed RB carbon with 30% Teflon.
            CoTMPP loading: 15% by weight of active layer. Elec-
            trolyte: 9 $\underline{M}$ NaOH. Temp: 85°C. Total pressure: 1 atm.

Another interesting development is the work of Mrs. S. Sarangapani (41) at CWRU on O_2 reduction on the bicobalt planar complex
shown in Fig. 10 adsorbed on pyrolytic graphite. Rotating disk-
ring electrode measurements in alkaline solution have shown the O_2
reduction to proceed by the 4-electron pathway on this adsorbed
complex. A likely prospect again is a bridged -O-O- adduct as the
precursor to the bond breaking step.

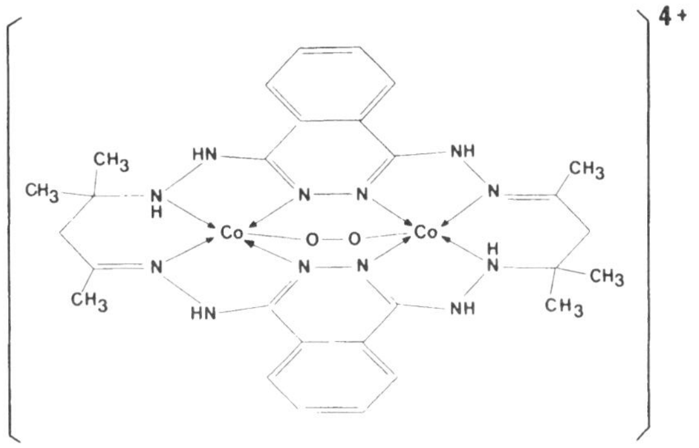

Figure 10.   Dimetal complex $[Co_2TAPH]^{4+}(NO_3)_4$. [TAPH = 6,7,8,9,
            12,19,20,21,22,25 decahydro-8,8,10,21,21,23-hexamethyl,
            -5,26:13,18-bis (azo) dibenz [1,2,6,7,12,13,17,18] ox-
            raaza cyclodocosine].

While these results with transition metal complexes are encouraging fundamental developments in O_2 electrocatalysis, it is difficult to translate them into practical electrodes, principally because of catalyst stability problems. The cobalt tetramethoxy phenyl
porphyrin catalyzed gas-fed carbon electrodes have shown high performance for up to 500 hr at high current densities ($\sim$300 mA/cm^2)
under the conditions indicated in the legend of Fig. 9. Thereafter
the activity rather rapidly declines. While the failure mechanism

has not been fully identified, in general it is a stability problem.
In an attempt to achieve better stability while still retaining
reasonable catalystic activity, various research workers have partially pyrolyzed transition metal macrocyclics on carbon supports.
Just what happens during the pyrolysis is not fully clear but depends
on the temperature. The macrocyclics are thermally quite stable
materials. Operating times of 5000 to 10,000 hours have been achieved
with gas-fed O_2 cathodes using pyrolyzed cobalt porphyrins and
phthalocyanines at the Institute of Electrochemistry in Moscow, but
the polarization characteristics are not as low as in Fig. 8. The
pyrolysis temperatures used by this group were very high (up to 950°C).
Work in the author's laboratory at CWRU has shown that the porphyrin
structure is not retained and that much of the cobalt is converted
to the metal. Very complex changes can occur also in the carbon
support structure because of the catalytic effect of the cobalt on
the carbon transformations at these high pyrolysis temperatures.
Perhaps some of the cobalt still is coordinated to nitrogen and this
influences the catalytic activity and stability.

In any event interest in the macrocyclics continues very high.

Acknowledgment: The author is pleased to acknowledge support of the
research at CWRU described in this paper by the U.S. Office of Naval
Research and the U.S. Department of Energy.

References

1. D.W. Carnell and C.R.S. Needs, "Energy-Saving Catalytically
   Active Cathodes for Caustic-Chlorine Production," Paper No. 260,
   National Meeting, The Electrochemical Society, Boston, May 6-12,
   1979. Extended Abstracts 79-1, pp. 671-2.

2. W.W. Carlin and W.B. Darlington, "Activated Cathodes for Reduced
   Power Consumption in Electrolyte Cells," Paper No. 261, loc. cit.
   Extended Abstracts 79-1, pp. 673-5.

3. I. Malkin and J.R. Brannan, "Reduced Hydrogen Overpotential in
   a Chlorine Cell," Paper No. 262, loc. cit. Extended Abstracts
   79-1, pp. 676-7.

4. E. Yeager, "The Sodium Amalgam-Oxygen Cell," in Fuel Cells, W.
   Mitchell, ed., Chapt. 7, Academic Press, New York, 1963, pp.
   300-328.

5. M. Eisenberg, in Advances in Electrochemistry and Electrochemical
   Engineering, C. Tobias, ed., Vol. 2, Interscience, New York, 1962.

6. A. Damjanovic, M.A. Genshaw and J. Bockris, J. Electrochem. Soc.,
   114, 466 (1967).

7. E. Yeager, J. Zagal, B. Nikolic and R. Adzic, "Optical and Electrochemical Studies of Adsorbed Transition Metal Complexes and
   Their O_2 Electrocatalytic Properties," Paper No. 344, National

Meeting, The Electrochemical Society, Boston, May 6-12, 1979. Proceedings of the Third Symposium on Electrode Processes, The Electrochemical Society, Princeton, NJ, 1980, pp. 436-456.

8. J. Collman, M. Marrocco, P. Denisevich, C. Koval and F.C. Anson, J. Electroanal. Chem., 101, 117 (1979).

9. E. Yeager, C. Krouse and K. Rao, Electrochim. Acta, 9, 1057 (1964).

10. I. Morcos and E. Yeager, Electrochim. Acta, 15, 953 (1970).

11. J. Appleby and J. Marie, Electrochim. Acta, 24, 195 (1979).

12. A. Khutornoi, P. Bindra, R. Amadelli and E. Yeager, "Oxygen Reduction on Underpotential Deposited Metal Monolayers in NaOH," Paper No. 29, National Meeting, The Electrochemical Society, Boston, May 6-12, 1979. Extended Abstracts 79-1, pp. 79-81.

13. J.D.E. McIntyre, S. Gottesfeld and W.F. Peck, "Electrochemical Catalysis by Foreign Metal Adatoms," Paper No. 339, loc. cit. Extended Abstracts 79-1, pp. 864-5.

14. R.R. Adzic and A.R. Despic, J. Phys. Chem. N.F., 98, 95 (1975).

15. R.R. Adzic, A.V. Triphkovic and R.T. Atanasoski, J. Electroanal. Chem., 94, 231 (1978).

16. J. McGinnety, N. Payne and I. Ibers, J. Am. Chem. Soc., 91, 6301 (1969).

17. H. Behret, H. Binder, and G. Sandstede, in Proc. of the Symposium on Electrocatalysis, M. Breiter, ed., The Electrochemical Society, Princeton, NJ, 1974, pp. 319-338; H. Behret, H. Binder, and G. Sandstede, Electrochim. Acta, 20, 111 (1975).

18. A. Tseung, B. Hobbs, and A. Tantram, ibid., 15, 473 (1970).

19. L. Pauling, Nature, 203, 182 (1964).

20. See e.g., B. Hoffman, D. Diemente and F. Basolo, J. Am. Chem. Soc., 92, 61 (1970).

21. E. Yeager, "Mechanism of Electrochemical Reactions on Non-Metallic Surfaces," in Electrocatalysis on Non-Metallic Surfaces, NBS Special Publication 455, 1976, pp. 203-219.

22. F. Beck, W. Dammert, J. Heiss, H. Hiller and R. Polster, Z. Naturforsch. 29A, 1009 (1973).

23. F. Beck, Ber. Bunsenges, Physik. Chem., 77, 353 (1973).

24. E.I. Ochiai, Inorg. Nucl. Chem. Letters, 10, 453 (1974).

25. T.G. Traylor and C.K. Chang, J. Am.Chem. Soc. 95, 5810 (1973).

26. M.J. Bennett and P.B. Donaldson, J. Am. Chem. Soc., 93, 3307 (1971).

27. W.P. Schaefer, Inorg. Chem. 7, 725 (1968).

28. J. Huang, R.K. Sen and E. Yeager, J. Electrochem. Soc., 126, 786 (1979).

29. H.C. Stynes and J.A. Ibers, J.Am. Chem. Soc., 94, 5125 (1972).

30. W. Brimgar, C. Chang, J. Gerbel and T. Traylor, ibid., 96, 5597 (1974).

31. J.S. Griffiths, Proc. Roy. Soc. (A) 235, 73 (1956).

32. C. Oloman, "Trickle Bed Electrochemical Reactors," National Meeting, The Electrochemical Society, Seattle, Wash., May 1978. Extended Abstracts 78-1, No. 469.

33. M. Tarasevich, A. Sadkowski and E. Yeager, "Oxygen Electrochemistry," in Comprehensive Treatise of Electrochemistry, Electrodics: Kinetics, Vol 6, B.E. Conway, J.O'M. Bockris, E. Yeager and R. White, eds, Plenum Press, in press.

34. J.P. Collman, P. Denisevich, Y. Konai, M. Marrocco, C. Koval and F.C. Anson, J.Am. Chem. Soc. 102, 6027 (1980).

35. J. Zagal, P. Bindra, and E. Yeager, J. Electrochem. Soc. 127, 1506 (1980).

36. R. Amadelli, J. Molla, P. Bindra, and E. Yeager, J. Electrochem. Soc., submitted.

37. J. Zagal, R. Sen and E. Yeager, Inorg. Chem. 16, 3379 (1977).

38. B. Nikolic, R. Adzic and E. Yeager, J. Electroanal. Chem., 103, 281 (1979).

39. R. Kötz and E. Yeager, ibid., 113, 113 (1980).

40. J. Molla and E. Yeager, to be submitted for publication.

41. S. Sarangapani, F.Urbach and E. Yeager, to be submitted for publication.

42. R. Zurilla, R. Sen and E. Yeager, J. Electrochem. Soc. 125, 1103 (1978).

DISCUSSION

Prof. W. Vielstich, Bonn University: Raney metal catalysts in the non-sintered form are used in the Siemens H_2/O_2 fuel cell, the fine pore layer being an asbestos foil (so-called "supported electrodes"). They have shown a remarkable reliability and reproducibility in batteries up to 20 kw. At the operating current density of 400 mA/cm^2 (6 M KOH, 80°C) the voltage of the single cells stays at 730 ± 10 mV. On the other hand, the cost of these electrodes is higher than that of the Teflon bonded carbon type electrodes.

E. Yeager: Such electrodes could probably be adapted to the chloralkali cell application described in my lecture.

Prof. W. Vielstich: Oxygen electrodes for use in both modes (reduction and evolution) should avoid carbon or graphite in the oxygen
evolution region. Good performances have been obtained with separated
catalysts for each mode. In our institute at 50 mA/cm^2 more than
2000 cycles have been obtained with plastic bonded metal electrodes.

E. Yeager: I agree with your comment concerning the desirability of
avoiding carbon in bifunctional O_2 electrodes. On the other hand,
industrial groups in the U.S. and Germany have obtained reasonable cycle life with bifunctional electrodes in which relatively stable carbons
are used, for iron-air secondary batteries. These electrodes have
multilayer structures which tend to minimize the O_2 evolution reaction on the high area carbon in the electrode during charging.

Prof. F. Goodridge, University of Newcastle: Speaking as a user of
air electrodes, what are the chances of obtaining a high performance
reversible O_2 electrode?

E. Yeager: Near reversible performance is obtained for O_2 cathodes
in high temperature fuel cells such as the molten carbonate H_2-air
fuel cell and even in the Apollo H_2-O_2 fuel cell with a molten KOH
hydrate melt at $\sim$200°C. Unfortunately the O_2 cathode suffers from
large polarization in acid electrolytes even at 200°C and in alkaline
electrolytes at lower temperatures. We are still lacking a sufficiently active catalyst for near reversible behavior at reasonable
current densities under these conditions. At very low current densities, some of the perovskites (e.g., strontium lanthanum cobaltic
oxide) and the pyrochlores (e.g., lead ruthenate) show near reversible behavior in alkaline electrolytes. Perhaps with further effort,
the current densities can be increased. I have "guarded optimism".

Dr. Sankar Das Gupta, HSA Reactors, Ltd., Toronto:
1. Can the macrocyclic compounds be used as anodes for oxygen pro-
    duction?
2. If they could be used as anodes, what would be the lifetime of
    such electrodes in the severe oxidative environment that exists
    for oxygen evolving anodes?
3. Would pyrolyzed macrocyclics be more stable for oxygen evolution
    reactions?

E. Yeager: In our laboratory, we have not had any experience with
the transition metal macrocyclics as catalysts for the anodic generation of O_2. I understand from Dr. M. Savy of the CNRS Laboratories
on Interfacial Chemistry in Bellevue, France, that Mo phthalocyanine
has reasonable activity for O_2 generation and some stability. Over
extended periods I would expect that the macrocyclic ligand would
undergo oxidation and degradation. Further, if carbon is the support,
it could be attached as pointed out by Prof. Vielstich.

I believe it would be worthwhile to try the pyrolyzed macro-cyclics. If the pyrolysis is at high temperatures as mentioned in my lecture, it may well be that the catalytic agent is no longer the macrocyclic but rather the transition metal which would be in the form of an oxide at the potentials involved in O_2 evolution.

PROSPECTS AND PROBLEMS IN ELECTROCHEMICAL ENGINEERING [*]

Charles W. Tobias
Materials and Molecular Research Division
Lawrence Berkeley Laboratory and Department of Chemical
  Engineering
University of California
Berkeley, California 94720

ABSTRACT

Developments in electrochemical engineering science over the
past 20-30 years have led to introduction of quantitative techniques for scale-up, design, and optimization of cell-processes.
At the interface between the complex and heterogeneous field of
electrochemistry, and chemical engineering, the electrochemical
engineer faces a lack of basic information on phenomena occurring
at electrode surfaces. This is especially evident in the case of
simultaneous and consecutive component processes which involve
changes in phase at electrodes: gas evolution, and electrocrystallization from the electrolyte or from another solid phase. New
efforts are needed in the direction of characterizing processes,
in the quasi steady state, which continue over long periods of
time. Direct physical observation and interpretation of key
phenomena, such as crystal growth, bubble dynamics, change in state
of aggregation of catalysts, and, in general, the characterization
of changing surface properties in situations corresponding to industrial process conditions, are needed if we are to further improve
material- and energy-efficiency, and to reduce capital and labor
costs.

---

[*] Full manuscript not available

A CHEMICAL ENGINEER'S PERSPECTIVE OF CELL DESIGN AND SCALE UP

Francis Goodridge

University of Newcastle upon Tyne
Department of Chemical Engineering
Newcastle upon Tyne, NE1 7RU, England

INTRODUCTION

May I start by expressing my appreciation to be asked to take part in this centennial celebration. In this context I thought it appropriate to look back at some of the cell designs of the recent past, consider briefly some scale up problems and to finish up by trying to identify worthwhile areas of activity in the near future. I must emphasize that examples of cell design have been chosen not because they are the most successful or important but because either I have been personally connected with their development or have been involved in research on them.

It is fair to say that at the turn of the century electrolytic cell design was, with the exception of relatively isolated processes such as barrel plating, synonymous with plates dipping into a tank. It is my thesis that the use of a chemical engineering approach to the design of cells has led to a situation where electrolytic reactors can now be competitive with catalytic ones. To be able to assess the performance of reactors, quantitative criteria must be used. Three possible ones are the space time yield, Y_{ST}, the chemical yield, Y_C, and the energy yield, Y_E, as defined by:

$$Y_{ST} = \frac{\text{Amount of product obtained}}{\text{Unit time x Unit cell volume}}$$

$$= a \times i \times Q_E \times C_E \tag{1}$$

$$Y_C = \frac{\text{Actual amount of product obtained}}{\text{Maximum amount obtainable for a given conversion}} \tag{2}$$

61

$$Y_E = \frac{\text{Amount of product obtained}}{\text{Energy consumed}} = \frac{Q_E \times C_E}{V} \tag{3}$$

Of these it has been the space time yield of electrolytic cells that has in the past been inferior to that of liquid phase catalytic reactors. How some design features affect these three parameters is shown in Table 1.

TABLE 1

REQUIRED FEATURES	PARAMETER AFFECTED		
(a) High electrode area per unit cell volume		Y_{ST}	
(b) Uniform electrode potential	Y_C	Y_{ST}	Y_E
(c) Low internal ohmic resistance		Y_E	
(d) Good mass and heat transfer	Y_C	Y_{ST}	

Let me now discuss four cell designs to illustrate my point of a chemical engineering approach.

SOME RECENT CELL DESIGNS

The four examples I am going to give are the fixed bed Nalco cell (1) which produces tetraalkyl lead in large tonnages, the fluidised bed cell (2) which was first developed in our laboratories in the seventies and has now reached industrial application, the falling bed (3) and the sieve plate cell (4) both in an early state of development.

The requirements for the Nalco cell were the ability to cope with the gradual use of the anode in a continuous manner as well as points (a) and (d) in Table 1. The solution as shown in Fig. 1 is an anode in the shape of a packed bed of lead spheres with steel cathodes covered in a diaphragm dipping into the bed. This is a greatly simplified version of the original cell design (5) but is still a courageous and brilliant solution to a difficult problem using what is termed a three-dimensional electrode. It is not difficult to model potential and current distribution for such an electrode (6) but the indications are that in the present case the reaction is mass transfer controlled and hence feature (b) is of no impact under these conditions.

The second design, again based on a well known chemical engineering unit operation, is the fluidised bed cell which in the application discussed here is used for the removal of metal from dilute solutions. Required features are again (a) and (d) together with

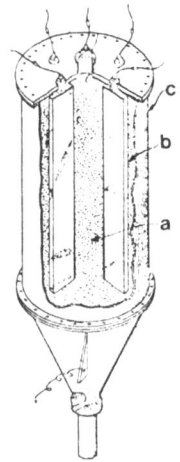
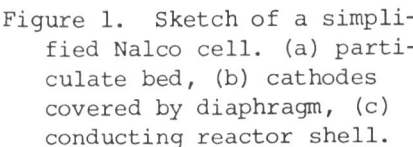

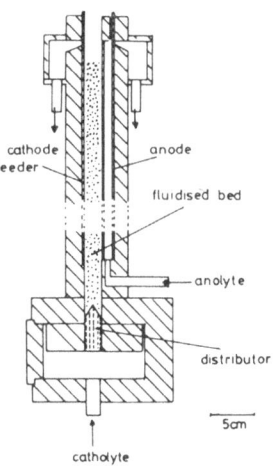

Figure 1. Sketch of a simpli-
    fied Nalco cell. (a) parti-
    culate bed, (b) cathodes
    covered by diaphragm, (c)
    conducting reactor shell.

Figure 2. Section through a
    fluidised bed cell of rec-
    tangular geometry.

the ability to operate in a continuous fashion. A fluidised bed
electrode using roughly spherical particles of say, 300-1000 μm in
diameter has all these properties. Figure 2 shows a cross-section
through our pilot plant cell (2), which is 0.25 m^2 in diaphragm
cross-section, the width of this bed being 0.025 m in the direction
of current flow. Size segregation allows the larger particles to
be elutriated from the bottom and smaller ones to be returned to
the top of the cell, thereby permitting continuous operation. This
electrode is more difficult to model than the packed bed but two of
my co-workers have achieved some success in doing so on the basis
of a bipolar charge transfer process (7). For example, one practical feature explained by this model is the unexpectedly high purity
of copper when deposited from relatively impure solutions. Unfortunately, time does not permit to discuss this model in any detail,
sufficient to say that it postulates bipolar aggregates where a
reversible metal system dissolves at one end of the aggregate and
deposits at the other. In other words, a refining action takes
place explaining the high degree of purity of the metal deposit.
The industrial Aczo unit (8) used for effluent treatment and shown
in figure 3 has a cylindrical geometry and in terms of overall
design is not too distant in concept from the Nalco cell in figure 1.

    The partial bipolar nature of the fluidised electrode leads
me to my next design. It was found that certain metals, such as
manganese, could not be deposited in a fluidised bed cell. Figure
4A shows a typical particle circulation pattern in a liquid

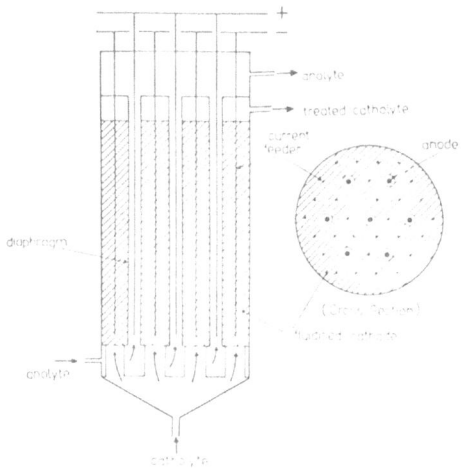

Figure 3.   Cylindrical Aczo cell.

fluidised bed.   By inclining the cell to the horizontal (4B) a so-
called spouting bed is obtained where a relatively narrow dilute
particle phase moves up rapidly past the diaphragm and a wider
dense particle phase falls slowly past the current feeder.   Because
of good particle contact bipoles would be absent from the dense
phase.   The falling bed cell takes this one step further by, as
shown in figure 5, removing the rising phase from the potential
field and passing it up behind the current feeder.   Now, all deposition occurs in the monopolar dense phase and so far manganese has
been deposited with a current efficiency of 40%.   This is quite
encouraging for a first attempt, since current efficiencies for the
conventional commercial process at a considerably lower current
density are only in the range of 65-70% (9).

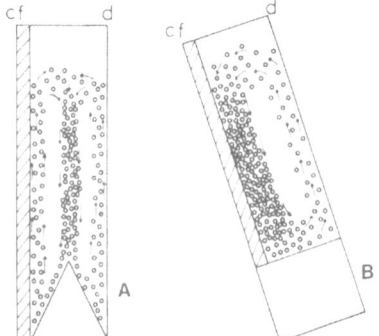

Figure 4.   Modes of fluidisation. (A) conventional fluidisation,
             (B) spouting bed.   c.f. = current feeder. d = diaphragm.

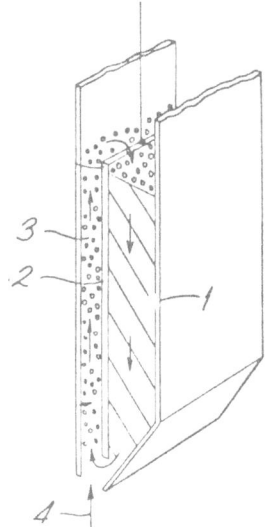

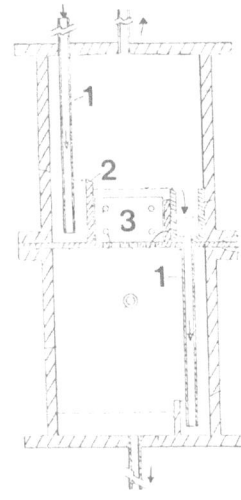

Figure 5. Falling bed cell.
(1) anode, (2) cathodic
current feeder, (3) rising
phase, (4) inlet electro-
lyte.

Figure 6. Elevation of sieve
plate cell. (1) downcomer,
(2) weir, (3) sieve plate
electrode.

My final example is the sieve plate cell which as the name
implies is based directly on the design of a sieve plate absorption
tower, well known in chemical engineering applications. The cell
is shown schematically in figure 6. Electrolyte flows down the
downcomer, over a weir, across a series of bipolar rectangular
plates, down another downcomer and on to a second group of electrode
plates (not shown). Reactant gas passes up through the electrolyte

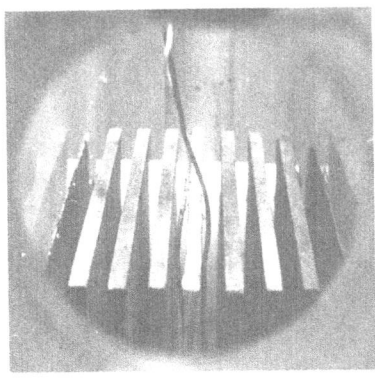

Figure 7. Detail of bipolar sieve plate electrodes.

between the plates. The cell acts as a very efficient absorption
tower and thus it is possible to produce propylene oxide with a
high current efficiency, space time yield and a minimum of byproducts.
It is significant that the very sparingly soluble butylene can be
expoxidised with a current efficiency of better than 60%.

SOME PROBLEMS OF SCALE UP

    In view of the limited time available, let me confine my
remarks to the scale up of the fluidised bed reactor shown in figure 2. One of the main problems is hydrodynamic rather than electrochemical, namely, electrolyte distribution. Up to a bed
of, say, 0.3 m, a relatively simple distributor will suffice. Since
commercial size is of the order of 1 m, then the design philosophy
should be to use a sectional cell. After all, this is analogous
to the scale up of some reforming processes, where the original
work was done in tubular reactors 3" in diameter and the industrially
sized version employed tube bundles of the same diameter. Returning
to the cell in figure 2, since the distance along the flow of current is limited to about 0.025 m, this will also be the thickness
of the industrially sized module.

    Another problem is the virtual absence of data on mass trans-
fer rates to particles in the bed itself as distinct from the wall
of the reactor. Work in our laboratory on mass transport to fixed
and moving particles in a fluidised bed of glass beads gave some
interesting results. Figure 8 shows an apparently successful correlation between the mass transfer factor j_D and a particle Reynolds
number Re_D, where (also figures 9 & 10) particle and probe diameters
are denoted by ▲ = 1.4 mm, ● = 2.1 mm and ■ = 3.5 mm, Sc = 1280 and

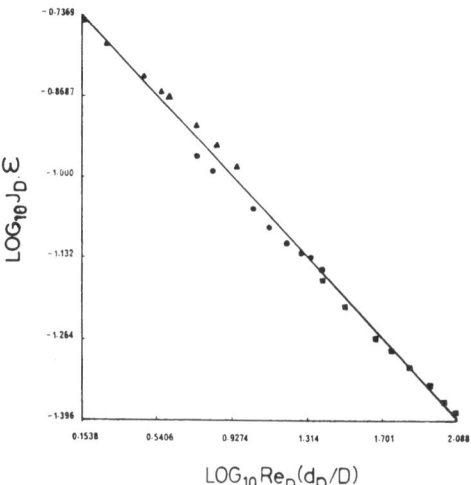

Figure 8. Plot of j_D versus Re_D.

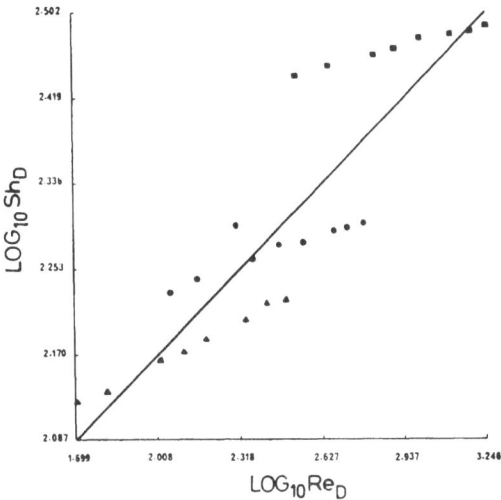

Figure 9.   Plot of Sh_D versus Re_D.

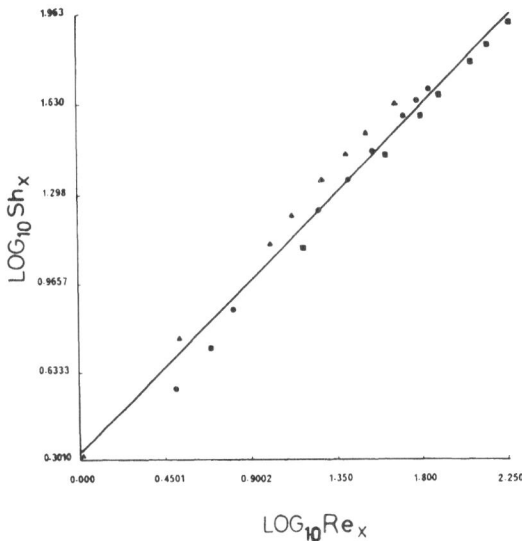

Figure 10.   Plot of Sh_x versus Re_x.

bed expansion ranges from 5-100%. This is in fact the most common
way mass transfer coefficients are correlated in the literature.
That all is not as well as it appears is demonstrated in figure 9,
where it is clear that the same data are not correlated by:

$$Sh_D = constant \ x \ Re_D \qquad\qquad (4)$$

One cannot help speculating whether the apparent success of the correlation in figure 8 is due to the fact that the superficial velocity u, which appears on both sides of equation (1) varies by 200% as against the mass transfer coefficient k_L that only changes by a modest 20%. It is, however, possible to get a correlation if d_p, the particle diameter, is replaced by x, the mean distance of separation between particles. Interestingly the resulting equation:

$$Sh_x = C\ Re_x^b \qquad \begin{array}{l} C = 1.9 - 2.2 \\ b = 0.73 - 0.77 \end{array} \qquad (5)$$

correlates mass transfer coefficients not only for fixed and moving particles, but also for transport to the vessel wall. Clearly, there is a limit to the range of application of correlation (5) since as x increases Sh_x should not approach infinity. Furthermore, correlation (5) suffers from the same disadvantage as j_D against Re_D, since in the case of (5) x appears on both sides of the equation and can mask trends in the value of k_L. It can, however, be regarded (just as the plot of j_D) as a way of obtaining values of k_L, provided care is taken not to exceed the imposed limits of applicability. Neither of these correlations can really be regarded as a design equation since neither values of x nor ϵ are immediately available to the design engineer. A correlation that can be used as a design equation and in the limit reduces to the behavior of a single particle has been developed by us (10) but unfortunately time limitations do not permit a detailed discussion. Difficulties arise when particle size distribution has to be taken into account, but that is another story.

OUTLOOK FOR THE FUTURE

Let me spend the remainder of my time in speculating about the immediate future of cell design. In my opinion, we are rapidly approaching the situation chemical reactors reached at the turn of the century where a relatively small number of properly engineered designs were almost invariably used in a process. I therefore do not foresee a future for a large number of additional novel cell designs. Instead I envisage three important areas of activity:

(a) Consolidation and engineering of a small number of cell designs.

(b) A search for additional constructional materials, particularly electrode surfaces and diaphragms.

(c) The application of recently developed control techniques to improve cell performance.

Since (c) is an area of research engaging our attention I will say a little bit more about it. With the advent of cheap micro-processors and minicomputers it is now viable to dedicate these to cope with interactive control parameters. One of the topics we are

investigating is the control of the outlet concentration of metal
effluents from the fluidised bed cell shown in figure 2. Figure 11
depicts a flow sheet of the pilot plant. Provision is made to be
able to produce flow and concentration perturbations and there are
two control loops, one for flow, the other for controlling the outlet concentration by a change in current. Since variations in flow
affect bed expansion and alter the current due to the change in
resistance, the flow loop is interactive with the current loop but
the reverse is not the case. The plant is just being commissioned
but figure 12a and 12b show a simulation (11) of the expected effects
as a result of flow disturbances, the former without control action,
the latter on applying a well tuned digital PI controller. Since
these disturbances represent practical difficulties, economic viability will decide on the extent of application of these techniques.

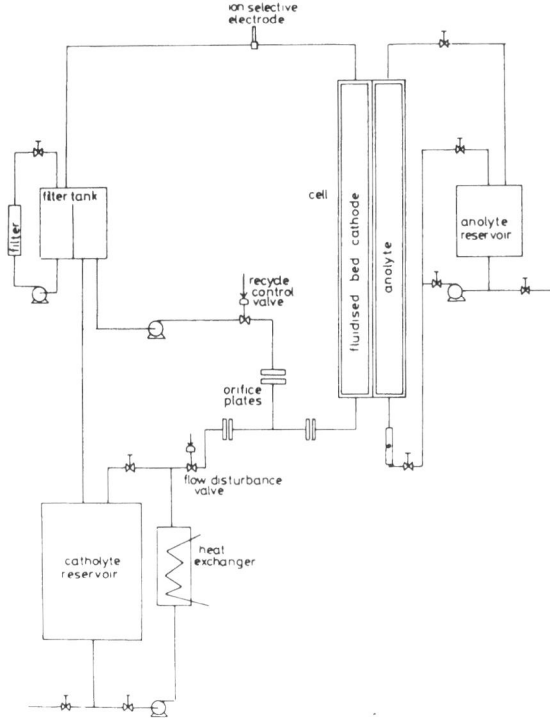

Figure 11. Flow sheet of fluidised bed pilot plant for the treat-
            ment of metal effluents.

     May I conclude by saying that I am very optimistic about the
increase of electrochemical techniques in selected areas of industry.
Among these I would include synthetic methods for the production of
fine chemicals and pharmaceuticals and the treatment of metal effluents.

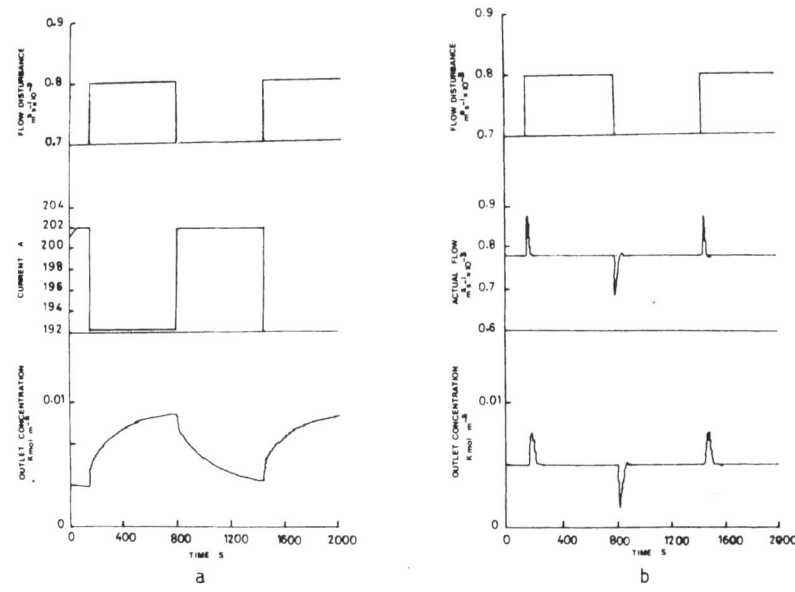

Figure 12.   Simulation of plant behavior on experiencing a flow
             disturbance.   (a) without control action,   (b) with
             digital PI control action.

NOMENCLATURE

a   Specific electrode area m^{-1}

b   Exponent of Reynolds number   dimensionless

C   Constant in equation (5)   dimensionless

C_E Current efficiency   dimensionless

d_p Particle diameter   m

D   Bed diameter   m

D_L Diffusivity in the electrolyte $m^2 \, s^{-1}$

i   Current density Am^{-2}

j_D Mass transfer factor $= \dfrac{k_L}{u} Sc^{2/3}$ dimensionless

Q_E Electrochemical equivalent   kmol coulomb^{-1}

Re_D Reynolds number $= \dfrac{\rho u d_p}{\mu (1-\varepsilon)}$ dimensionless

Re_x Reynolds number $= \dfrac{\rho u x}{\mu \varepsilon}$ dimensionless

Sc   Schmidt number $= \mu/\rho D_L$ dimensionless

Sh_D Sherwood number = $k_L d_p/D_L$ dimensionless

Sh_x Sherwood number = $k_L x/D_L$ dimensionless

u   Superficial electrolyte velocity ms^{-1}

x   Mean distance of separation between
         particles   m

Y_C Chemical yield   dimensionless

Y_E Energy yield $kmol\ KJ^{-1}$

Y_{ST} Space time yield $kmol\ m^{-3}\ s^{-1}$

ε Bed voidage   dimensionless

μ Electrolyte viscosity $Ns\ m^{-2}$

ρ Electrolyte density $kg\ m^{-3}$

REFERENCES

1. Nalco Chemical Co., U.S. Patent 3,573,178 (1971)

2. F. Goodridge and C.J. Vance, Copper deposition in a pilot-plant-
    scale fluidised bed cell, Electrochim. Acta, 24, 1237 (1979).

3. K. Scott and A.R. Wright, U.K. Patent application 7908039 (1979).

4. F. Goodridge, K. Lister, R.E. Plimley and K. Scoffham, The
    Production of alkene expoxides using a novel cell design.
    Paper presented at the Spring meeting of the Electrochemical
    Society, St. Louis, Missouri, May 1980. Abstract No. 431.

5. Nalco Chemical Co., British Patents 1,048,987 and 1,071,322
    (1966).

6. F. Goodridge and M.A. Hamilton, The behaviour of a fixed bed
    porous flow-through electrode during the production of p-amino-
    phenol, Electrochim. Acta 25, 481 (1980).

7. R.E. Plimley and A.R. Wright, A bipolar mechanism for charge
    transfer in a fluidised bed electrode. Paper presented at the
    Spring meeting of The Electrochemical Society, St. Louis, Mis-
    souri, May 1980. Abstract No. 400.

8. C.M.S. Raats, H.F. Boon and G. van der Heiden, Fluidised bed
    electrolysis for the removal or recovery of metals from dilute
    solutions. Paper presented at the International Symposium
    "Chloride Hydrometallurgy," Brussels, Belgium, September 1977.

9. M. Harris, D.M. Meyer and K. Auerswald, The production of elec-
    trolytic manganese in South Africa, J. of South African Inst.
    Min. and Met., 77, 137 (1977).

10. F. Goodridge, D. Nassif and R.E. Plimley, Mass transfer studies
     in liquid fluidised beds, Trans. Instn. Chem. Engrs. (In press).

11. A.J. Morris, A.R. Wright, S.C. Denis and Y. Nazer, Micropro-
     cessor control of an electrochemical process for metal removal
     from an aqueous effluent. Paper presented at a meeting on
     "Energy considerations in electrolytic processes," sponsored
     by the Society of Chemical Industry and the Institution of
     Mining and Metallurgy at the University of Newcastle upon
     Tyne, England, July 1980.

DISCUSSION

Dr. W. E. Haupin, Alcoa Technical Center, New Kensington, PA:

    Please explain the mechanism of charge transfer in a fluid
bed electrode.

Professor F. Goodridge: Our initial idea was to try and explain
the charge transfer process in terms of a collision mechanism.
One can show, however, that in order to account for the high currents sustained by a fluidized bed electrode one would require something like 10^4 to 10^5 collisions per second between particles and
1 cm^2 of the current feeder. This is not in accord with the relatively lazy particle movement observed in a working fluidized bed.
Furthermore, there are experimental findings, such as the presence
of anodic potentials in a cathodic bed, which cannot be accounted
for in terms of a collision mechanism.

    Our present view is that a major part of the charge is trans-
mitted in the bulk of the bed by bipolar aggregates of particles
acting like a bipolar electrode in, say, a refining process. This
would also explain the unexpectedly high purity of deposit from a
relatively impure electrolyte. It follows that at any given moment
net deposition would only occur on particles actually in contact
with the current feeder which therefore acts as an extended area
electrode. Calculations using a Monte Carlo technique have shown
that this charge transfer model predicts observed high frequency
potential distributions in a surprisingly detailed manner.

NAFION® MEMBRANE AND ITS APPLICATIONS

Walther G. Grot
E.I. du Pont de Nemours & Co., Inc.
Experimental Station, 323
Wilmington, Delaware 19898

The interest in electrolytic processes has increased considerably in recent years. In many of these processes, a divided cell is required and the success or failure of the operation often hinges on the availability of a separator with the required properties. These properties include mechanical strength and long-term stability in the operating environment of the cell. Oxidative stability is of particular importance because a major advantage of electrochemical processes is the generation of powerful oxidizing agents at the anode. Similarly, in fuel cells or batteries, high cell voltages require the use of strong oxidizers.

With the introduction of NAFION®, cell separators of exceptional chemical and thermal stability are now available. With the exception of elemental fluorine and molten alkali metals, NAFION® is stable against chemical attack at temperatures of up to 150°C. In addition, it has high ionic conductivity which means that for most applications the maximum current density obtainable is not limited by the membrane but by other factors such as gas release in the electrolyte. Current transport is predominantly cationic.

The development of NAFION® started about 18 years ago when, based on exploratory work in our Central Research Department, the synthesis of the following polymer was proposed. (Figure 1)

The polymer is obtained in the form of a meltfabricable precursor containing $-SO_2F$ functional groups. After fabrication to the desired shape, these groups are converted to $-SO_3H$ or other ionic groups by a suitable chemical treatment.

73

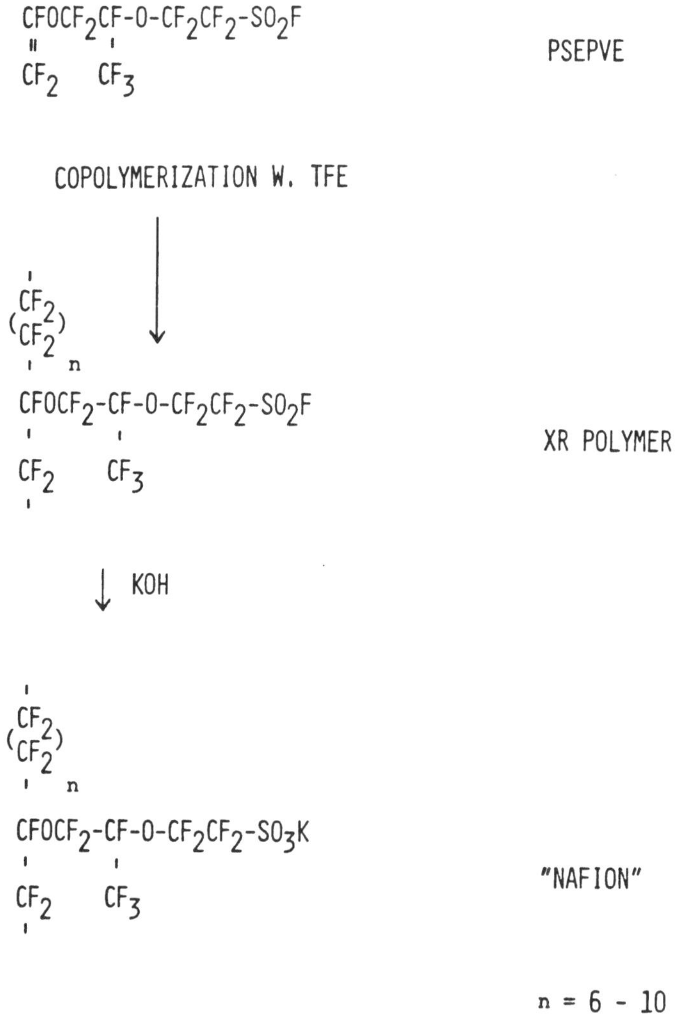

CFOCF$_2$CF-O-CF$_2$CF$_2$-SO$_2$F   PSEPVE
‖   ‖
CF$_2$ CF$_3$

COPOLYMERIZATION W. TFE

│
▼

│
CF$_2$
(CF$_2$)
│ n
CFOCF$_2$-CF-O-CF$_2$CF$_2$-SO$_2$F   XR POLYMER
│   │
CF$_2$ CF$_3$
│

↓ KOH

│
CF$_2$
(CF$_2$)
│ n
CFOCF$_2$-CF-O-CF$_2$CF$_2$-SO$_3$K   "NAFION"
│   │
CF$_2$ CF$_3$
│

n = 6 - 10

Figure 1

For many industrial applications the strength of the membrane is improved by reinforcing the film with a fabric made of Teflon®.[1,2] The properties of such fabrics are shown in Table 1.   For maximum conductivity, it is important to have a large open area. Using a plain weave, structures more open than T-12 are not sufficiently stable to permit handling in commercial production. Leno weave permits more open structures because of the better anchoring of the weft. A very open construction can be achieved through the use of rayon as a sacrificial fiber. In T900G, for instance, one strand of Teflon® alternates with 4 strands of rayon resulting in a fabric of fairly good stability. In use, the rayon is destroyed leaving a very open fabric of Teflon® behind.[3]

Table 2 lists Nafion® separators which have been sold commercially and their suggested applications. Chlor-alkali electrolysis is by far the most important from a commercial standpoint. I will comment later on a new high performance chlor-alkali membrane not shown in Table 2.   It is called Nafion® 901X and will be commercially available in 1981.

In the chlor-alkali application, the membrane cell separator has to meet a number of very demanding requirements. The schematic of a membrane chlor-alkali cell is shown in Figure 2:

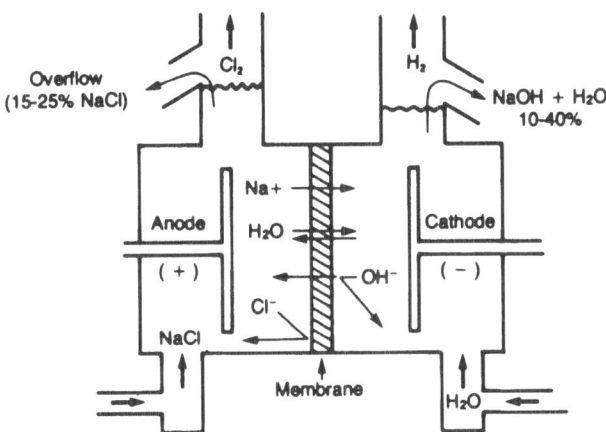

Figure 2.   Membrane Chlor-Alkali Cell.

There is a strong economic incentive to produce a highly concentrated sodium hydroxide solution as the catholyte. This is particularly true for a large merchant plant where shipping cost for dilute caustic are prohibitive.

While cation exchange membranes in principle reject anions by the Donnan exclusion, the requirement of high caustic concentration combined with the high electrical mobility of the OH ion

Table 1.   Properties of Fabrics Made of Teflon®

	T-12	T-24	T-900G
Weave Pattern	Plain	leno	plain
Thread Count Per cm	16 x 16	10 x 10	6 x 6 Teflon®
			24 x 24 Rayon
Denier	400	400	200 Teflon®
			50 Rayon
Thickness, mils	10.8	7	4.6
% Open Area (optical)	25	32	70 (after removal of the Rayon)

Table 2.

Type	Structure*	Application
117	7-1100	Fuel cells,
125	5-1200	batteries, dialysis, etc.
214	1.5 EDS/7-1150/T-24	Chlor-alkali electrolysis
295	1.5 EDA/7-1150/T900G	up to 30% NaOH
315	2-1500/4-1100/T-12	
316	1.5-1500/4-1100/T-12	Chlor-alkali electrolysis
390	1.5-1500/4-1100/T-900G	up to 20% NaOH
324	1-1500/5-1100/T-24C	
415	5-1100/T-12	Chromic acid regeneration,
417	7-1100/T-12	Electroplating Chemicals,
425	5-1200/T-12	General purpose membrane

*Thickness (in mils) and equivalent weight (EW) of the polymer layers are shown. For instance, Nafion® 214 (1.5 EDA/7-1150/T24) consists of a 7 mil thick film of 1150 EW polymer treated with ethylene diamine (EDA) to a depth of 1.5 mil and laminated to T-24 fabric .1 mil $\hat{=}$ 25.4µm.

makes the chlor-alkali electrolysis one of the most demanding in terms of membrane selectivity.

In order to meet these stringent requirements of selectivity, we developed the concept of a barrier layer on the catholyte side of the membrane.[4] The polymer composition of this surface layer is selected to give the anion rejection required for high current efficiency and the consequences of the inherently poorer conductivity are minimized by keeping the barrier layer very thin. The composition of support layer is selected to give highest conductivity. The barrier layer can be a separate layer of a different composition or it can be created by a chemical modification of one surface of the NAFION®. The sulfonyl fluoride precursor is a convenient starting point for such modifications. For example, one surface of a sulfonyl fluoride film can be converted to a sulfonamide by treatment with ammonia or an amine. Treatment with ethylene diamine (EDA) has been particularly effective probably because it causes a small amount of crosslinking of the surface layer. EDA modified membranes are indicated by a first digit of 2 in the NAFION® code.[5,6]

NAFION® codes with a beginning digit 3 indicate membranes containing a sulfonic acid barrier layer of higher equivalent weight. They are ideally suited to the production of dilute caustic up to 20% concentration.

The performance of representative examples of different types of membranes in laboratory chloro-alkali cells is shown in Table 3. It should be noted that laboratory cell voltages may not represent accurately those found in commercial size units.

The performance of NAFION® 417 is shown only to illustrate the effect of the absence of a barrier layer; this membrane is not intended for chlor-alkali applications.

The performance of NAFION® 324 vs. time is shown in Figure 3. This product is suitable for making intermediate strength caustic and is most often used when the caustic is consumed on-site. As I mentioned before, Du Pont has developed a new high-performance chlor-alkali membrane called NAFION® 901X suitable for higher caustic concentrations.

The results obtained from extensive testing in Du Pont laboratories, chlor-alkali manufacturer laboratories and commercial cells throughout the world indicate the new product will be superior in energy utilization, operating costs and overall investment requirements to any other chlor-alkali membrane available today.

Table 3. Membrane Development

NAFION® MEMBRANE	CLOTH % OPEN AREA	RESISTANCE Ω-cm IN 0.6 N KCl	LABORATORY CELL VOLTAGE	CAUSTIC STRENGTH %	CURRENT EFFICIENCY %
214 (EDA/7-1150/T24)	32	5.3	4.2	28	90+
315 (2-1500/4-1100/T12)	25	5.3	4.4	20	80-85
324 (1-1500/5-1100/T24)	32	3.9	4.1	20	80-85
390 (1.5-1500/4-1100/T900G)	70	3.8	3.9	20	80-85
417 (7-1100/T12)	25	4.2	4.2	20	58

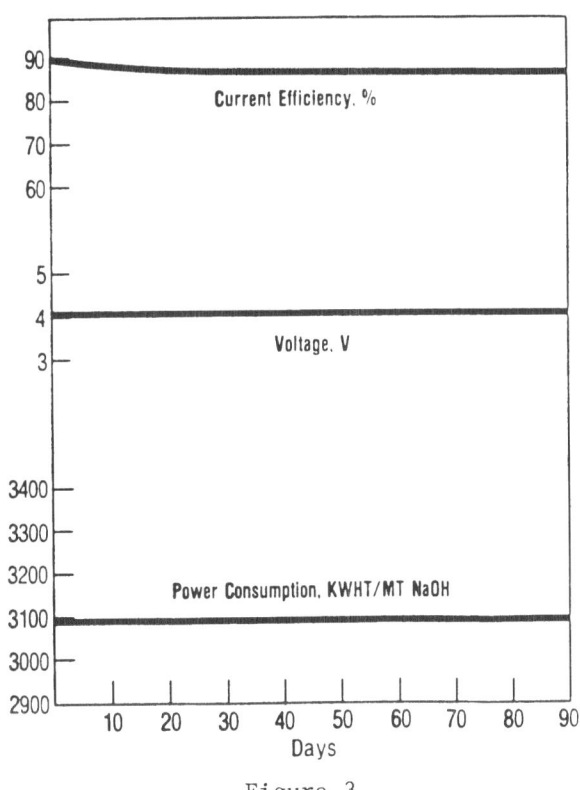

Figure 3

Performance of NAFION® 324
31 A/dm^2, 20% NaOH, 80°C, pH 1 Brine, 30ppm H$_3$PO$_4$
Boiled 30 Minutes at Room Temp.Under Vacuum

Du Pont is confident that the new membrane will operate at efficiencies that will average 94 percent or greater for the first two years of operation. It produces caustic at a concentration of 33 percent and at a very high purity.

NAFION® 901X is capable of operating at minimum voltage over a wide range of current densities. This flexibility permits cycling of current density in order to take advantage of low off-peak power costs.

It will be commercially available in 1981.

The 3 chlor-alkali processes are compared in Figure 4. The amalgam process is highest in cell voltage and therefore in electrical power consumption. The extra energy input to the cell is dissipated as waste heat in the denuder. This cell produces pure, concentrated caustic directly. None of the water in the anolyte is transferred to the catholyte. This requires the use of solid salt for the resaturation of the depleted anolyte. The amalgam cell is now unacceptable for environmental reasons in many parts of the world.

The diaphragm cell has the lowest cell voltage of the 3. This is in part due to the lower current density made possible by the lower cell investment. All the water in the anolyte is transferred to the catholyte eliminating the need for anolyte resaturation with solid salt. This cell is actually a net producer of solid salt if brine is used as a starting material. The biggest drawback of this cell is the low quality of the caustic produced, requiring a rather expensive evaporation and separation process. Even after this process the purity of the product obtained is not suitable for some applications.

The membrane process is in respect to cell voltage, water transport across the cell and caustic concentration, intermediate between amalgam and diaphragm process. The amount of brine that needs to be recirculated is reduced compared to the amalgam process, both because of higher permissible brine depletion and because some water is transported through the membrane.

Compared to the diaphragm process, the membrane cell produces a catholyte of sufficient concentration and, more importantly, purity to be directly suitable for most applications. The membrane process therefore offers particularly attractive economics for on site generation of chlorine and caustic.

But even when evaporation is required for shipment of the caustic, the amount of steam required is greatly reduced and the salt separation is eliminated. This combined with higher product purity results in a favorable comparison with the diaphragm process.

Amalgam

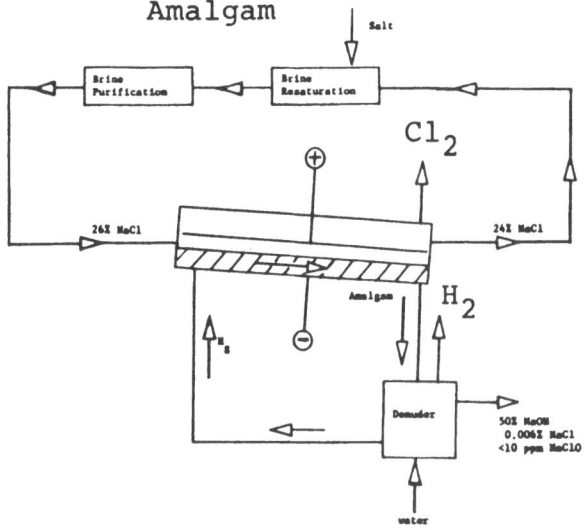

Diaphragm

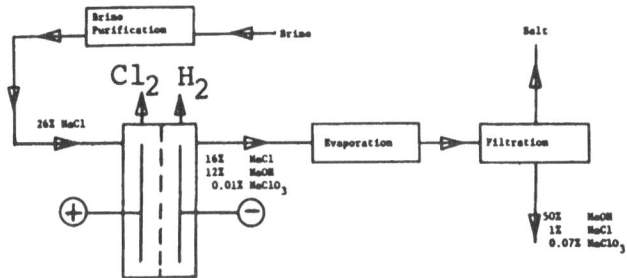

Membrane

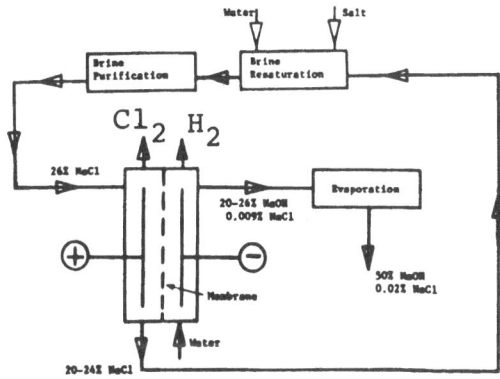

Figure 4

Continued improvements in membrane performance are expected to result from major Du Pont research effort aimed at developing improved ion-exchange membrane products.

The evolution of cell design is equally important. Just as the design of diaphragm cells has been refined over many years, we will see a similar evolution of membrane cell design.

The efforts to develop uses for NAFION® in the past have been concentrated in the chlor-alkali field because of the enormous scale of this application. In the future I would expect the development of new applications in the fields of electroorganic chemistry because of NAFION®'s relatively high conductivity in solvent/water mixtures. In inorganic electrochemistry I would expect further developments in the field of fuel cells and batteries and particularly in the recovery and regeneration of chemicals in the electroplating and metal finishing industry.

I would like to discuss the regeneration of chromic acid as an example because it again utilizes the chemical stability of NAFION®. The chromic acid to be regenerated may originate from a number of different uses.

The most straightforward case is the regeneration of chromic acid which has been used in the oxidation of some organic material, for instance, the etching of plastic parts prior to metallizing. In addition to 3 valent chromium and sulfuric acid, this solution usually contains considerable amounts of unreacted chromic acid. Electrolytic reoxidation of such a solution is accomplished readily using lead anodes and dilute sulfuric acid as the catholyte.

A cylindrical cell geometry is attractive for chromic acid regeneration because it simplifies the cell design, particularly with respect to gasketing. Chromic acid regenerators of such a geometry, using a tube of NAFION® of about 80 mm diameter as a separator, are being sold commercially and are used widely in the plating industry.[7]

Most important perhaps from a commercial standpoint is the regeneration of a chrome plating bath which has accumulated an unacceptable amount of metal cations (Cr^{+++}, Fe^{+++}, Cu^{++}, etc.). Cr^{+++} is readily converted to chromic acid by anodic oxidation while the other cations are removed by electrodialysis through the membrane. The efficiency of this electrodialysis is fairly low because most of the current in the strongly acidic solution is carried by H^+ ions, but is still economically attractive because the amount of metal ions that need to be removed is small.

In this application a catholyte must be selected which will

not contaminate the anolyte with undesirable anions. In a chrome
plating bath, the ratio of H_2SO_4 to CrO_3 has to be controlled quite
carefully usually at about 1 to 100. Even a small amount of back
migration of sulfate ions from the catholyte upsets this ratio.
One interesting solution to this problem is the use of an organic
anion as a conductor in the catholyte. Such anions are destroyed
by the chromic acid in the anolyte.[8]

Another application in which the removal of metal ions is
necessary, in addition to the anodic oxidation of chromium ions,
is the regeneration of chromic/sulfuric acid solutions used for the
etching of printed circuits. In this case, however, the quantity
of copper ions to be removed is substantial and the economics of
the process depend very much on the efficiency of copper ion transport through the membrane.

In order to determine this we electrolized a solution (anolyte)
containing:

$CuSO_4 - 5H_2O$	138 gm/l
$Cr_2(SO_4)_2 - 12H_2O$	108 gm/l
CrO_3	53 gm/l
H_2SO_4	96 gm/l

using a lead anode, a platinum wire screen cathode in 20% sulfuric
acid as the catholyte, and a NAFION® 120 cell separator. After
passage of the theoretical amount of current, the anodic current
efficiency was determined by the titration of the chromic acid
content of the anolyte while the efficiency of the copper ion
transport was determined by the weight increase of the cathode.
The anodic current efficiency was close to 100% in all cases, i.e.,
the Cr^{3+} content of the anolyte was almost completely converted to
chromic acid, while the efficiency of copper ion transport ranged
between 8% at a current density of $15A/dm^2$ and 13% at a current
density of $30A/dm^2$.[9]

The removal of copper ions from the solution was therefore
incomplete at a point where all of the original chromic acid had
been regenerated. The economics of this process would depend on
the amount of copper ion build up which could be tolerated in the
bath. Complete removal of copper is possible by first passing the
spent etch bath through the cathode compartment of the cell to
reduce the remaining chromic acid, and deposit copper on the
cathode. The effluent of the cathode compartment is then fed into
the anode compartment of the same cell in order to regenerate
chromic acid. (Figure 5)

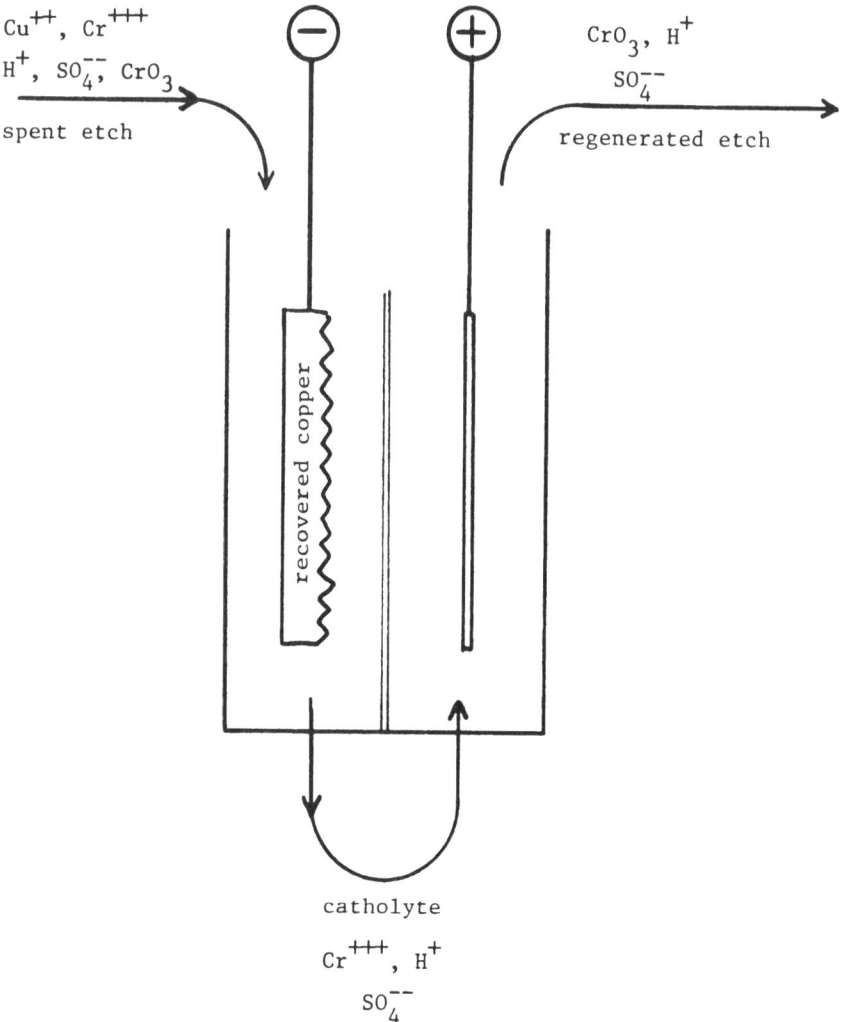

Figure 5.   Electrochemical system using a Nafion® membrane to
regenerate chromic acid.

Other examples of chromic acid solutions that can be regenerated electrolytically using a NAFION® membrane include solutions
that have been used in the surface treatment of brass, aluminum
and other metals. As the concern about the release of metal ions
into the environment increases, more applications of this nature
will develop.

REFERENCES

1. E.I. du Pont de Nemours & Co.(W.G. Grot), U.S. 3,770,567
    (November 8, 1971).

2. E.I. du Pont de Nemours & Co.(W.G. Grot), U.S. 3,849,243
    (August 1, 1973).

3. E.I. du Pont de Nemours & Co.(W.G. Grot), USP 4,021,327
    (May 3, 1977).

4. E.I. du Pont de Nemours & Co.(W.G. Grot), U.S. 3,784,399
    (September 8, 1971).

5. E.I. du Pont de Nemours & Co.(P.R. Resnick and W.G. Grot)
    U.S. 4085071 (April 18, 1978).

6. E.I. du Pont de Nemours & Co.(W.G. Grot), U.S. 3,969,285
    (July 13, 1976).

7. Betty A. Rose, Industrial Finishing, May 1979, p. 44-47.

8. John L. Raymond and Robert Z. Reath (Robert F. Ehrsam)
    USP 3,909,381 (Sept. 30, 1975).

9. W.G. Grot, Chem. Ing.-Tech. 50 (1978) Nr. 4 p. 299-301.

ACKNOWLEDGEMENTS

Table 1   -   From reference 9

Figure 2 -   From T. Berzins, "Electrochemical Relationships in
              Chlor-Alkali Cells Employing NAFION® Membranes"
              Paper presented at the 152nd Meeting of the Electro-
              chemical Society, Atlanta, Georgia, October, 1977.

Table 3   -   Adapted from E.H. Price "The Commercialization of
              Ion-Exchange Membranes to produce Chlorine and
              Caustic Soda" ibid

Figure 3 -   From D.E. Maloney "Fluorocarbon Membranes for Chlor-
              Alkali Industry" Paper presented at the Symposium
              of the Electrochemical Technology Group SC1, London,
              June, 1979.

Figure 4 -   Adapted from D. Bergner "Elektrolytische Chlorerzeugung
              nach dem Membranverfahren" Chemiker. Zeitung 101,
              No. 10, p. 434, (1977).

DISCUSSION

Dr. Sankar Das Gupta, HSA Reactors, Ltd., Toronto:

1. During operation, can a cation exchange Nafion® change into an anion exchange, perhaps by losing its sulphonic acid group via some exchange mechanism?

2. Nafion® is a dimensionally unstable diaphragm (contracts by 15% when it dries), so how would the Cr systems operate, when they need to operate in environments of non-continuous operations? (e.g., the unit only operates for 10 hours/day, 5 days/week.)

3. What is the transport rate of chlorine from the anode chamber to the cathode chamber? For example, it is well known that water molecules travel readily through the Nafion® membrane with about 4 molecules travelling with every H^+ ion.

Dr. W. Grot:

1. I cannot think of any mechanism which would result in the formation of an anion exchange group.

2. If the membrane was installed and clamped in the wet state, the shrinkage as a result of drying could result in stretching of the membrane and therefore a loose fit on rewetting.

   Shut-down of a chromic acid regeneration unit would not necessarily require discharge of the electrolytes. In many cases, maintaining a trickle current for the protection of the electrodes would be sufficient. If discharge of the electrolytes is necessary, the cell should be refilled with distilled water.

3. Any chlorine diffusing through the membrane would result in the formation of sodium hypochlorite as the primary product. This could be subject to cathodic reduction as well as thermal decomposition (formation of sodium chlorate and oxygen).

   Typical level of total oxidizing salts is less than 50 ppm (as sodium chlorate) indicating very low levels of chlorine diffusion.

Dr. Floyd L. Ramp, B.F. Goodrich Research and Development Center, Cleveland: Some time ago we observed, in a cell using a KCl anolyte and a phosphate catholyte - pH ∿5, a lower cell resistance if the process was started with very dilute KCl. The KCl concentration was then brought to operating conditions. We attributed the persistence of low cell resistance to phosphate ions trapped in the membrane which was highly swollen in the initial stages. Have you had any related experience?

Dr. W. Grot: Your observation may be related to preconditioning of the membrane. We recommend conditioning of the membrane in the H^+ form for 30 minutes in boiling water in order to obtain lowest resistance. Without this treatment the membrane would exhibit higher resistance when used in a KCl solution of modest concentration.

Operating the membrane in a very dilute KCl solution for a period of time, particularly at elevated temperatures, may precondition the membrane in a similar fashion.

I believe it would be unlikely that phosphate ions would be trapped in the membrane for any length of time.

Jeff C. Cole, Consultant, The H. K. Ferguson Company and Department of Energy: With membrane 901 X, 33% caustic and 94% current efficiency, what is the corresponding voltage drop?

Dr. W. Grot: Since this is a new development, I am unable to provide any additional information.

Prof. Francis Goodridge, University of Newcastle upon Tyne: As Professor de Nora indicated this morning, one of the future important applications of electrochemistry would be copper winning from dilute leach liquors. Using a fluidized bed electrode it would be necessary to have a cationic membrane to allow the acid to enter the anolyte and allow the latter to be used again as leach acid. Is there any fundamental reason why an anionic membrane of a performance equivalent to that of Nafion® should not be produced?

Dr. W. Grot: Use of an anion exchange membrane would indeed permit recovery of the leach acid as the anolyte. Copper removal from the catholyte would have to be nearly complete to permit discharge of the catholyte into the environment. Using a cation exchange membrane such as Nafion® would recover the leach acid as the catholyte. Since this stream would be reused for leaching, complete removal of the copper would not be necessary. The relative merits of the two approaches are debatable.

The fundamental difficulty in developing chemically and thermally stable anion exchange resin is the inherent instability of commonly used anion exchange groups and the reduction of base strength associated with fluorine substitution. Perfluorinated tertiary amines, for instance, have no basic character.

THEORETICAL MODEL FOR THE STRUCTURES OF IONOMERS: APPLICATION TO

NAFION[®] MATERIALS

K.A. Mauritz and C.J. Hora

Diamond Shamrock Corporation
T.R. Evans Research Center, P.O. Box 348
Painesville, Ohio 44077

A.J. Hopfinger

Department of Macromolecular Science
Case Western Reserve University
Cleveland, Ohio 44106

INTRODUCTION

Considerable progress has been made in the solution theory of polyelectrolytes. However, for the condensed phase analogs of polyelectrolytes, ionomers, this is not the case. Eisenberg (1) has put forth an initial theory of ionomer structure which contains conceptual formalisms of general usage. His theory has been consulted extensively in the work reported here. Ponomarev and Ionova (2) have attempted to construct a sophisticated statistical mechanical model to describe the thermodynamics of ionomers. Recently, Gierke (3) has described a theory of ion transport in the Nafion[®] ionomer based upon a specific molecular organization.

To be complete, three physicochemical states of the ionomeric material need to be considered in a structural theory: a) dry, b) in aqueous solution, and c) in aqueous solution containing mobile ions. Obviously, considering only aqueous solution is a limitation. However, this is the solvent of interest in the majority of applications of ionomers. Ionomers, irrespective of physicochemical state, tend to phase separate. That is the polar groups, along with associated solvent and ions if present, cluster together to the exclusion of the nonpolar units. This molecular picture is the one which emerges from the detailed studies of a relatively limited

number of ionic polymers. All of these are hydrocarbon-based
materials and include polyethylene (4-12), polystyrene (13-18),
polybutadiene (19,20), and some polar polymers containing low molecular weight salts (21). In all cases, the bound ionic species has
been a carboxyl group.

The goal of the work reported here has been to devise a theory
which predicts the polar/nonpolar phase separation as a favorable
thermodynamic process. In addition, the effects of each of the
physicochemical forms on the thermodynamic and structural characteristics of the biphasic material have been sought. A pseudo
quantitative application of the theoretical formalism has been made
for Nafion®. The values for the requisite molecular parameters
were estimated from a combination of experimental bulk thermodynamic
data and molecular structure calculations employing both molecular
and quantum mechanics (22,23).

THEORY: (1-GENERAL MECHANISM OF CLUSTER FORMATION)

We begin by restricting ourselves to an equilibrium representation of ionomer structure. Then, on a time average basis, the
polar groups on the ionomer sidechains are able to "see" one another
through their dipolar interactions. From Monte Carlo simulation
calculations we have been able to show that the configurational
dipole-dipole interaction free energy for unrestricted dipolar
motion is of the form

$$F = C \ln(n_c - 1) \qquad (1)$$

where C is a constant characteristic of;
a) life-time of the dipoles,
b) strength of the dipoles, and
c) dielectric medium.

The, n_c, corresponds to the number of interacting dipoles.

In the ionomer, mutual localization of polar groups through
their dipolar behavior necessitaties expansion and contraction of
the associated polymer chains. If we assume this process is random
to the extent that chain expansions and contractions are equal in
number, then the elastic deformation energy per dipole is

$$W = \frac{3kT}{2\langle h^2 \rangle} (\Delta d^2) \qquad (2)$$

where T is temperature, Δd the average chain expansion/contraction,
and $\langle h^2 \rangle$ the meansquare end-to-end chain dimension. Equation (2)
is strictly valid only for a Gaussian chain. However, we can extend
is utility by defining Δd in terms of the physicochemical state of

the ionomer, and by generalizing the definition of $<h^2>$ to

$$<h^2> = N \left[(\frac{n-1}{2})\ell\right]^\Delta \left(\frac{1 - \cos\alpha_0}{1 + \cos\alpha_0}\right) \quad (3)$$

N = the number of monomer units/chain;
n = the number of CX_2 - units/monomer backbone chain;
ℓ = trans-distance for the CCC backbone groups
α_0= the CCC backbone valence bond angle;
Δ = chain ordering parameter; fixed = 2 for a random distribution
   of chain segments in space.

In the most simple representation, the size of a cluster, in terms
of n_c, could be found by solving

$$F = W + T\Delta S(n_c) \quad (4)$$

where $\Delta S(n_c)$ is the change in the entropy of the ionomer due to
clustering. $\Delta S(n_c)$ can reasonably be equated to the loss in the
configurational entropy of a typical chain due to clustering, which
when normalized to a per dipole scale is,

$$\Delta S(n_c) = - (\frac{n_c}{N}) R\ell n [N! (N!n_c)!] \quad (5)$$

where N = (Chain molecular weight)/(Monomer molecular weight) and
the right hand side of eqn. (5) can be simplified using Sterlings'
approximation. In practice, additional structure and energy terms
must be considered for each physicochemical state. We now discuss
each physicochemical state in this regard.

(2-DRY STATE)

   The sidechain polar groups, neutral or charged, serve as the
dipoles. The only additional energy term which needs to be added
to eqn. (4) is a destablizing term on the right to account for the
"surface tension" between the cluster and nonpolar phases. Equation
(4) then take the form,

$$F + \varepsilon r^2(n_c) = W + T\Delta S(n_c) \quad (6)$$

where $r(n_c)$ is the radius of the cluster and $\varepsilon = 4\pi\varepsilon_0$, which, in
turn, ε_0 is the surface energy per unit area.

(3-IN AQUEOUS SOLUTION)

   When the ionomer is placed in aqueous solution, water molecules
diffuse into the ionomer and hydrate about polar groups. Depending
upon the history of the ionomeric material, as well as the relative rates
of aqueous diffusion and cluster formation, clusters may or may not

have formed. We have considered the case in which clusters are
not formed prior to solvating. In this situation we first must
consider the polar group-hydrate structure which is termed a
hydration shell. The size of the hydration shell is again postulated to result from a free energy balance, Polar group ... H_2O
interactions coupled with H_2O ... H_2O intrahydration shell interactions are balanced against the elastic deformation energy needed
to move chains out of the way of the growing hydration shell. In
addition, a surface tension term must again be included to account
for the aqueous-nonpolar interface generated by the growing hydration shell. The general form of this energy balance is,

$$K_e(r_h(n_h)^m - r_p^m) + \varepsilon' r_h^2 = n_b A_1 + (n_h - n_b) A_2 \qquad (7)$$

where the first term on the left in eqn. (7) is a generalized,
spherically symmetric, deformation energy in which K_e is the
elastic force constant, $r_h(n_h)$ is the radius of the hydration shell
for n_h water molecules in the hydration shell, and r_p is the
unsolvated radius of the polar sidechain unit. The second term on
the left is the interfacial surface tension term with ε' being
the H_2O-nonpolar surface tension energy. The first term on the
right is the free energy of hydration of the polar sidechain unit
with n_b being the effective number of first layer hydrating water
molecules, and A_1 is the average free energy of hydration per water.
The second term on the right is free energy gain due to H_2O ... H_2O
hydration shell interactions with A_2 being the average free energy
of interaction per H_2O.

For the assumed temporal basis, the hydration shells now come
together through dipolar interactions to form a hydrated cluster.
The form of eqn. (6) can be used once a term on the left is added
to account for a potential loss in free energy due to the explusion
of water molecules from hydration shells overlapping. This loss
in energy can, however, be compensated by interhydration shell
H_2O ... H_2O interactions and a net decrease in the total H_2O-
nonpolar interfacial area. In total eqn. (6) now becomes,

$$F' + \varepsilon' r^2(n_c) + 0(n_c, n_h, F') = W' + T\Delta S(n_c) + 0'(n_c n_h, A_2) \qquad (8)$$

where the explusion overlap function is of the form

$$0(n_c, n_h, F') = \phi_1 \Delta R^2(n_c, n_h, F') r_h(n_c) + \phi_2 \Delta R^3(n_c, n_h, F') \qquad (9)$$

in which ϕ_1 and ϕ_2 are molecular constants characteristic of groups
present, and ΔR is the radial overlap distance of two hydration
shell. $0'(n_c, n_h, A_2)$, the interhydration H_2O ... H_2O energy term,
is of the same form as eqn. (9), but with different ϕ's which
depend upon A_2.

The primes associated with F and W in eqn. (8) are to remind one that the numerical forms of these expressions differ from those of eqn. (6). The effective dipole of a hydration shell is now due to the sidechain group (irrespective of neutral of charged form) plus the net dipolar arrangement of the water molecules. Hence F' is different from F. The elastic work, W', requires moving hydration shells plus polymer chains. Thus W' is not equal to W.

(4-AQUEOUS SOLUTION CONTAINING IONS)

We will assume that for,

$$\text{MMG}^{(-)} \to I^{(+)} \; \underset{k_2}{\overset{k_1}{\rightleftharpoons}} \; G^{(-)} \to H^{(+)} \quad k_2 > k_1,$$

where $\text{MM}\,G(-)$ is the anionic form of the polar sidechain unit and $I(+)$ is the mobile solution cation. Under these conditions the size of the polar clusters can be found by solving eqn. (8). However, the numerical constants in the equation will be different from those of the purely aqueous solution situation. The effective hydration shell dipole will now depend also upon the type of mobile cation and its solution concentration. The size of the ion-dipole hydration shell will also be altered by the presence of the solution cation. It must be stressed that $(\text{MM}\,G(-) \to I(+))$ is not considered to be a rigid form possessing a discrete bond length. Rather, on a time average basis this dipolar assembly is realized as a significant state of the system.

APPLICATION TO NAFION$^{\circledR}$ (1.-MOLECULAR PARAMETERS)

The general theory has been applied to Nafion$^{\circledR}$ ionomers whose general repeat structure is given as (I):

$$\begin{array}{c} +\!\!\!\!-(CF_2)_Y - CY -\!\!\!\!+ \quad \text{where X is small} \quad (I) \\[1mm] | \\ (O - CF_2 - CF)_X - OCF_2CF_2SO_3H \\ | \\ CF_3 \end{array}$$

In figures 1a through 1c are schematic representations of the molecular organization of the hydrated structures for Nafion$^{\circledR}$ as postulated in the theory.

In order to begin to quantitatively apply the theory, it is necessary to express explicitly the relationship between r_h and n_h, r_c and n_c, and the force constant K_e. We have assumed a four parameter model to accomplish this goal. First, we define three packing factors ρ_1, ρ_2 and ρ_3 as;

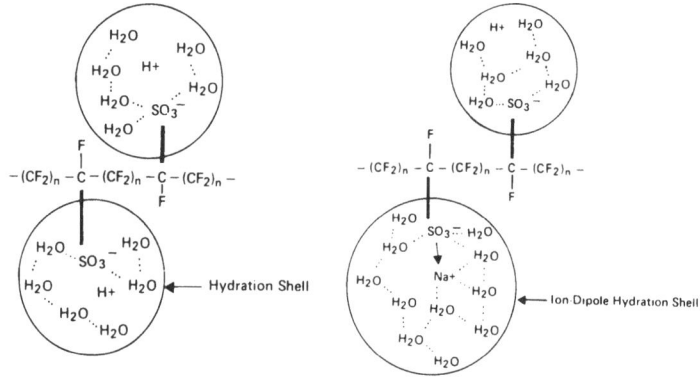

a. **Wet (Hydrated) Material
   — Sulfonic Acid Form**

b. **Wet Material (Na+ Ions Present)
   — Ion-Dipole Formation**

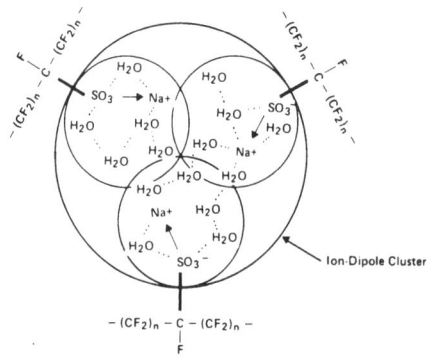

C. **Wet Material (Na$^+$ Ions Present)
   — Ion—Dipole Cluster Formation**

Fig. 1.   Local Schematic Representations of Individual Nafion$^®$
          Chain Structure.

ρ_1 = packing factor for water molecules in a purely aqueous
     hydration shell,

ρ_2 = packing factor for water molecules in hydration shells con-
     taining a cation, e.g. ion-dipole hydration shells, and

ρ_3 = packing factor of hydration shells in a cluster.

This allows us to establish the needed relationships. For
example the relationship between r_h and n_h is

$$n_h = \rho_1(r_n^3 - r_{so_3^-}^3)/r_w^3 \qquad\qquad (10)$$

where $r_{so_3^-}$ is the steric radius of an unhydrated so_3^- group, r_w is
the steric radius of a water molecule, and spherical symmetry is
assumed for all molecular entities.

The elastic force constant, K_e, is set equal to the functional form of the bulk elastic modulus (24) and normalized as the fourth parameter of the model. The remaining parameters were estimated using molecular and quantum mechanical calculations.

(2.-EXPERIMENTAL CALIBRATION)

A trial-and-error procedure has been used to calibrate the model, e.g. define ρ_1, ρ_2, ρ_3 and K_e. The molecular parameters for Na$^+$ cations are given in Table 1. Figure 2 contains plots of water absorption for the sulfonic acid form of Nafion® in contact with water as a function of ionomer equivalent weight, EW. Both experimentally measured (24) and theoretically predicted water absorption-EW curves are presented. Admittedly, the theoretical curve deviates from the experimental data, especially at increasing EWs. However, in the range of major commercial interest (i.e., EW = 1000-1200), the agreement between theory and experiment is quite good (±6%).

Table 1: Molecular parameters for Nafion® ionomers in aqueous solutions where Na$^+$ ions can be present.

$$K_e = \left[3.971 - 12.185 \left(\frac{EW}{1000} \right) + 14.225 \left(\frac{EW}{1000} \right)^2 \right.$$
$$\left. - 7.439 \left(\frac{EW}{1000} \right)^3 + 1.467 \left(\frac{EW}{1000} \right)^4 \right]$$

C (for ion-dipole hydration shell clusters) = -56 kcal/mole

P_1 = .74	A_2 ($H_2O \ldots H_2O$) = -1.8 kcal/mole
P_2 = .30	ϕ_1 = .26 (kcal/mole)/$\mathring{A}^3$
P_3 = 1.50	ϕ_2 = .29 (kcal/mole)/$\mathring{A}^3$
$r_{SO_3^-}$ = 2.4 $\mathring{A}$	ϕ_3 = .16 (kcal/mole)/$\mathring{A}^3$
r_w = 1.85 $\mathring{A}$	ϕ_4 = .19 (kcal/mole)/$\mathring{A}^3$
ε = .09 (kcal/mole)/$\mathring{A}^2$	m = 4
ε' = .19 (kcal/mole)/$\mathring{A}^2$	α_0 = 112°
A_1 = ($h_2O \ldots SO_3^-$) = -18.8 kcal/mole	ℓ = 2.30 A°

K. A. MAURITZ ET AL.

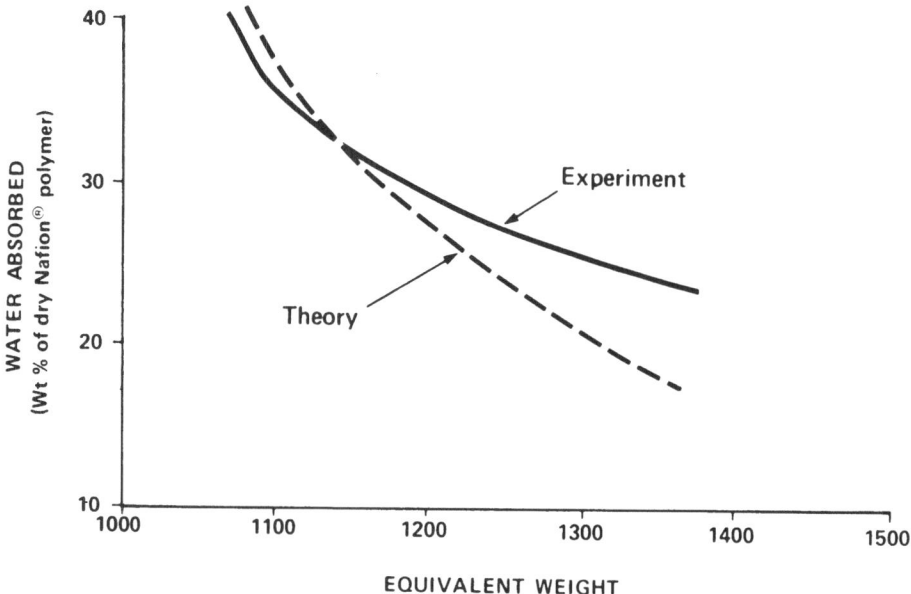

Figure 2:   Comparison of Experimentally Measured and Theoretically
            Predicted Water Absorption vs. Equivalent Weight for
            Standard Nafion®.

     Plots of water absorption and wet density of the sodium-
sulfonate form of the ionomer as a function of the solution molarity
of Na^+ ions are shown, respectively, in Figures 3 and 4 for 1100
and 1200 EW materials. As with Figure 2, both theoretical predictions and experimental data are presented in these latter figures
to demonstrate the ability of the calibrated model to quantitatively
describe the equilibrium solution behavior of standard Nafion®
ionomers.

     As demonstrated by the results presented in Figures 2-4, the
model does a fairly good job of describing the equilibrium density
and water absorption properties of Nafions®. The single most
important study supporting the basic cluster structure in Nafion®
is that of Yeo and Eisenberg (25). They found a small x-ray peak
in the "Nafion-Cs" complex at a Bragg distance of 51 Å (the EW of
the sample material was reported to be 1365).

     The presence of this peak must be attributed to the presence
of large scattering centers (the ion-clusters) in the polymer.
This combination of experimental and theoretical data lead us to
conclude that the conceptual picture of Nafion® is probably
correct. In addition, quantitative application of the conceptual
theory leads to reasonably well predicted macroscopic properties.

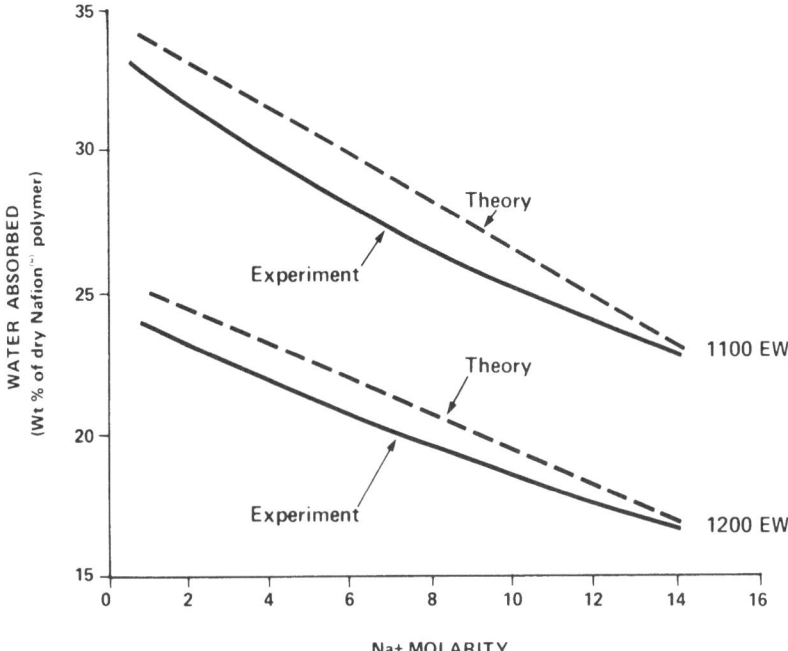

Figure 3:   Comparison of Experimentally Measured and Theoretically
            Predicted Water Absorption vs. Na$^+$ Molarity for Standard
            Nafion$^®$ in the Sodium Sulfonate Form.

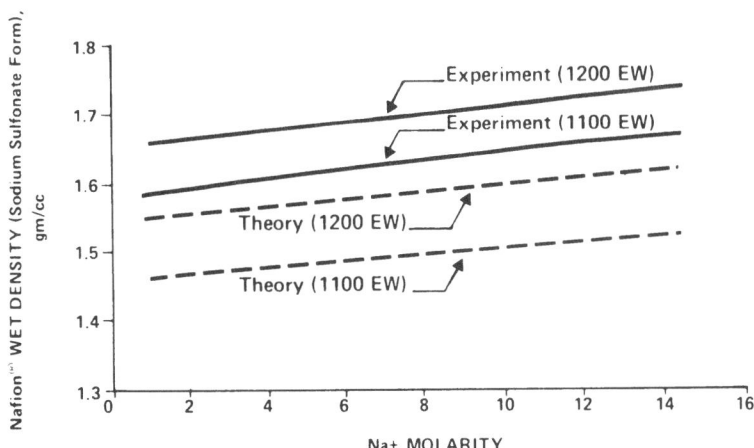

Figure 4:   Comparison of Experimentally Measured and Theoretically
            Predicted Membrane Wet Density vs. Na$^+$ Molarity for
            Standard Nafion$^®$ in the Sodium Sulfonate Form.

(3-PREDICTION OF MOLECULAR ORGANIZATION)

Using the calibrated model, it is possible to make a predictive description of the molecular organization in the ionomeric material. In particular we conclude this paper by reporting the effect of the external solution ion concentration upon the size of the hydration shell cluster. In figure 5 we have plotted $n_h = N$, the number of water molecules per ion-dipole hydration shell, as a function of n, which is proportional to EW, for three different Na^+ concentrations. It can be seen that the water retained in an ion-dipole hydration shell of a cluster increases slowly with increasing ion concentration, and drops off rapidly, in a roughly linear fashion, with increasing n (EW) for all three ionic concentrations. From this we conclude that n (EW) of the polymer, that is, more generally, polymer structure, plays a more crucial role in water domain formation than does external ionic concentration. This is an important observation in regard to design of Nafion® membrane separators.

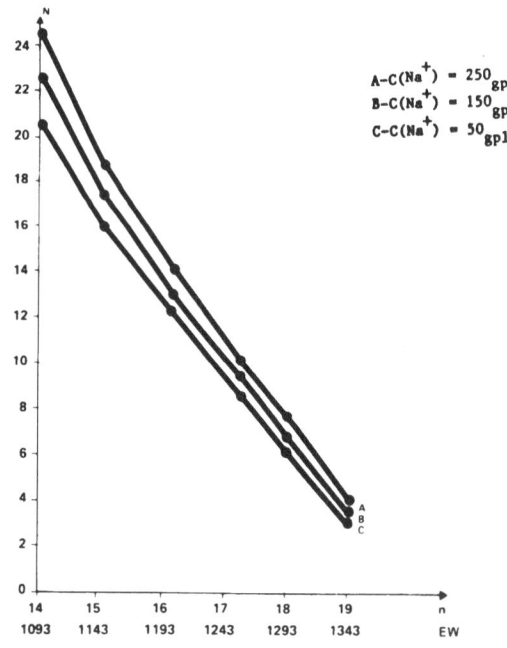

Figure 5: N versus n (EW) for three different external aqueous
            -Na⁺ concentrations.

ACKNOWLEDGEMENTS:

A.J. Hopfinger would like to thank Diamond Shamrock Corporation and his collaborators at Diamond for making this investigation possible.

REFERENCES:

1. A. Eisenberg, Macromolecules, 3, 147 (1970).
2. O.A. Ponomarev and I.A. Ionova, Vysokomol. Soyed. A16, 1023 (1974).
3. T.D. Gierke, "Ionic Clustering in Nafion® Perfuorosulfonic Acid Membranes and Its Relationship to Hydroxyl Rejection and Chlor-Alkali Current Efficiency", presented at 152nd National Meeting, The Electrochemical Society, Atlanta Georgia, Oct. 10-14 (1977).
4. R.W. Rees and D.J. Vaughan, Polymer Preprint, 6, 296 (1965).
5. T.C. Ward and A.V. Tobolsky, J. Appl. Polym. Sci., 11, 2403 (1967).
6. E.P. Otocki and T.K. Kwei, Macromolecules, 1, 401 (1968).
7. E.P. Bonotto and E.F. Bonner, Polym. Preprint, 9, 537 (1968).
8. W.J. MacKnight, L.W. McKenna, and B.E. Read, J. Appl. Phys., 38, 4208 (1968).
9. P.J. Phillips and W.J. MacKnight, J. Polym. Sci., A2, 8, 727 (1970).
10. K. Sakamoto, W.J. MacKnight, and R.S. Porter, ibid., A2, 8, 727 (1970).
11. S.R. Rafikov et al., Vysokomo, Soyed A15, 1974 (1973).
12. C.L. Marx and S.L. Cooper, J. Macromol. Sci., B9, 19 (1974).
13. W.E. Fitzerald and L.E. Nielson, Proc. Roy, Soc., A282, 137 (1964).
14. N.Z. Eril and H. Morawetz, J. Colloid Sci., 19, 708 (1964).
15. A. Eisenberg and M. Navratil, J. Polym. Sci., B10, 537 (1972).
16. A. Eisenberg and M. Navratil, Macromolecules, 6, 604 (1973).
17. M. Navratil and A. Eisenberg, Macrolecules, 7, 84 (1974).
18. A. Eisenberg and M. Navratil, Macromolecules, 7, 90 (1974).
19. E.P. Otocka and F.R. Eirich, J. Polym. Sci., A2, 6, 921 (1968).
20. E.P. Otocka and F.R. Eirich, J. Polym. Sci., A2, 6, 933 (1968).
21. J. Moacanin and E.F. Cuddihy, J. Polym. Sci., C14, 313 (1966).
22. A.J. Hopfinger, "Conformational Properties of Macromolecules", Academic Press, New York (1973).
23. G.A. Segal (ed.) "Semiempirical Methods of Electronic Structure Calculations". Parts A and B, Plenum-New York (1977).
24. W.C.F. Grot, G.E. Munn, and P.N. Walmsley, "Perfluorinated Ion Exchange Membranes", presented at the 141st National Meeting, The Electrochemical Society, Houston, Texas, May 7-11 (1972).
25. S.C. Yeo and A. Eisenberg, Polymer Preprints, 16(z), 104 (1975).

DISCUSSION

Dr. P. K. Subramanyan, Gould Labs, Cleveland: What would be the response of the system Nafion® + H_2O when subjected to pressure, both expansive and compressive? Would it obey a thermodynamic relationship?

I think if your initial assumption with respect to the balance of an expanding pressure by water molecules and a counter-acting pressure by the matrix is correct, then the water molecules within the matrix will respond in a thermodynamic sense to the application of pressure. It will be possible to calculate the partial molar volume of water within the Nafion® matrix from this response.

Prof. A. Hopfinger: It is obviously difficult to quantitatively predict the effect of pressure on the phase separation structure of an ionomer. Qualitatively, compressive (external) orthotropic pressure should decrease water absorption and, therefore, cluster size. The expansive pressure is already modelled in our theory by considering the free energy of hydration. The effect is to swell the material.

With regard to your second comment, we do in fact determine the molar volume of water within the Nafion® matrix. It is part of the energy balance theory.

Dr. Floyd L. Ramp, B. F. Goodrich Research and Development Center, Cleveland: Will annealing Nafion®, in the ionic form with water at high temperature and pressure, or in the SO_2F form by heat before hydrolysis influence the properties of Nafion®, i.e., allow cluster growth?

Prof. A. Hopfinger: I do not know the experimental results of annealing Nafion® although such experiments have, I believe, been done at Diamond Shamrock and Du Pont. Our model cannot handle the possibility of annealing, but my feeling as a polymer physical chemist is that annealing at high temperature and pressure could lock the ionomer into a metastable structure of limited phase separation. That is, the number, size, and relative ordering of clusters will be reduced.

RECENT AND ANTICIPATED DEVELOPMENT OF

INDUSTRIAL ORGANIC ELECTROCHEMICAL SYNTHESIS

Manuel M. Baizer

School of Engineering and Applied Science
University of California
Los Angeles, California 90024 U.S.A.

A history of industrial organic electrochemical synthesis could be constructed from existing books (1) and review articles (2). It would show that until relatively recently the efforts put forth to develop this technology did not compare with the massive assaults on other technological problems. It would show sporadic successes and failures rather than smooth progress from era to era. It would show that certain electrochemical processes that were advantageous at the time they were introduced were later replaced by other "conventional" technologies. Since in the non-communist world the preferability of one process over another depends almost exclusively on considerations of economy of production, one must conclude that until the recent past organic electrochemical processes were too expensive compared to alternative processes leading to the same product(s).

The uneven rate of development in "normal" times of organic electrochemical synthesis (and its required concomitant advances in electrochemical engineering) would per se have made predictions concerning future prospects hazardous. And in addition a series of events have occured since 1965 which were not predicted from past history and whose consequences cannot be extra-polated in a straight line into the future.

1. Monsanto puts its electrochemical production of adiponitrile on stream and Nalco its commercial tetraalkyllead process. Both are large-scale processes and were reported in the world technical and commercial literature. These visible successes encourged many others, who might otherwise not have done so, to initiate programs in organic electrochemical synthesis. Some of these programs were too hastily mounted and ill thought out and had to be abandoned. Nevertheless very much more remains in progress now than before 1965. It can confidently be predicted

that each new announcement of commercialization of a process on a
substantial scale will draw more and more companies into at least
a consideration of the capabilities of this technology. Thus each
firm that introduces a process advantageous to itself spurs the
further development of the field and thereafter the further accrual
of second-order benefits to itself.

Probably more important than the broadening of the scope of
laboratory organic electrochemical synthesis has been the acceleration of the development of the corresponding engineering. Many
new cell designs, electrode materials and forms, and economies of
operation have been reported and adopted in the last fifteen years.
The present-day ability to up-scale a given laboratory process to
pilot plant stage and beyond is orders of magnitude better than it
was just a few years ago.

Another development which followed in the wake of renewed
industrial and academic interest in the field was the publication
of a number of up-to-date books on organic electrochemistry (3)
and engineering (4). In addition, many new national and international forums sprang up at which chemist/engineers could discuss
their and others' work and could become personally acquainted with
many members of the relatively small "brotherhood" of scientists
working in this area. Very often the chemist previously unac-
quainted with the capabilities of organic electrochemistry, who
early scored even a laboratory success by applying electrochemical
techniques became an enthusiastic exponent of the method and
spread his "message" by lecture tours to universities and industries.

2. Since about 1973 a number of events which perhaps could
have been foreseen by a not-yet-developed breed of wise futurolo-
gists have severely jolted the organic chemical industry:
increasing cost and decreasing availability of petroleum feedstocks
which had for a couple of generations supplanted practically all
other sources of the building blocks for large tonnage products;
increasing demand by citizens' representatives (ultimately
government) that industry drastically reduce the concentration of
pollutants it was releasing to air and water.

Now existing processes and new ones considered for installation had to be cost-estimated to take into account not only the
usual items but the cost of disposing properly of unwanted by-
products. The search for cheaper raw materials intensified.
Technologies were and are being reassessed. Processes which for-
merly had seemed too expensive to install and operate could now,
applying the new criteria, become very attractive. The opportunities to introduce organic electrochemical processes designed so
as to be clean, energy efficient, and capable of utilizing, at
least in part, non-petroleum-derived starting materials are increasing.

One consequence of the imposition of the above constraints
upon the chemical industry has been the attempt to convert challenge
into opportunity. E.g., the noxiousness of sulfur dioxide stack

fumes has led to the invention of processes for recovering the gas
and converting it to saleable sulfuric acid; the need to dispose
safely of excess by-product HCl has spurred the adoption of processes for converting the HCl to chlorine or a desired chlorinated
compound. Just so, organic electrochemical synthesis can and will
provide responses to newly developed challenges. This point will
be discussed again later.

The above matters are of concern to individual companies, of
course, but also to governmental agencies of industrialized
countries. E.g., the U.S. Department of Energy has sponsored a
study (5) on the potentialities for energy savings via organic
electrochemical processing; the U.K. has conducted a questionnaire
survey under the direction of Dr. Derek Pletcher, University of
Southampton, on the needs of British industry in this field; the
West German Ministry of Research and Technology has commissioned
a Dechema study (6) which considers among other factors, raw
materials supplies, energy saving, pollution abatement, and the
benefits of introducing new technology; the West German Deutsche
Forschungsgemeinschaft sponsored a Schwerpunktprogramm in applied
electrochemistry 1971-1977 whose results were summarized in a
colloquium (June 1978) and a report (W. Jaenicke, G. Sandstede,
and H.J. Schäfer, editors)

Within the limitations of interpretation mentioned early
above, it is instructive to examine some of the organic electrochemical syntheses which were commercialized or considered for
commercialization even before 1973 (Table 1). The following should
be noted:

1. There is a variety of direct and indirect additions
   and reductions.
2. The largest scale and the largest variety of processes
   stem from very large companies (Monsanto and Asahi,
   BASF respectively).
3. Advantage was taken of the achievement of a particular
   chemical transformation (Monsanto) or particular
   applicability of proprietary technology (BASF) in
   developing the syntheses in question.
4. Sustained and still-continuing effort over a number
   of years has been required.
5. A large number of moderate-sized processes can be
   based on exploitation of expertise in a given
   technology (rotating electrodes, India) and of
   particular economic conditions (again India).

There is no easy way to obtain information on what processes
were taken to an early stage of development and then dropped and,
if so, whether the abandonment should now be reconsidered in view
of the socio-economic developments of the last seven years. We
are forced by the large number of patents that have issued but
have not been consummated to face provocative questions: Were the
inventions ultimately deemed to be uneconomical in the local context? Were they cost-estimated to death by hostile "objective"

Table 1.   Some Reported Practical Electroorganic Syntheses[a]

Product	Raw Material	Co. (Country)	Scale (Status)[b]	Type of Process
Acetylenedicar-boxylic acid	Butynediol	BASF(FRG)	(P.P.)	Oxidation of functional group
Adiponitrile	Acrylonitrile	Monsanto(US) Monsanto(UK) Asahi(Japan)	10^8kg/yr. (Comm) 10^8kg/yr. (Comm) 2×10^7kg/yr(Comm)	Reductive Coupling
o-Aminobenzyl alcohol	Anthranilic acid	BASF (FRG)	(P.P)	Reduction of functional group
1-Amino-4-metho-xynaphthalene	1-Nitronaphtha-lene	BASF (FRG)	(P.P)	Reductive rearrangement
o-Aminophenol	Nitrobenzene	(Japan) Holliday(UK)	(Comm) (Comm)	Reductive rearrangement
Aniline	Nitrobenzene	(India)		Indirect reduction(Ti▪)
o-Anisidine	Nitrobenzene	BASF(FRG)	(P.P)	Reductive rearrangement
Anthraquinone	Anthracene	Holliday(UK)	(Comm)	Indirect oxidation
Benzaldehyde	Toluene	(India)	(P.P.P.)	Indirect oxidation (Mn▪)
o-Carbomethoxy-benzyl alcohol	Dimethyl terephthalate	Hoechst(FRG)	(P.P.)	Reduction of functional group
1,4-Dihydronaph-thalene	Naphthalene	Hoechst(FRG)	(P.P.)	Reduction
1,4-Dihydronaph-thylethers	Naphthyl ethers	Hoechst(FRG)	(P.P.)	Reduction
Dihydrophthalic acid	Phthalic acid	BASF(FRG)	(Comm?)	Reduction
2,5-Dimethoxy-dihydrofuran	Furan	Japan/ BASF/FRG	(Comm) (Comm)	Oxidative addition
Fluorinated organics	Hydrocarbons, aliphatic car-boxylic acids	Dai Nippon (Japan)	(Comm)	Anodic substi-tution
Gluconic acid	Glucose	(India)	3×10^5kg/yr. (Comm)	Oxidation of functional group
Glyoxylic acid	Oxalic acid	(Japan)	(Comm)	Reduction of functional group
Hexahydrocar-bazole	Tetrahydrocar-bazole	BASF(FRG)	(Comm)	Reduction
Hydroquinone or Quinone	Benzene	Several	(P.P.P.)	Paired synthesis or anodic oxi-dation + chemi-cal reduction
Maltol	Furfuryl alcohol	Otsuka(Japan)	(P.P.P.)	Oxidation
Metanilic acid	m-Nitrobenzene-sulfonic acid	BASF(FRG)	(P.P.)	Reduction of functional group
2-Methyl-1-naphthyl acetate	2-Methylnaph-thalene	BASF(FRG)	(P.P.)	Anodic substi-tution
Pinacol	Acetone	(Japan) BASF(FRG)	(P.P.P.)	Reductive coupling
the Pinacol	o-Hydroxypro-piophenone	Sorapec(France)	(P.P.)	Reductive coupling
Piperidine	Pyridine	Robinson Bros. (UK)	1.2×10^5kg/yr. (Comm)	Reduction
Propylene oxide	Propylene	BASF(FRG) Others in UK & FRG	(P.P.P.) (P.P.P.)	Paired synthesis
4,4'-Bis-pyridi-nium salts	Pyridinium salts	(Japan)	(P.P.P.)	Paired synthesis
Salicylaldehyde	Salicylic acid	(India)	(P.P.P.)	Reduction of functional group
Sebacic acid diesters	Adipic acid half esters	BASF(FRG) (Japan) (U.S.S.R.)	(P.P.P.) (P.P.P.) (Comm?)	Crum Brown-Walker
Succinic acid	Maleic acid	(India)	6×10^4kg/yr. (Comm)	Reduction
Tetradecanedioic acid	Monomethyl azelate	Soda Aromatic Co. (Japan)	(Comm)	Crum Brown-Walker
Tetraethyllead	Ethylmagnesium halide	Nalco(U.S.)	(Comm)	Anodic

a) A much more complete compilation is given in Ref. 6)
b) Comm=commercial; P.P.=pilot plant; P.P.P.=past pilot plant

cost estimators (not an uncommon phenomenon)? Or did conservative
Middle Management inflict a lingering death on the new and retain
the comfortable old by discussing and discussing (cf. government
commissions)?

The post-1973 era will continue for many years to present the
problems mentioned above and will therefore certainly witness the
acceleration of the rate of development and adoption of new organic electrochemical syntheses. The challenges (widely recognized) and opportunities (i.e., suggested research and development
areas) are summarized briefly in what follows.

Challenge: Increasing cost of energy

Response: The cost of electrical energy is projected to increase
less rapidly than that of other forms. There is a need to develop
better electrodes in terms of specificity, durability, overvoltage,
etc.; there is an invitation to design better cells which minimize
cell voltage and maximize space-time yields. Processes should be
adapted to operation in undivided cells wherever possible. The
development of suitable paired syntheses obtained in a single
undivided cell will minimize energy requirements.

Challenge: Increasing cost of hydrocarbon raw materials

Response: Take further advantage of specificity and control
possible electrochemically to increase yields and therefore desired utilization of feedstock. Particular opportunities exist
in direct/indirect oxidation/functionalization of hydrocarbons.
Further syntheses from CO, $H_2(CH_3OH, CH_2O\ CH_3COOH)$, CO_2, CH_4, NH_3
and from renewable resources, particularly carbohydrates.

Challenge: Increasing need to control plant effluents

Response: In-cell recycling of by-products, e.g., HX. High yields.
At worst, H_2 or O_2 as sacrificial by-products. Paired reactions
Use of catalytic rather than stoichiometric quantities of all
chemical reagents now used. Use of electrochemistry to destroy
noxious effluents (e.g., CN^-) or to recover toxic trace metals.

Challenge: Barriers to innovation

Response: Possibility of government support of a new technology,
without sacrificing proprietary interests, to minimize initial
risk-taking. Increasing availability of information, expertise,
books, consultants, intensive university courses, sponsored outside work on cell design and operation, licensing arrangements to
reduce initial cost of entry into field.

Challenge: Current research of undefined applicability

Response: Largely unexploited are applications to organic synthesis of (a) newer electrodes such as DSA's, varieties of carbon,
alloys, "chemically modified" electrodes (C.M.E.'s), semi-con-
ductors, oxides, (b) newer electrode forms such as packed and
fluidized beds, carbon fibers, carbon felt, carbon cloth, (c)
newer cell design such as the capillary gap and its variants, the
colloid mill, the "Swiss Roll", the pump cell, (d) newer membranes
such as du Pont's Nafion and Asahi Glass' polymeric carboxylic
acid, (e) techniques such as vibrating electrodes, current reversal, pulsing, wiping electrode surfaces in situ, (f) a widened

electrochemical domain by use of e.g., BF_4^-, PF_6^-, FSO_3^-, and $CF_3SO_3^-$
anions, (g) new reactions such as functionalization of hydrocar-
bons and paired syntheses, (h) new mechanistic probes such as
electrochemistry-cum-spectroscopy, (i) electrocatalysis which may
mean (at least) modification of electrodes to reduce overvoltage
for a given reaction or electro-regeneration of reagents, (j)
biological electrochemistry (7), as, e.g., in electroregeneration
of co-enzymes, (k) photoelectrochemistry (8) which may mean (at
least) combining electrochemical redox with a photo-chemically
promoted reaction, and (1) glow discharge as gas phase electrolysis (9) which has aroused renewed interest after a lapse of
several decades.

Industrial and academic organic electrochemists have well-
defined tasks whose accomplishment will aid in the accelerated
development of industrial organic electrochemical synthesis:

1. The capabilities of practical synthetic organic electrochemistry are more effectively conveyed to mission-oriented
organic chemists and engineers. To all that has already been
done must be added the attempt to introduce the discipline as
part of undergraduate and graduate courses in organic chemistry
and engineering.

2. The horrendous costs to industry of high risk research
and development are abated by academic-industrial cooperation and,
if necessary, by initial government support. This type of activity has been of great benefit in the U.K.

3. There are, as we expect there will be, continued advances
in the much discussed obvious technical aspects of the field:
cell designs of higher space-time yields, more desirable membranes
(if needed) and electrodes, novel electrode materials (e.g., metal
oxides) and forms (e.g., carbon fibres), more confident predic-
ability of upscaling, etc.

4. There is convincing demonstration that most, if not all,
of the present processes which use stoichiometric quantities of
reductants and oxidants (except air) can be replaced by electrochemical processes which use only catalytic amounts of these
reagents.

5. There is greater emphasis on the type of synthesis which
electrochemistry can accomplish uniquely or exceptionally well
rather than in a "me-too" fashion.

6. The manufacturers of medicinals, fine chemicals, and
specialties, long considered obvious candidates for adoption of
electrochemical processes, are targeted for special attention.

7. Research and development programs are much better thought
out than in the past. A top echelon executive must agree that if
a laboratory process is technically and economically sound, if the
product involved is of enduring interest to the company, and if
no completely unforeseen circumstances intervene, he will continue to provide economic and moral support until success is
achieved.

8. As a corollary of some of the points above: chemists and

engineers will work together on a given project from the time of the first successful laboratory demonstration of a synthesis.

ACKNOWLEDGMENT

The author is very grateful to the following, who graciously supplied him data and expert opinion relevant to Table 1: Drs. H. Wendt, Technische Hochschule, Darmstadt, F.R.G.; Udupa, Central Electrochemical Research Institute, Karaikidi, India; W.J.M. van Tilbogr, Koninklijke/Shell Laboratories, Amsterdam, The Netherlands; M. Seko, Asahi Chemical Industry Co., Ltd., Tokyo, Japan; Jansen and Lohaus, Hoechst, Frankfurt a.M., F.R.G.; T. Shono, Kyoto University, Japan; J.M. Savéant, University of Paris VII, France; Reif and Pape, BASF, Ludwisghafen, F.R.G.; K. Junghans, Schering A.G., Berlin (West); W. Wintermeyer, Merck, Darmstadt, F.R.G.

REFERENCE

1. S. Swann, Jr. and R.C. Alkire, "Bibliography of Electro-Organic Syntheses, 1801-1975," The Electrochemical Society, Inc., Princeton, N.J. 1979; G. Dupernell and J.H. Westbrook (Eds.) "Selected Topics in the History of Electrochemistry", Proceedings Volume 78-6, The Electrochemical Society, Inc., Princeton, New Jersey, 1978.
2. M.M. Baizer, J. Electrochem. Soc., 124, 185C (1977); S Swann, Jr., ibid., 99 125C (1952); H.J. Creighton, ibid., 99, 127C (1952).
3. A.P. Tomilov et al., "The Electrochemistry of Organic Compounds", Halsted Press, New York (1972) translation of 1968 book); A.J. Fry, "Synthetic Organic Electrochemistry", Harper and Row, New York, 1972; M.M. Baizer (Ed.), "Organic Electrochemistry", Marcel Dekker, Inc., New York, 1973; F. Beck, "Electroorganische Chemie", Verlag Chemie, Weinheim, F.R.G., 1974; N.L. Weinberg (Ed.), "Techniques of Electroorganic Synthesis", Wiley-Interscience, New York, Part I, 1974, Part II, 1975; D.T. Sawyer and J.L. Roberts, Jr., "Experimental Electrochemistry for Chemists", Wiley-Interscience, New York, 1974; S. Ross, M. Finkelstein, and E.F. Rudd, "Anodic Oxidation", Academic Press, New York, 1975. "Electroorganic Chemistry" (Ed. T. Shono), Japan, 1980 (in Japanese).
4. A.T. Kuhn (Ed.), "Industrial Electrochemical Processes", Elsevier, Amsterdam-London-New York, 1971; P. Gallone, Trattato di Ingegneria Electrochimica", Tamburini Editore, Milano, 1973.
5. Electrochemical Technology Corporation, "Final Report on a Survey on Organic Electrolytic Processes" ANL-OETM-79-5. Propared for the Office for Electrochemical Project Management, Argonne National Laboratories, under contract No. 31-109-38-4209 (1979).

6. DECHEMA, "Forschung and Entwicklung zur Sicherung der Rohstoffversorgung," Part III, Volume 3, "Electrochemical Processes", Frankfurt a.M., 1976.

7. E.g., M.J. Eddowes and H.A.O. Hill, J. Am. Chem. Soc., 101, 4461 (1979); M. Aizawa et al., Biotech. Bioeng., 18, 209 (1976); H. Jaegfeldt et al., Anal. Chem. Acta, 97, 221 (1978); T.C. Wallace and R.W. Coughlin, Anal. Biochem., 80, 133 (1977).

8. For recent application in synthesis, see e.g., J.M. Bobbitt and J.P. Willis, J. Org. Chem., 42, 2347 (1977), H. Inoue et al., Nature, 277 (5698) 637 (1979); P. Yeh and T. Kuwana, Chem. Lett., 1145 (1977).

9. E.g., papers by H. Suhr.

We wish to thank the editors of the Journal of Applied Electrochemistry and of Electroorganic Chemistry for allowing us to reprint substantial portions of this paper.

ELECTROCHEMICAL POWER GENERATION

Karl Kordesch

Technical University Graz
Institute for Inorganic Technology
Graz, Austria, A-8010

INTRODUCTION

Achieving efficient and economic energy conversion and storage
is the new direction of electrochemical technology. The problem of
supplying conveniently small amounts of electrical energy was simply
solved with throw-away batteries. This may not be desirable in the
future, even the cheap manganese dioxide cell should be made rechargeable ! In order to save precious oil improved batteries must
be built for electric vehicles with the additional effect of quietly
improving our city environments. "Refillable" fuel cells may be
the answers instead of rechargeable secondary batteries. They may
be of the alkaline type like the cells built for space applications
or acidic types as used for stationary powerplants more successfully.
The availability of hydrogen as energy carrier and fuel will become
decisive for a future electrochemical power scenario. However,
other elements like lithium or aluminum may be important for the
design of portable electrical power sources with high energy density.

The technology of electrochemical power generation is too
large to be covered within one paper, therefore I will cover some
specific areas of progress in the field of small cells for the consumer and some advance in the knowledge about carbon-based fuel
cells which may become the power sources for the (hybrid-) electric
car in the next century.

The development of alkaline manganese dioxide cells, air-zinc
cells and carbon electrodes of many kinds has traditionally been
pursued in the Cleveland area, partly in the industrial laboratories,
partly here at the Case Western Reserve University, with new élan
now at the Case Laboratories for Electrochemical Studies.

Similar subjects have recently become research topics at the Technical University of Graz. Therefore, the good wishes of the technical personnel of my institute to the people of CLES, at the occasion of the Case Centennial Celebration are most sincere. May be that there are also some historical connotations in order: Cleveland had its Michelson and Morley, while Graz had its Hess and Schrödinger - that Graz had also a Copernicus is only a merit of its 850 years of history as a city.

THE ALKALINE MANGANESE DIOXIDE CELL, A RECHARGEABLE SYSTEM

The decade 1960 to 1970 started the rapid rise of the primary high-drain alkaline manganese dioxide - zinc cell. Table I shows the commercial success of this type of "Dry Cell" by comparing the yearly sales figures in the U.S.A. with those of the Leclanche-type cells.

TABLE I

Sales of "Dry Cells" in the U.S.A.
(in millions)

	1960	1965	1970	1975	1980
Total	151	221	248	480	910
Leclanche	150	210	219	384	546
Alkaline	1	11	29	96	364

Alkaline MnO_2 - Zn cells are rechargeable when the discharge of the manganese dioxide is limited to the extent of the homogeneous reaction phase (from MnO_2 to $MnO_{1.5}$) whereby the protons originating from the water molecules are introduced into the structure of the MnO_2, forming $MnOOH$. As the discharge proceeds, the concentrations of the Mn-III-ions and the OH^- ions in the lattice increase and both species become distributed throughout the metal oxide structure (1). The original MnO_2 lattice expands during the discharge, but has also the capability to contract again on charge. This phenomenon was checked in recent experiments where the limit of the recharge-ability was investigated (2). The expansion was measured with an electronic gage operating on the induction principle so that all mechanical movements during charge- and discharge cycling could be recorded electrically. Figure 1 shows the results: one set of cycles belongs to a freely mounted disc electrode, the other to a confined disc electrode. It is obvious that the behaviour of a cylindrical electrode, confined in a steel can must outperform a poorly supported flat plate electrode, unless it is bonded or put under pressure by other means. The patent literature has recognized these facts empirically by providing binders (3) and structural means of assuring coherence (4).

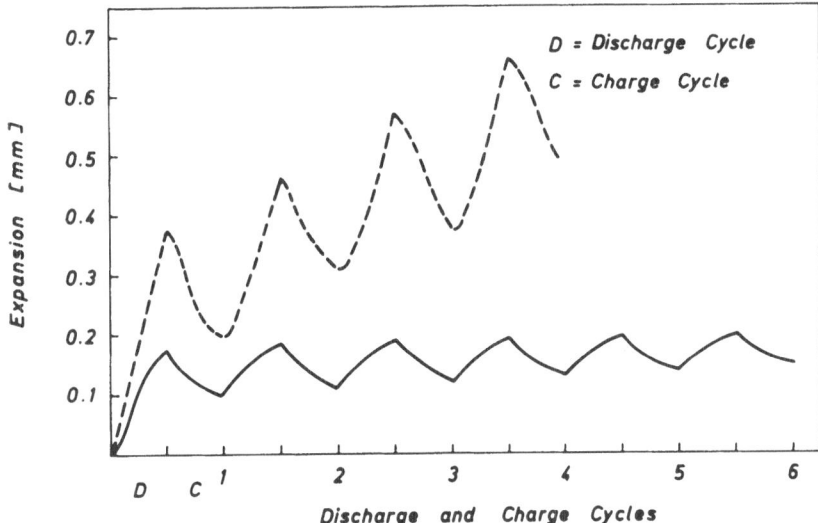

Fig.1: The expansion of a manganese dioxide disc-electrode
during discharge- and charge cycles for a confined
and a freely mounted test sample. The expansions and
contractions are accompanied by resistance changes.

The Rechargeability as a Function of the MnO_2-Type

It was of great interest to investigate the available International Common Samples (5) in respect to their rechargeability.
For this purpose a test cell had to be constructed which provided a
reproducible electrode confinement (preset pressures) and allowed
accurate reference measurements of the electrode potentials (6).
After initial experiments, which eliminated the natural ores
as poorly rechargeable materials, the remaining electrolytic and
chemically produced MnO_2 samples were tested under a pressure of
27 N/cm^2, discharged for 60 min. at a current density of ∼10 mA/cm^2
and then charged for five hours potentiostatically at 1.70 V vs. Zn.
These test conditions resulted in an accuracy of ± one cycle for
the different materials. The pressure was set to the lowest value
which produced accurate results (the test equipment had a range
from 1 to 1000 N/cm). In the following Figs. 2 and 3 the difference
between I.C. No.2, an electrolytically produced MnO_2 and I.C. No.5,
a chemically made MnO_2 is shown. All samples tested were discs
of 0.5 mm thickness and weighed 0.096 g (containing 20 % graphite
and 0.5 % acetylene black). The depth of discharge reached at the
first cycle was 35 % of the 1 e capacity of the MnO_2 sample.

The relationship between cycling behaviour and the capacity
of the MnO_2-sample under primary cell conditions was also investigated. No correlation was found. The surface had no influence either.

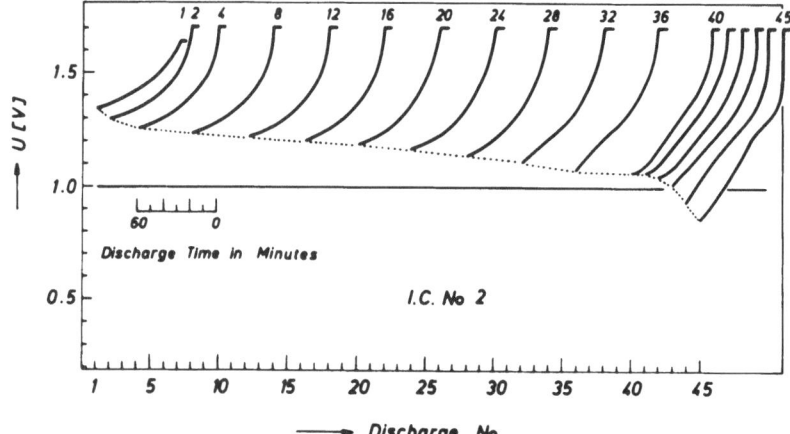

Fig.2: The discharge- charge cycling behaviour of the I.C.2,
International Common Sample of MnO_2, electrolytic
grade No. 2. Depth of discharge: 35 % of 1 e equiv.
Discharge: 10 mA/cm^2 for 1 hr. Charge: 1.70 V vs. Zn.
Pressure of confinement: 27 N/cm^2.

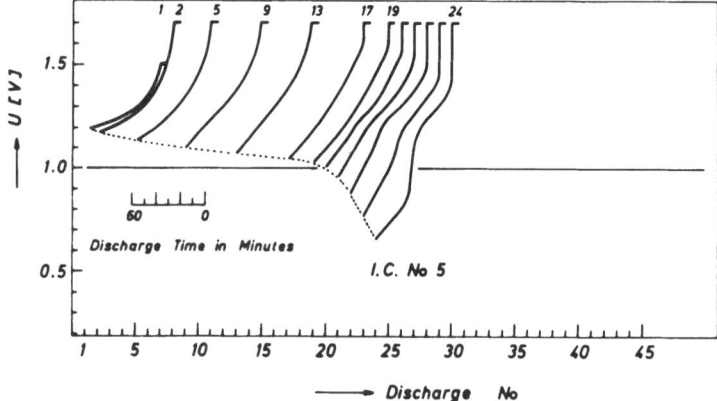

Fig.3: The discharge-charge cycling behaviour of the I.C.5,
International Common Sample of MnO_2, chemically
produced, No. 5. Depth of discharge: 35 % of 1 e equiv.
Discharge: 10 mA/cm^2 for 1 hr. Charge: 1.70 V vs. Zn.
Pressure of confinement: 27 N/cm^2.

Note: the charging of the MnO_2 was done potentiostatically
over a fixed time period of 5 hours. This corresponds
realistically to the "Constant voltage- taper charge
method" recommended for rechargeable MnO_2-Zn cells (7).

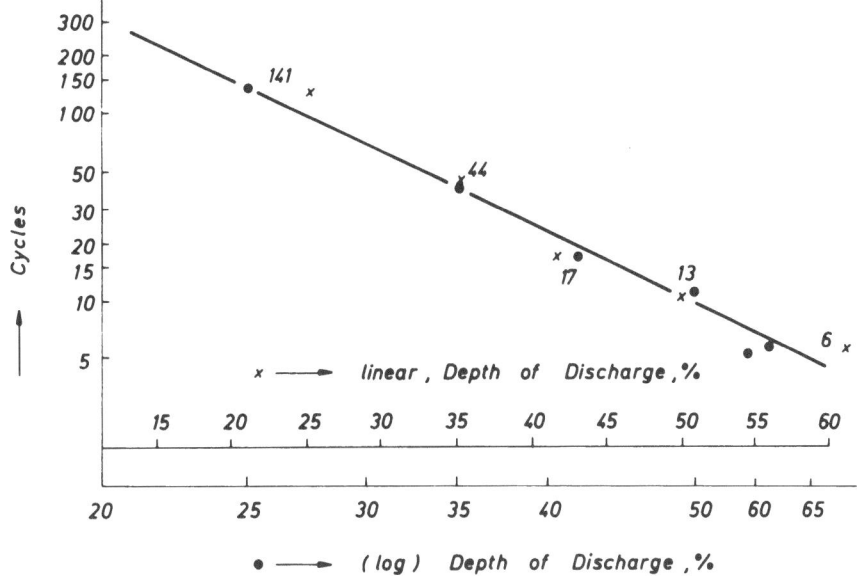

Fig.4: Cycle life as a function of the depth of MnO_2-discharge

Because of the limitation to the range of the homogeneous phase
of discharge, the cycle life is critically influenced by the depth
of discharge. Figure 4 shows a semi-log and a log-log plot of this
property of the manganese dioxide. Other systems show a similar
behaviour, for the lead-acid cell it was explained on statistical
grounds, e.g. gradual loss of active material (8).

The Outlook for a Practical Rechargeable MnO_2-Zn Cell

The good rechargeability of the zinc anode must be established
by providing a conductive structure onto which the zinc can be re-
plated efficiently (9). In a sealed, zinc-limited system an oxide
balance must be maintained to prevent gassing after discharge and
overcharge problems must be avoided if the cell is fully charged and
forgotten on the charge equipment (10). All these requirements of a
foolproof consumer battery must be fulfilled at reasonable cost.
fortunately the supply of manganese dioxide and zinc is not limiting.
As a result of optimization and rebalancing the present outlay
of primary alkaline MnO_2-Zn cells it can be expected that the final
rechargeable cell will have about half the capacity of the primary
cell (e.g. 3 Ah for an LR-14 or C-size cell), with a capability
of perhaps 100 recharges down to 50 % - economically a viable goal!

THE REDISCOVERY OF AIR-ZINC CELLS

Air-depolarized cells have the advantages of high capacity and low cost. However, the necessary air access to the cathode and to other internal components causes moisture losses or gains and also results in carbonation of the caustic electrolyte. The air-zinc cells of the past had therefore a short life after opening of the breathing channels. Modern electronic equipment, like hearing-aid devices or calculators, even watches, have now such a minimal need for current output that very tiny access-holes suffice (11). This mechanical restriction of vapour- and gasexchange can be made more reproducible by applying selective O_2- permeable membranes to control moisture and carbondioxide exchanges (12). To assure also a peak-current capability the air-cathode can be combined with MnO_2 (as catalyst and depolarizer) in hybrid-electrode systems (13).

It is surprising how well these, actually long-time known and abandoned air-zinc cells serve in the new field of microelectronics, establishing a low cost competition for the silver oxide-zinc cells and probably also for some of the non-aqueous Li-cells and other miniature cells where the life expectancy is not of utmost priority.

HIGH RATE PULSE DISCHARGE CELLS

Applications which require high rate discharge have become more frequent in recent years. One example is the starting of small engines , e.g. for lawnmowers. A lead-acid cell which features the spiral-wound thin-electrode design and is marketed for several years in the D-, X- and "beercan"-size" (14) is especially effective for heavy-duty service. Figure 5 illustrates a 10 C discharge rate for D- and X- cells which have a nominal C/10 capacity of 2.5 and 5 Ah when used at low drains (lantern service). With this type of cell it should be noted that the design of a completely sealed lead-acid cell has been successfully approached. The cell is overcharge-able at rates between C/10 and C/20 because the oxygen recombination cycle is effectively used in the large surface system employing a special fibre-glas separator wetted with sulfuric acid - there is no free electrolyte in the cell. From a technical standpoint this cell combines the desired characteristics of an extremely good charge acceptance on a constant voltage regime (2 C- fast charge initially dropping down to C/20 finally) with a sealed construction. The cycle life is not as good as that of the Ni-Cd cell, it varies, but good 200 and more cycles, depending on the depth of discharge are available; another example of low-cost, economy-minded technology.

It should be mentioned that the rechargeable alkaline MnO_2-Zn cell in a rolled (spirally-wound) design can compete with this lead-acid cell in power density and probably surpass it in capacity by a factor of two, considering the heavy weight of the lead system.

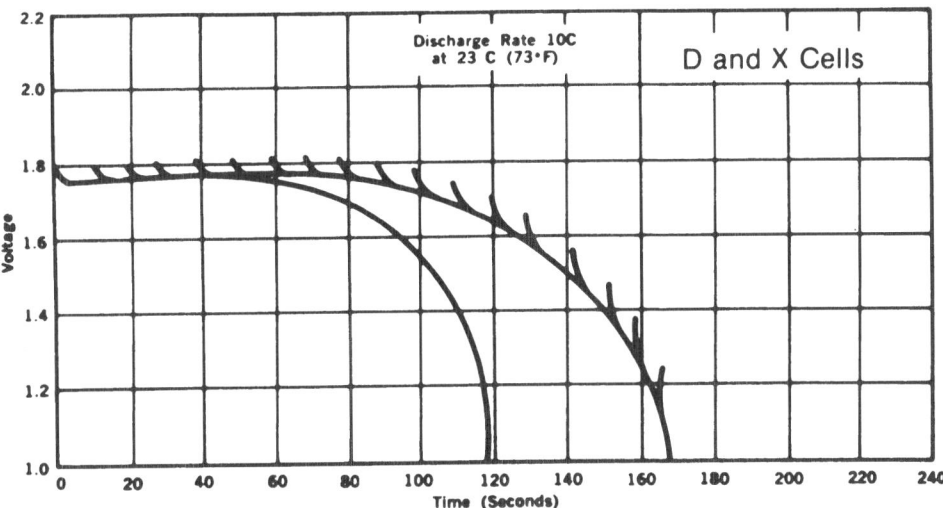

Fig.5: High-rate pulse discharge of spirally wound lead-acid
cells (14). Continuous discharge and 16.7% dutycycle
(10 second pulse of 25 A and 50 A for D and X cells,
respectively, followed by 50 seconds rest).

A primary cell which shows exceptionally high pulse currents
powers the Polaroid SX-70 camera. It is again a thin plate design
which makes use of the large surface available to improve the drain
capability of a very low-cost system: the MnO_2-Leclanche-Zinc Cell.
As a commercial 6-Volt battery for many uses it is now marketed
in a special holder (15). Remarkable is the special gas-venting
system which uses a gas-permeable membrane. The bipolar construction
avoids the cell interconnections which would be required with the 4
standard AAA-cells (alkaline L-30) approximately competing with it
in pulse-drain discharge but surpassing the flat cell in capacity
by a factor of two. An alkaline flat cell battery has not been
put on the market yet. The replacement of the four AAAA-size cells
in the camera battery (Kodak) 7 K 67, or 539 with a thin-plate
alkaline 6 V stack should be the next step to expect. See Fig. 6.

THE COMPETITION BETWEEN AQUEOUS AND NON-AQUEOUS BATTERIES

Lithium batteries with aprotic electrolytes, giving two or up
to three times the voltage of aqueous systems will certainly conquer
the miniature cell market. One of the following papers will give an
exhaustive review of the progress in this field. However, it should
be considered that the lithium price has skyrocketed and even cells
using Li and MnO_2 are getting expensive. A replacement of lithium
by sodium is possible under certain conditions (16).
High power output and rechargeability are still "aqueous" domains.

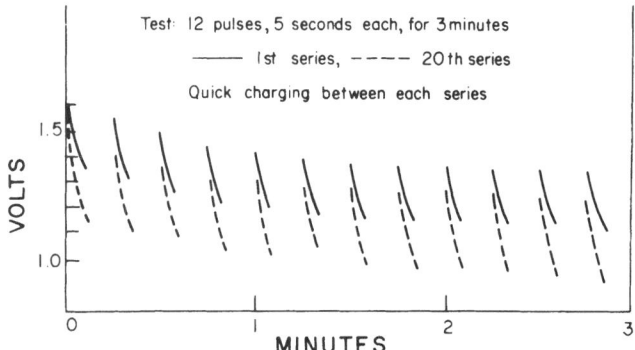

Fig. 6: 20 Ampere pulse-testing of a thin-plate alkaline MnO_2-
zinc battery. Each cell contains 21 g of MnO_2. (4)
Electrode area 30 cm^2, 0.5 mm thick. Recharged for
15 minutes at 1.7 V const. voltage. The continuous
20 A capacity of this cell is 2 Ah (6 min- or 10C rate)

FUEL CELLS AS POWER SOURCES FOR ELECTRIC VEHICLES

Fuel cells are electrochemical powersources which produce
electrical energy as long as they are supplied with fuel and air.
With that capability they fullfill the first requirement for a car
which must operate over long distances. In addition, fuel cells
have a high efficiency, high specific power output and have no noise
or emission problem. In stop-and-go traffic they use fuel only when
power is produced.

The internal combustion system depends entirely on oil products
the need for switching to a transportation system which uses non-
petroleum primary fuels is obvious. If fuel cells are used instead
of secondary batteries then it is possible to avoid the powerplant losses of 60 %, the transmission and the recharging losses
and save the time necessary for recharging the batteries also.
The practically unlimited range of a fuel cell powered vehicle
is probably most important. Only the size of the tank and the fuel
availability determine the range.

The "Hydrogen Economy" based on converted coal-, solar- and
nuclear power is expected to arrive around the year 2000. (17)
Hydrogen or hydrogen carriers (hydrides, ammonia, alcohols, etc.)
can be chosen for converter or direct systems to be selected on the
basis of economic considerations.

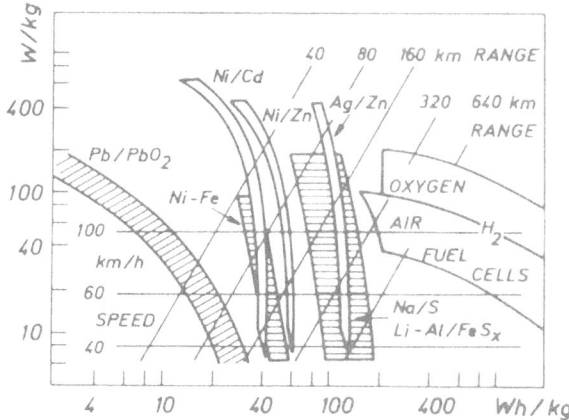

Fig.7: Comparison of different battery systems. The energy
       density is shown as a function of the power output/kg
       needed to propel a ∼1000 kg electric vehicle at a fixed
       certain speed over a required range. Battery-wgt.: 300 kg

The diagram in Fig.7 compares different galvanic systems suitable for electric vehicle propulsion. It shows that a commercially
available lead-acid battery has a capacity of 40 Wh/kg at the 2 W/kg
level(corresponding to the 20 hour discharge rate) but indicates a
drop to 10 Wh/kg when the power demand rises to about 40 W/kg.
At the same time that means that a car travelling at 80 km/hr would
only have a range of 20 km - very marginal, but reflecting the real
situation (18) of todays electric automobile technology, not count-
ing battery and car improvements which are only in the research- and
early development stage. Of course, for the moderate requirements
of fleet vehicles (postal vans, delivery trucks) not exceeding
50 km/h the range becomes 80 km - which is definitely useful in the
city, relieving environmental problems and reducing oil consumption.

Fuel cells have a long history (19) and have come a long way
even in the last decade, especially the power level has increased
and air-operation of a hydrogen fueled battery will be approaching
a peak at 200 W/kg but the better efficiencies (55-60%) will be in
the 100 W/kg range. Oxygen operation has been recommended repeatedly
but the advantages of the high power output are offset by the need
to transport oxygen in some form. To remedy this situation, a peak-
power source (secondary battery) may be used in a hybrid system.

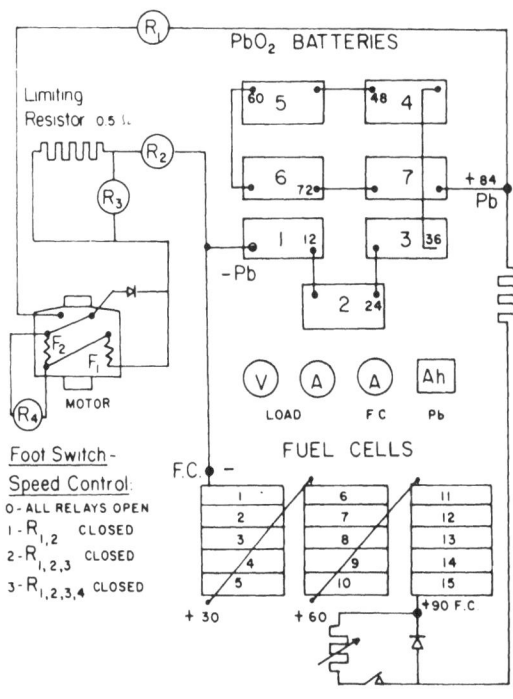

Fig.8: The lead-acid / hydrogen-air hybrid system used in the
Kordesch four passenger car (Austin A-40, 1000 kg). (20)

Several benefits accrue from a hybrid power plant: the rechargeable
battery is always ready to drive the car and will start the fuel cell
system's accessories on demand. A lead-acid battery has satisfying
properties (starter, SLI-battery) and sufficient cycle life if it
is not subject to frequent deep discharges (the rule under electric
car operation with a rechargeable battery alone). Of course, a Ni-Zn
battery would be better – but presently it is still in the development stage. A test vehicle using a lead-acid / hydrogen-air hybrid
system was built by the author in the early 70's (20) and extensive
experiments were performed in respect to the driving capabilities
of such a system. The principal circuit diagram of the Kordesch-
fuel cell hybrid system is shown in Figure 8. A nominal 90 V fuel
cell battery of the alkaline type (Union Carbide Corp.) with a max.
continuous output of 6 kW (enough for operating the car at 60 km/h)
was connected to a 3.4 kWh lead-acid battery with a peak output of
16 kW. The lead battery weight: 150 kg. The fuel cell system weighed
170 kg and had a capacity of 33 kWh. The hydrogen, which was supplied
from compressed-gas steel cylinders (80 kg) was utilized with an
efficiency of better than 50 %. The energy density : 200 Wh/kg.
The driving range of the vehicle was 300 km with each refilling,
which took 2 minutes. The car operated for 4 years over 20,000 km.

Phosphoric Acid Fuel Cells

At present the phosphoric acid fuel cell is technologically the most advanced system. The main reason for this is that it has many attractive engineering features. It is compatible with reformed fuels and the temperature of operation (200 °C) allows a thermal feedback to increase the overall efficiency. For a stationary system there is little incentive to reduce the weight and volume of the battery, but for vehicular applications it would be necessary to optimize all components in respect to power density and cost.

All the phosphoric acid matrix fuel cells developed in the United States (21,22) use noble metal catalysts, however, the amount applied to carbon electrodes has been lowered to less than 0.5 mg/cm^2 and recovery processes are considered feasible. European studies (23) resulted in hydrogen catalysts without Pt-metals but the current densities achieved are relatively low.

Solid Polymer Electrolyte Fuel Cells

These cells are based on the use of perfluorsulfonic acid membranes and are an outgrowth of the Gemini Space Program (24) The SPE fuel cell is attractive for vehicular applications because of its fast start-up capability and its tolerance to shock and vibrations. Compatible with reformed fuels it can also operate at high power densities. The difficulty is water-management, there is a tendency to dry out which can be countered by humidifying the reactants.

Alkaline Electrolyte Fuel Cells

Alkaline fuel cells had been developed to a very high level of sophistication for space craft application (Apollo-Program) and Space-Shuttle power supply testing (25). Unfortunately, these NASA-programs did not concentrate on the saving of noble metal catalyst. Industrial programs aimed at the development of alkaline hydrogen-air fuel cells succeeded to reduce the requirements for expensive catalysts (19) and today it is possible to operate air electrodes at very high current densities even without noble metal catalysts.

The main problem with alkaline fuel cells is the carbonate formation when CO or CO_2 containing fuels are used. The 0.03 % CO_2 in the air can be removed by replaceable absorbers, but the large amounts in reformed hydrocarbon fuels require a rather intricated and uneconomical processing not easily amenable for moving vehicles, at least at the present state of the art. Simple catalytic shift converters which suffice for acidic systems are not good enough for the alkaline system, the remaining % CO is too high.

DEMONSTRATION VEHICLES WITH FUEL CELLS

One of the first vehicles demonstrating the capabilities of hydrogen-oxygen fuel cells was the tractor built by Allis-Chalmers Mfg. Co. in 1959. It was powered by a 15 kW alkaline Battery using porous nickel electrodes and asbestos as separator to immobilize the KOH electrolyte. The same company showed later a golf cart and a lift truck powered by a 3 kW hydrazine-oxygen battery. Probably the most extensive study of a fuel cell powered vehicle was the "Electrovan" of General Motors Corp. built in 1965 - 66 , equipped with a 32 module hydrogen-oxygen battery constructed by Union Carbide Corp.. The nominal output was 32 kW, the peak power 160 kW. The fuel cell power plant had a weight of 1500 kg, with a total car weight of 3230 kg. Max. speed was 115 km/h and the acceleration (0-100 km in 30 sec.) was satisfactory. The driving range with cryogenic gases was 250 km, with compressed gases: 150 km. Fig.9 shows a phantom view of the "Electrovan" (26).

As a response to the results obtained with the GM "Electrovan" the following improvements were considered necessary:
    (1) The use of a fuel cell - secondary battery hybrid
        system to optimize the performance / weight ratio
    (2) Use air instead of oxygen
    (3) Use compressed hydrogen instead of cryogenic hydrogen
    (4) Increase the life of the fuel cell system by operating
        it only when the car is in use - complete shut down
        over longer idle periods or garage time.
    (5) Lower catalyst cost, use of a more active carbon base.

Kordesch (20) built such a vehicle with Union Carbide Corp. electrodes and operated it from 1971 to 1975. The outlay of the hybrid system was shown in Fig. 8. The principal diagram of the fuel cell system is shown in Fig. 10. Lead-acid battery and fuel cell system worked in a load-sharing mode which was achieved by directly connecting the terminals and supplying the motor from the combined output, with no electronic intermediate circuit (except a protective diode to prevent electrolysis in the fuel cells) between.

The hybrid principle was generally adopted for fuel cell cars and further proven in a vehicle of the U.S. Army, a truck fitted with a 20 kW hydrazine-air battery built by Monsanto Research Corp. and a lead-acid battery. Shell Research Ltd. equipped a DAF-44 car with a hydrazine-air/lead battery hybrid in 1972 (27).

What Fuels to use ?

Alcohol looks at the moment as the most promising fuel, it is also contemplated to be the result of efforts to convert coal into an automobile fuel. It is easily converted to hydrogen in a reformer. Hydrocarbons (propane, etc.) are more difficult to convert.

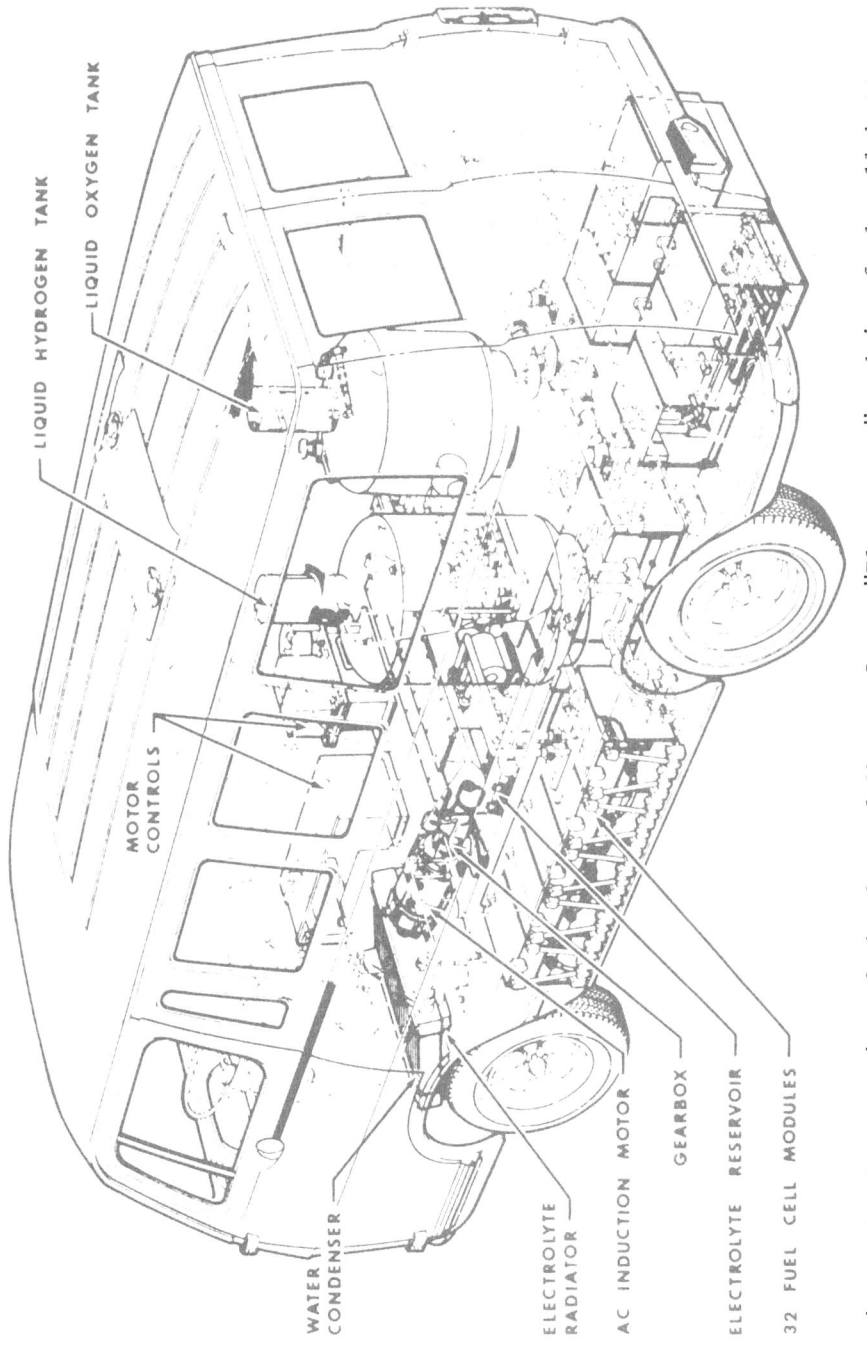

Figure 9: Phantom view of the General Motors Corp. "Electrovan" and its fuel cell battery.

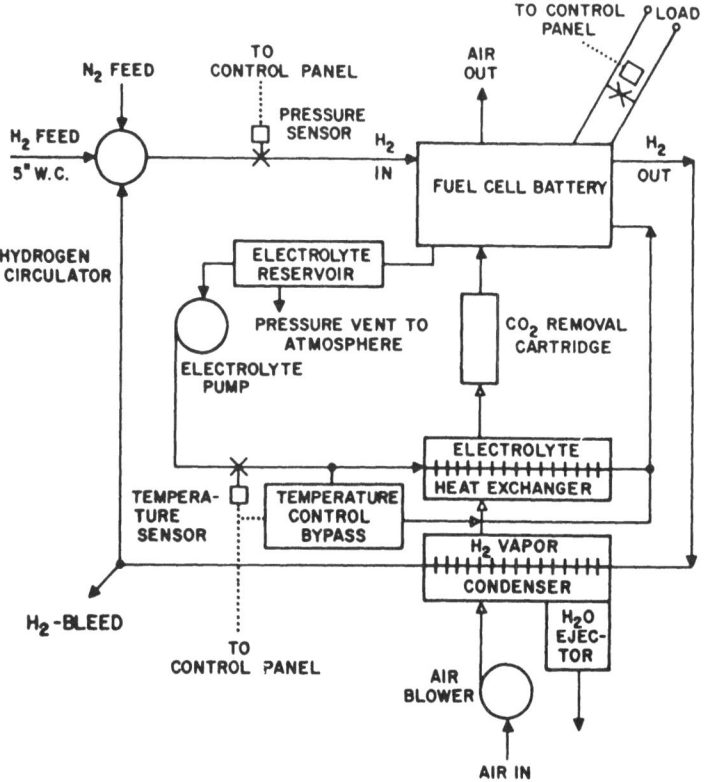

Fig.10: Hydrogen – Air Fuel Cell System (Union Carbide Corp.)

Hydrogen is the ideal fuel and can be used in any type of fuel cell with the best efficiency possible. Methods of storing hydrogen that have been considered include metal hydrides, alkali-metals and also ammonia in liquified form. The latter method has the advantage of a low pressure storage but the disadvantage of needing a cracking unit. In view of the wide availability of liquid ammonia for farm uses and the low temperature of catalytic decomposition it seems to be a good choice – however, it depends on hydrogen production, the same way as hydrogen itself : in a future hydrogen economy there will be plenty. It is interesting to note that liquid ammonia fuel cell systems are reported to be in use in the Peoples Republic of China, serving as power sources in rural areas (28). Unfortunately, 60 % Hydrazine can not be considered a common fuel for public use.
   For the immediate future compressed hydrogen in special light-
   weight tanks looks most promising.

This situation limits the choice of fuel cell systems to only two:
   phosphoric-acid fuel cells for converted fuels and
   alkaline fuel cells for hydrogen only

RECENT CONSIDERATIONS ABOUT FUEL CELL POWERED VEHICLES

A large effort was made in the last 10 years to develop improved secondary batteries. Lead-acid batteries may reach 50 Wh/kg energy density, advanced Ni-Zn or Ni-Fe batteries may store 60 - 70 Wh/kg. The only rechargeable systems over 100 Wh/kg will be the Na-S, the $Li-FeS_2$ and the $Zn-Cl_2$ systems. The latter ones are high temperature batteries with objectionable features which perhaps can be overcome, but the main disadvantage stays: they need recharging.

These considerations have awakened the hopes for fuel cell-powered vehicles. The suggestions to operate automobile engines on H_2 gas have helped the fuel cell aspects as well as the world production figures for ammonia : 50 million tons (made via hydrogen).

The company ELENCO (Electrochemical Energy Conversion) formed in 1976 as a part of Bekaert (Belgium), DSM (The Netherlands) and SCK-CEN (Belgium) are working on the development of alkaline hydrogen air fuel cell systems for vehicles, mainly busses are considered in their program (29). The concept calls for hybrids between 50 kW fuel cells and lead-acid batteries. Cost estimates for the fuel cells were around $ 200 per kW output projected to 1990. Operating costs for the busses are estimated to match those of diesel busses at that time. The availability of nuclear power plays an important role, one must consider that Europe is not as rich on coal resources as the United States.

At Brookhaven National Laboratory (30) efforts were made to check the possibility for building improved alkaline cells. Tests showed that Union Carbide Corp. electrodes manufactured recently give a far higher current density on air than the electrodes used in the Kordesch vehicle (20). The new electrodes combine the old principle of hydrophobicity with that of controlled surface wetting and increased air pressure operation made possible by a dual porosity nickel - carbon electrode. With these electrodes the same size fuel cell system could give 12 kW instead of the previous 6 kW.

The easiest way of convincing people of the merits of a fuel cell vehicle system is a demonstration: at Los Alamos Scientific Laboratories (31) experiments are going on with phosphoric-acid fuel cell systems - with and without reformers - to optimize the components for vehicle applications. A golf cart is operational.

At the Siemens A.G. in Erlangen, Germany, the work on hydrogen oxygen systems is continuing, a 7 kW alkaline battery operating at 80 °C at a pressure of 2 bar weighs only 70 kg (complete). It has the dimensions 24,5 cm x 24 cm x 100 cm, would fit a car easily.
Also in Germany, the studies at AEG-Telefunken aiming at non-noble metal hydrogen catalysts are continued. From Russia recent News: The Moscow Power Institute reports about a H_2-air powered car.

The Improvement of Fuel Cell Electrodes

     To produce electrodes with high performance characteristics, long life and low cost is the goal of industrial developments. In many instances fundamental knowledge turns out to be missing, carbon especially needs a better product definition and research in respect to the electrochemical and catalytic properties is necessary. An extensive literature search on carbon and a patent review on process-technology was made for Brookhaven National Laboratory by a study group from the Technical University Graz (32). The finding was that electrode development work is still a matter of trial and error with only a few important guidelines existing:

(1) Carbon materials suitable for electrodes are commercially produced for other purposes (e.g. the rubber industry). High chemical stability, electrical conductivity, a medium large surface are points for selection.

(2) Carbon blacks and pyrolytically deposited carbon are the preferred choices .

(3) Gas activation increases the low surface of some stable carbons which have had undergone heattreatments. Oxidizable groups ("hydrogen content") are removed by gas activation.

(4) Catalyst deposition must be uniform, extreme thin active layers are highly efficient in Pt-utilization. The decrease of catalyst surface can be slowed down in several ways.

(5) The binding agent PTFE which provides also the electrode repellancy must be carefully used : aqueous suspension and dry powder techniques need very different processing steps.

     Fig. 11 allows the visual comparison of carbon-black material before and after steam activation at 1000 °C. The magnification of the electron microscope pictures is 65,000 X. The activation with CO_2 produces similar results, only the exposure time is longer. (33) Recently electron microscope pictures showing the actual graphite structure and the Pt-catalyst on it have been produced (34). There is no doubt that the knowledge about carbon is increasing!

     A tool for the exploration of carbon-material reactivity under various conditions is the microcalorimetry. Different types oxidize faster, react with catalysts differently, but everything is expressed in the heat evolution -cumulative, unfortunately- and can at least be judged by comparison. It will take some time to unravel the reaction rates, partial contribution of side reactions, etc., but the results so far are very encouraging. The project is supported by the Austrian "Fonds zur Förderung der wissenschaftl. Forschung". The Fig. 12 shows two microcalorimeter curves, on the left the behaviour of Vulcan XC 72, a product of Cabot Carbon Co., frequently used in fuel cell electrodes of the alkaline and acidic types, on the right Shawinigan Acetylene black, a product of Gulf Oil Co. The curves of Pt-carbon interaction differ very much, the heat equilibrium after 15 days is significantly higher with Vulcan XC 72.

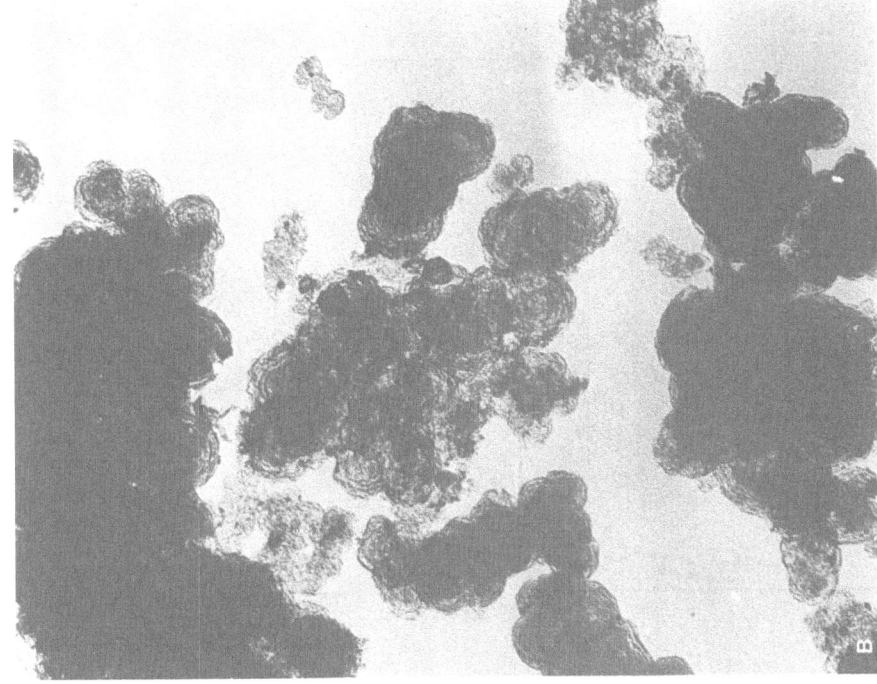

Fig.11: Lampblack-carbon before (left side) and after (right side) steam activation (65,000 X)

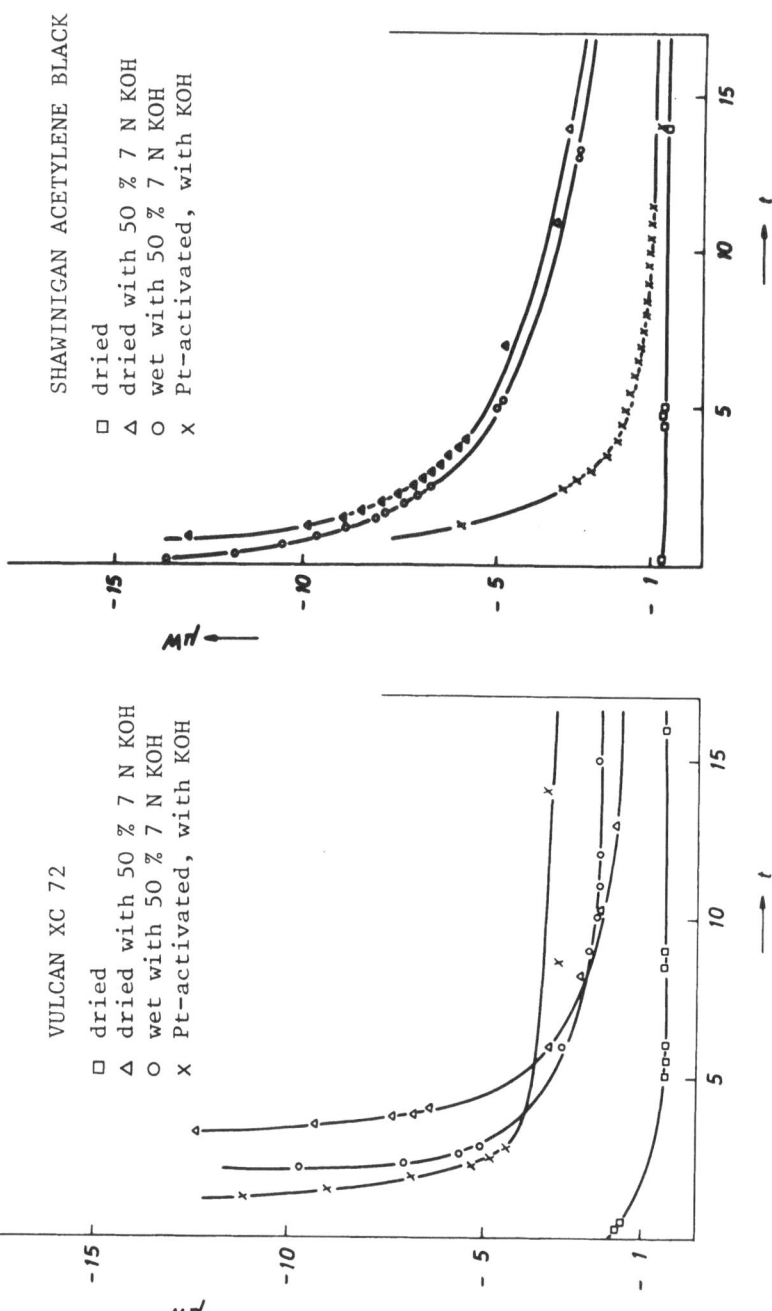

Fig.12: Left: Microcalorimeter curves of Vulcan XC 72, right: curves of Shawinigan Black

The same series of carbon materials has been investigated by cyclic voltammetry in respect to corrosion resistance in phosphoric acid (35). All these investigations are made in connection with the observation that the stability of carbon electrodes and catalysts seems to be less good at higher voltages (open circuit conditions) than under load conditions.

TESTING THE LEAD-ACID BATTERY BEHAVIOUR UNDER HYBRID CONDITIONS

The part of the secondary battery in a hybrid system is not without problems. Especially the charge acceptance at better than 70 % state of charge is a matter of concern (36). Since it is difficult to assess the performance of a hybrid system on the bench it was decided to operate the electric vehicle which has carried the fuel cell system until 1976 (20) as a hybrid vehicle with a 7 kW motor generator as power supply. Driving tests and battery tests were performed by the NASA Lewis Research Center under sponsorship of the Division of Transportation & Energy Conservation (37).

Similar tests are now continued at the Technical University of Graz with the purpose of studying the efficiency of other fuels than gasoline (for the motor-generator) and arriving at test results on lead-acid batteries which describe the behaviour under partial state of charge. Surprisingly, battery manufacturers do not have available such informations as e.g. the peak current obtainable from less than fully charged batteries at different (low) temperatures. Such data are important for hybrid operations of any kind, including mechanical (e.g. flywheel-) combinations.

CONCLUSIONS AND OUTLOOK FOR FUEL CELL OPERATED VEHICLES

The major task is still in the future: electrodes must be improved and the cost of the systems have to become lower. The biggest impetus for fuel cell progress will be the successful completion of the 4.8 Megawatt powerplant presently constructed in New York. A second powerplant (with improvement) is being built for Tokyo. For alkaline cells the important technological push may come from the successful use of air electrodes in chlorine-alkali cells for lowering the cell voltage. Once electrodes are mass-produced, the experimentations in the electric vehicle field will flourish.

REFERENCES

1. A. Kozawa, "Electrochemistry of MnO_2", in: Batteries, Vol. 1, Manganese Doixide, K.V. Kordesch, ed., Marcel Dekker, N.Y.,1974
2. R. Chemelli, J. Gsellmann, G. Körbler, K. Kordesch, Second International Manganese Dioxide Symposium, Tokyo, Oct. 27-29,1980

3.   K.V. Kordesch and R.E. Stark, U.S. Pat. 3,113,050 (1963)
4.   K.V. Kordesch and A. Kozawa, U.S. Pat. 3,945,847 (1976)
5.   Interntl.Common Sample Office, P.O.Box 6116, Cleveland, O.,44107
6.   K. Kordesch, J. Gsellmann and K. Tomantschger, 5 th Australian
      Electrochem. Meeting, Aug. 17-22, Perth, Australia, 1980
7.   EVEREADY Battery Engineering Data, Union Carbide Corp., 1976
8.   E. Voss and G. Huster, Chemie Ing. Technik 38,1966, p. 623
9.   K.V. Kordesch, U.S. Patents 3,042,732 (1962) and 3,288,642 (1966)
10.   K. Kordesch and J. Gsellmann, 11 th Power Sources Symp. Brighton
      in : Power Sources 7, J.Thomas, ed., Academic Press, 1979, p.557
11.   J. W. Cretzmeyer, H.R. Espig and R.S. Melrose, ibid.,p. 269
12.   K.V. Kordesch, U.S. Pat. 4,105,830 (1978)
13.   K.V. Kordesch, U.S. Pat. 3,883,368 (1975)
14.   GATES Energy Products, Battery Application Manual 1980
15.   Polapulse P-100 Six Volt Battery, Polaroid Corp., Kit No. 4155
16.   K.V. Kordesch and S.J. Cieszewski, Union Carbide Res.Rep.1975-8
17.   J.O'M. Bockris, The Solar-Hydrogen Alternative, J. Wiley, 1975
18.   K.V. Kordesch, Batteries, Vol.2, Lead-Acid Batteries and
      Electric Vehicles, Chapter 2, pp201-430, Marcel Dekker,N.Y.1977
19.   K.V. Kordesch "25 Years of Fuel Cell Development", Journ. of the
      Electrochem. Soc. 125, March 1978, pp. 77 C - 91 C
20.   K.V. Kordesch, Journ. Electrochem. Soc. 118, May 1971, pp.812
21.   L. Handley, United Technologies Corp., Natl. Fuel Cell Seminar
      June 26 - 28, Bethesda, M., 1979, Department of Defense, DOE.
22.   H. Marn, L. Christner, S. Abens, B. Baker, Energy Res.Co., ibid.
23.   "From Electrocatalysis to Fuel Cells", G. Sandstede, ed., 1972,
      Battelle Institute, Seattle Res. Center, Univ. of Wash. Press.
24.   J.F. McElroy, General Electric Co., National Fuel Cell Seminar
      July 11-13, San Francisco, Calif.,1978, Deptmt. of Defense, DOE
25.   Pratt & Whitney Aircraft Co., L.B.J. Space Center Houston,1973.
26.   SAE-Congress, Detroit, 1967, Paper Nos. 670176, 670181, 670182.
27.   M.R. Andrew, et al., SAE-Congress, New York, 1972, Paper 720191.
28.   Cha Chuan-sin, et al., Wu-han Univ., Power Sources 7, pp. 769.
29.   H. Van den Broeck, Progress in Batteries and Fuel Cells, Vol.2
      JEC Press Inc., 1979, P.O.Box 42041, Cleveland, O., 44142
30.   J. McBreen, G. Kissel, K.V. Kordesch, F. Kulesa, E.J. Taylor,
      E. Gannon, and S. Srinivasan, 15 th IECEC, Seattle, Wash.,1980
31.   "Fuel Cells in Transportation", B. McCormick, J. Huff, S. Srini-
      vasan, R. Bobbett, LASL-Report7634-MS, 1979.
32.   K. Kordesch, Survey about carbon and its role in phosphoric acid
      fuel cells, Final Report, Contract BNL 464459-S, Dec. 31, 1979.
33.   Ch. L. Mantell, Carbon and Graphite Handbook, Interscience,1968
34.   P. Stonehart, Stonehart Assoc.,Inc., National F.C. Seminar 1980
35.   J. McBreen, H. Olender, K.V. Kordesch and S. Srinivasan, Abstr.
      No. 21, Electrochem. Soc. Meeting, Hollywood, Fl., Oct.5, 1980
36.   K.V. Kordesch, Performance of Lead Batteries in a Generator-
      Hybrid Vehicle, 28 th Meeting of ISE, Sept.18-23, Varna, 1977
37.   R.F. Soltis, J.M. Bozek, R.J. Dennington and M.O. Dustin,
      "Baseline Tests of the Kordesch-Hybrid Passenger Vehicle" U.S.
      DOE, Deptm.of T. & E.C., CONS/1011-14,NASA-TM 73769, June 1978

DISCUSSION

Prof. John O'M. Bockris, Texas A & M University: I express surprise
that Professor Kordesch has not mentioned in his survey of electrochemical power generation the new fuel cells formed with sulfonic
acid electrolytes. These give a considerable increase in performance and are being tested in cars at Los Alamos.

    John Appleby of EPRI recently told me that he thought that
1 kW per kg would be obtained with these cells. This would then be
about the same as an internal combustion engine, and if Appleby's
statement can be confirmed in practice, I would see it resulting in
a direction in the effort in secondary batteries and a transfer to
concentrating on fuel cells for automotive transportation.

Prof. K. Kordesch: The evaluation of solutions of trifluoromethane
sulfonic (triflic) acid as fuel cell electrolyte is still going on.
Performance increases of more than 50 mV (compared with phosphoric
acid) have been achieved at similar load conditions. The new electrolytes make it also possible to operate at 60°C. Preliminary
studies in 340 cm^2 multicell units (at Energy Research Corporation)
have had good results after changing the wetting characteristics of
the electrode surface. The incoming air must be presaturated. Heat
removal problems exist at the low temperatures with matrix cells.

    At ECO, Inc., difluoromethane diphosphonic and -disulfonic
acid have been studied as electrolytes. At operating temperatures
in the 230°C range they seem to be stable.

Prof. W. Vielstich, Bonn University:

1. What is the main step forward in improving the energy efficiency
of iron air secondary batteries from 40 to 50%?

2. Is an energy efficiency of 50% acceptable for E.V. application?

Prof. K. Kordesch:

1. Both electrodes, the air cathode and the iron anode still need
improvements. The iron electrode self-discharge problems and the
gassing on charge are the main reasons for the low overall cycling
efficiency.

2. An energy efficiency of 50% is poor for a rechargeable battery;
it is marginally acceptable for a fuel cell system operating on
hydrogen and air. If one adds necessary reformer losses then the
efficiency is not far from that of a diesel engine.

BATTERIES OF THE FUTURE FOR VEHICLE APPLICATIONS

Elton J. Cairns

University of California, and
Lawrence Berkeley Laboratory
Berkeley, California   94720

I. INTRODUCTION

In consonance with the theme of this centennial symposium, it is appropriate to consider how electrochemistry in the automotive industry can make important future contributions to energy independence and a cleaner environment. Electric vehicles powered by rechargeable batteries offer the opportunity to shift a portion of the transportation energy demand away from petroleum, and toward other energy sources such as coal, nuclear, and hydroelectric. The importance of this opportunity can be understood by examination of Figure 1, (1) which is an energy flow diagram of the U.S. energy economy. Note that the transportation sector consumes an amount of oil equal to the total U.S. production (and more than all of the imports*). A reduction in the petroleum consumption by the transportation sector would allow a corresponding reduction in oil imports. This could be achieved by the use of electrical energy for at least part of our transportation. Note (Figure 1) that only a small fraction of the U.S.'s electrical energy is produced from oil.

The largest single segment of the oil consumption in the transportation sector is that of automobiles (about 55%). Therefore the largest impact with regard to shifting the petroleum demand can be made by the development of electric automobiles (rather than vans, trucks, etc.) that are attractive in the marketplace. It is well known (and will be clear from the discussion below) that the major impediment to the development of electric vehicles is the lack of batteries which have adequate

---

*Actual imports have been less than shown in Figure 1.

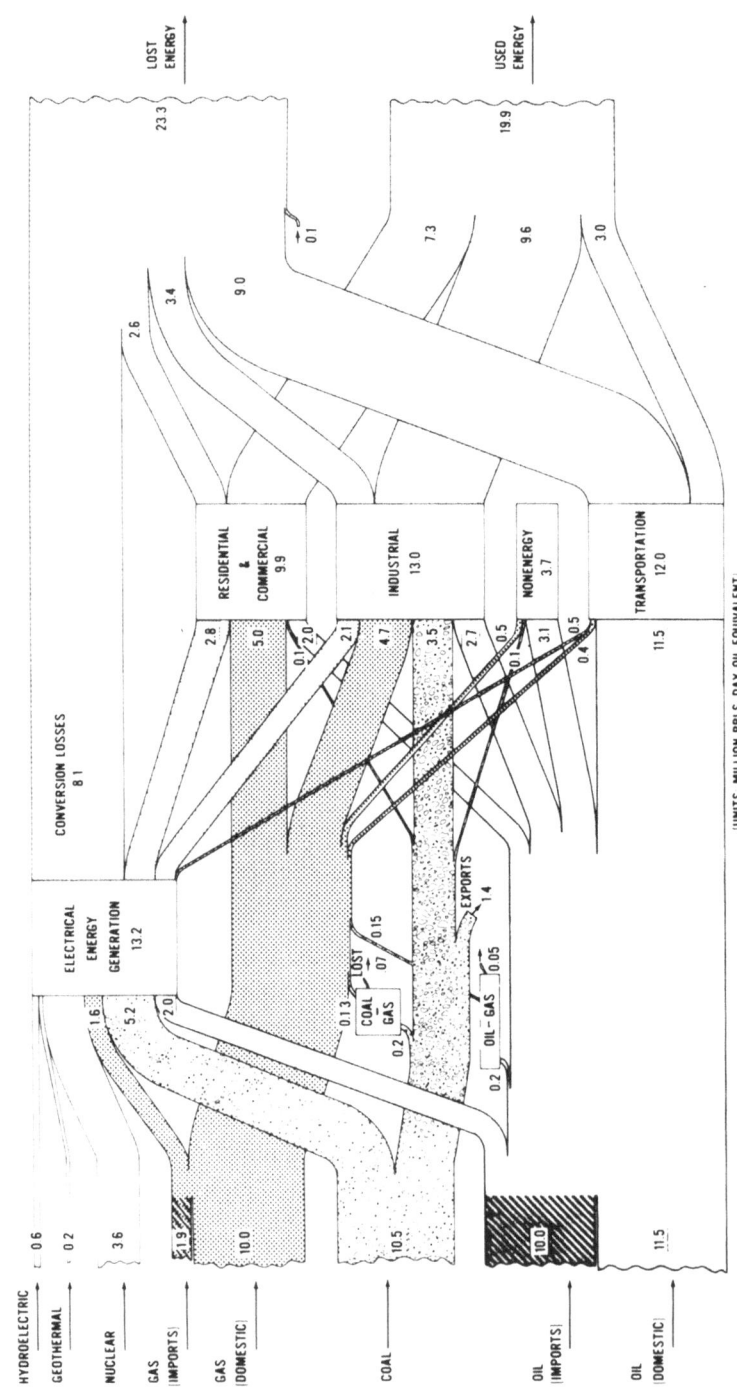

Figure 1.   Diagram of the energy economy of the U.S., projected to 1980. The units are expressed in millions of barrels of oil equivalent per day. (1 barrel of oil = 5.8×10^6 BTU)

performance (specific power and specific energy), durability, and cost. The main performance limitation has been inadequate specific energy (Wh/kg), proportional to vehicle driving range between recharges. Because of this, it is most appropriate to consider urban and suburban electric automobiles, rather than general purpose (including cross-country) electric automobiles. An urban electric automobile is shown in Figure 2. The gasoline version of this vehicle weighs about 1 tonne (1000 kg); electric versions weigh somewhat more, depending on the type of battery.

In addition to shifting the transportation energy demand somewhat away from petroleum, there is also the opportunity to increase the overall energy efficiency of automobiles. The energy efficiency for a spark-ignition automobile vs. that of an electric automobile, starting with coal as the primary fuel is shown in Figure 3. There is an efficiency ratio of about 1.7, favoring the electric automobile, even with the optimistic assumption of syn-fuel production at 70% efficiency. This efficiency advantage can be realized only if the mass of the electric vehicle is not so great as to counterbalance it. Of course, the relative effectiveness of energy use is the product of the efficiency ratio and the mass ratio of the vehicles.

A final advantage of electric vehicles is that of minimal air pollution. The pollutants are not released at the vehicle, but at

Figure 2. Photo of General Motors Electrovette, an urban electric automobile.

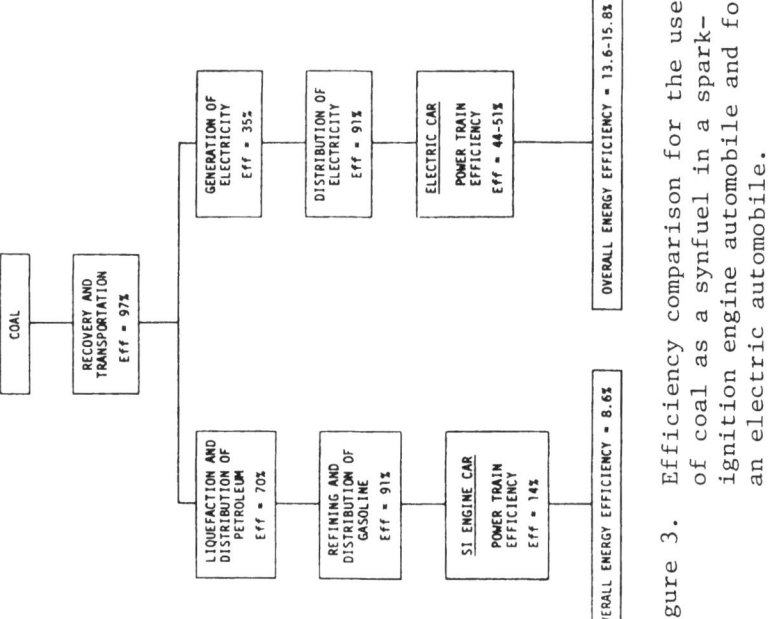

Figure 3. Efficiency comparison for the use of coal as a synfuel in a spark-ignition engine automobile and for an electric automobile.

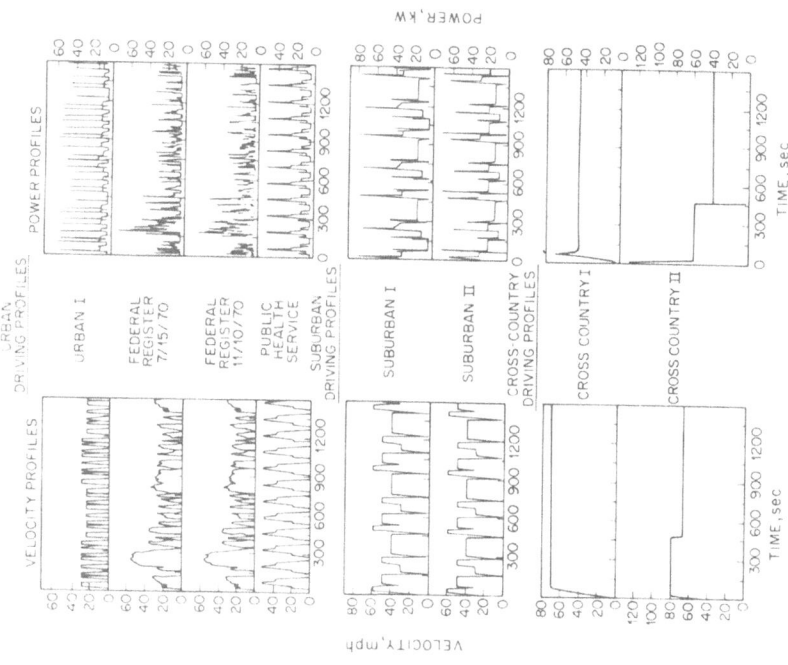

Figure 4. Sample driving profiles and corresponding power profiles (for a 1700 kg automobile). (3)

the power plant where they are more amenable to effective monitoring and control.

II. POWER AND ENERGY REQUIREMENTS FOR ELECTRIC VEHICLES

In order to devise appropriate batteries for electric vehicles, it is necessary to know the power and energy requirements of
the vehicle. These can be computed in a straightforward manner
from the equations of motion of the vehicle, and typical driving
profiles (velocity vs. time). The equations of motion are: (2,3)

$$P_b = \frac{P_r}{E_m \cdot E_e} + \frac{P_a}{E_{ae}} \tag{1}$$

$$P_r = V(F_r + F_w + F_g + F_a) \tag{2}$$

$$F_r = M_v gK \ (1 + 0.022 \ V) \tag{3}$$

$$F_w = \rho_a C_d A_f \ \frac{V^2}{2} \tag{4}$$

$$F_g = M_v g \ \sin \Theta \tag{5}$$

$$F_a = 1.1 \ M_v \ \frac{dV}{dt} \tag{6}$$

where P_b = power required from the battery, watts

P_r = power required at the rear wheels, watts

E_m = mechanical efficiency of the drive train (transmission,
          differential)

E_e = electrical efficiency of the drive train (power elec-
          tronics, motor)

P_a = power required for the work of the accessories

E_{ae} = efficiency of the accessories

V = vehicle velocity, meters/second

F_r = rolling resistance of the tires, newtons

M_v = vehicle test mass, kg

g = gravitational constant, 9.807 meters/second^2

K = coefficient of rolling resistance for the tires (0.008-0.015)

F_w = wind resistance, newtons

ρ_a = density of air, kg/m^3

C_d = air drag coefficient for the vehicle, 0.3-0.7, depending on streamlining

A_f = frontal area of the vehicle, m^2

F_g = gravitational force, newtons

Θ = angle of inclination of road

F_a = acceleration force, newtons.

Several driving profiles are presented in Figure 4. (3) These were used in conjunction with Equations 1-6 to yield the power profiles shown in Figure 4 for a vehicle mass of about 1700 kg. (3) It has been found that the energy consumption per unit distance, per unit vehicle mass is not very sensitive to the details of the driving profile, so that the results of the many computations can be summarized in the simple form of Table I. (4) A convenient set of numbers for battery design use is an energy consumption of 0.15 kWh/T·km, or 0.15 Wh/kg·km, and a power of 35 kW/T for acceleration.

For good vehicle design, a maximum of 30% of the vehicle mass may be assigned to the battery. (2) This allows an estimate of the vehicle range to be made, if the specific energy of the battery is known. Conversely, for a desired vehicle range, the required specific energy of the battery can be calculated:

$$R = \frac{S_pE \times f_b}{0.15}$$

where R = vehicle range under urban driving conditions, km

S_pE = specific energy of the battery, Wh/kg

f_b = fraction of the vehicle test mass assigned to battery.

Table I.   Energy and Power Requirements
for Urban Electric Vehicle

Energy Consumption*
   At Axle   0.10-0.12 kW·h/T·km
   From Battery   0.14-0.17 kW·h/T·km
   From Plug   0.18-0.23 kW·h/T·km

Peak Power Required
(0 to 50 km/h, $\leq$ 10 s)
   At Axle   25 kW/T (Test Wt.)
   From Battery   35 kW/T (Test Wt.)

Average Power Required

	At Axle	From Battery
Urban Driving		
(Avg. 32 km/h)	3-3.5 kW/T	4-5 kW/T
50 km/h Cruise	3-3.5 kW/T	4-5 kW/T

*These energy consumption figures correspond to urban
driving profiles such as the Federal Register driving
profile, and represent an average speed of about 32
km/h.

For a vehicle range of 150 km, and f_b = 0.3, the specific
energy of the battery must be 75 Wh/kg. Any battery with a lower
specific energy cannot achieve a 150 km range in a well-designed
electric vehicle. This is the major problem for battery development. No currently-available rechargeable batteries have a
specific energy of 75 Wh/kg or more, as will be seen below.

III.   ADVANCED BATTERIES FOR ELECTRIC VEHICLES

During the last decade, there has been an increasing interest
in the development of batteries having performance characteristics
attractive for electric vehicle use. The first characteristic
viewed as important has been the specific energy. Raising the
specific energy of a battery requires the use of reactants with
greater (negative) free energies of reaction, and lower equivalent weights. Some of the candidate reactants are compatible
with ordinary aqueous electrolytes, but others, such as the
alkali metals, require the use of nonaqueous electrolytes such as
molten salts, at elevated temperatures. First, the cells using
aqueous electrolytes will be discussed, then those with non-
aqueous electrolytes at high temperatures. To place the recent
developments in perspective, it is useful to review the status of
the only type of electric vehicle battery that is commercially
available now. Table II summarizes the situation for the Pb/PbO_2

Table II

$$Pb + PbO_2 + 2H_2SO_4 \rightarrow 2PbSO_4 + 2H_2O$$
$$E = 2.095 \text{ V}; \; 175 \text{ W·h/kg Theoretical}$$

Status
      Specific Energy   22–40 W·h/kg @ 10 W/kg
      Specific Power   50–100 W/kg @ 10 W·h/kg
      Cycle Life   300+ @ 10 W/kg, 60% DOD
      Cost $50–100/kW·h

Recent Work
      Replace Sb with Ca in positive current collector
      Maintenance-free cells
      Use $4PbO \cdot PbSO_4$ instead of $PbO + Pb_3O_4$ in positives
      Low-resistance current collectors
      Circulating electrolyte

Problems
      Sealing of cells
      Positive current collector corrosion
      Cohesion and adhesion of PbO_2
      High internal resistance
      Heavy

system. Notice that the highest specific energy achieved by a
Pb/PbO_2 module of any significant cycle life is 40 Wh/kg, or 23%
of the theoretical value (175 Wh/kg). Recent work to achieve this
value has included special computer-designed low-resistance current collectors, and provision for circulating the electrolyte to
promote rapid mass transport of electrolyte to the reaction sites.
A well-designed electric automobile can be expected to have an
urban driving range of 80 km with a 40 Wh/kg Pb/PbO_2 battery.

A modern version of the old Edison cell (Fe/KOH/NiOOH) has
been under development recently, and shows higher specific energy
than the Pb/PbO_2 cell, as might be expected on the basis of the
theoretical specific energies (267 vs. 175 Wh/kg). Table III
shows the status of this system. Full-sized vehicle batteries
are under test, and show the expected levels of performance. An
urban range of 100 km should be available. Lower-cost Fe and
NiOOH electrodes are being developed. One continuing set of
problems with this system is the low efficiency (60+%) and resultant high heat generation rate, caused by excessive gas evolution
primarily H_2 at the Fe electrode during recharge. This means
that a sealed, maintenance-free system is not feasible, and frequent electrolyte maintenance is required.

Table III

$$Fe + 2NiOOH + 2H_2O \rightarrow Fe(OH)_2 + 2Ni(OH)_2$$
$$E = 1.370 \text{ V}; \ 267 \text{ W·h/kg Theoretical}$$

Status
Specific Energy	40-50 W·h/kg @ 10 W/kg
Specific Power	50-100 W/kg @ 10 W·h/kg
Cycle Life	300-500 @ 10-25 W/kg, 80% DOD
Cost	>$100/kW·h

Recent Work
Improved Fe and NiOOH electrodes
200-300 Ah cells and modules
Vehicle-size batteries

Problems
H_2 evolution during recharge; can't be sealed
Heat evolution
Low efficiency ~60%
Capacity loss at low temperature

Another alkaline-electrolyte cell under development is the zinc/nickel oxide cell (Zn/KOH/NiOOH), which has a theoretical specific energy high enough to permit a 75 Wh/kg cell to be developed, for a vehicle range of 150 km. Cells currently available in small quantities have specific energies of 60-75 Wh/kg. Batteries of these cells have been tested in electric vehicles, and have yielded the expected doubling of the range with Pb/PbO_2 batteries. (2) Although Zn/NiOOH cells have some very attractive features (high specific energy, high efficiency, capability of being sealed), they have a cycle life of 100-200 deep cycles, compared to a minimum goal of 300 deep cycles (100% depth of discharge). The cycle life is limited by the behavior of the combination of the zinc electrode and the separator. The zinc electrode suffers from problems of 1) dendrite formation during recharge, causing cell shorting, 2) zinc redistribution during cycling, 3) densification during cycling, and 4) passivation during discharge. Any combination of these four problems can cause cell failure. Improvements in the zinc electrode have been a popular subject of research, but progress has been relatively modest. There have been some improvements in separators for the zinc electrode, so that dendrite problems are no longer the main cause of cell failure.

Significant cost reduction of zinc/nickel oxide cells will be necessary to allow use in electric vehicles. Polymer-bonded electrodes are being developed as one cost-reduction measure. The

status of this system is shown in Table IV, and a photo of a
vehicle-size battery is shown in Figure 5.

The zinc/chlorine ($Zn/ZnCl_2/Cl_2$) system has been under development for several years for possible use in stationary and electric vehicle applications. This system uses readily-available
materials, mostly carbon and polymers. The negative electrode is
zinc deposited on a dense carbon substrate. The chlorine is
stored outside the cell stack in a refrigerated storage area, in
the form $Cl_2 \cdot 8H_2O$, which is a yellow, ice-like solid. Chlorine
is supplied to the porous carbon positive electrodes via circulating warm aqueous $ZnCl_2$ electrolyte, which melts and dissolves
the $Cl_2 \cdot 8H_2O$, as shown in the schematic diagram, Figure 6. This
system is rather complex, but has been demonstrated in sizes up to
50 kWh, some of them as electric vehicle power plants.

The theoretical specific energy for the $Zn/Cl_2 \cdot 8H_2O$ cell is
405 Wh/kg, offering a possibility of a practical specific energy
approaching 100 Wh/kg. Specific energy values close to 70 Wh/kg
have been demonstrated. A 1.7 kWh system has yielded a cycle life
of about 1400 cycles, with periodic electrolyte maintenance.
Because of the fact that the zinc electrode potential is a few
tenths of a volt below the reversible potential of the hydrogen
electrode, there is a strong tendency for hydrogen evolution, and
low concentrations of transition metal (e.g. Fe, Co, Ni)

Table IV

$$Zn + 2NiOOH + H_2O \rightarrow ZnO + 2Ni(OH)_2$$
$$E = 1.735 \text{ V}; \ 326 \text{ W} \cdot \text{h/kg Theoretical}$$

Status
    Specific Energy   60-75 W·h/kg @ 30 W/kg
    Specific Power   100-150 W/kg @ 35 W·h/kg
    Cycle Life   100-200 @ 25-50 W/kg, 80% DOD
    Cost   >$100/kW·h

Recent Work
    Inorganic separators (e.g. ZrO_2, $Ni(OH)_2$, $Ce(OH)_3$,
        others)
    Microporous organic separators
    Sealed cells
    Nonsintered electrodes
    400 Ah cells and modules

Problems
    Sealing of cells - O_2 evolution and recombination
    Zn redistribution
    Separators

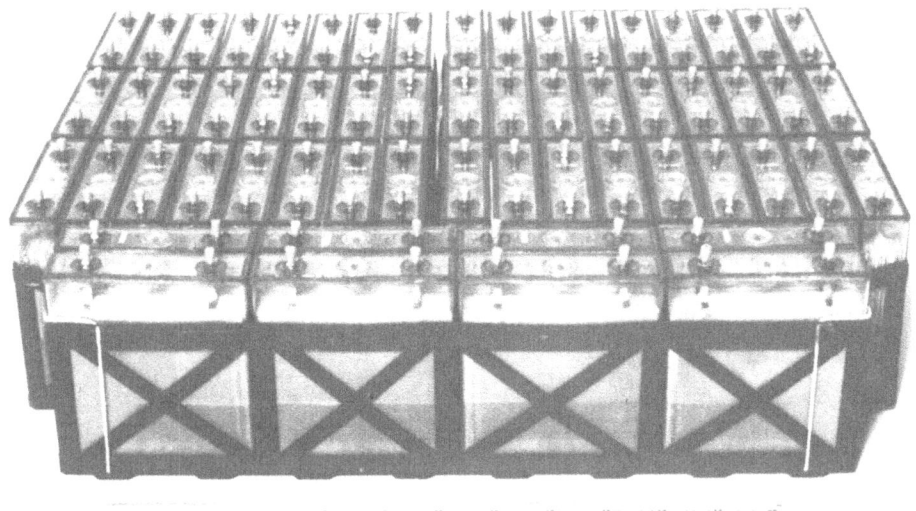

Figure 5.   Photo of a zinc/nickel oxide battery for an electric
            automobile.

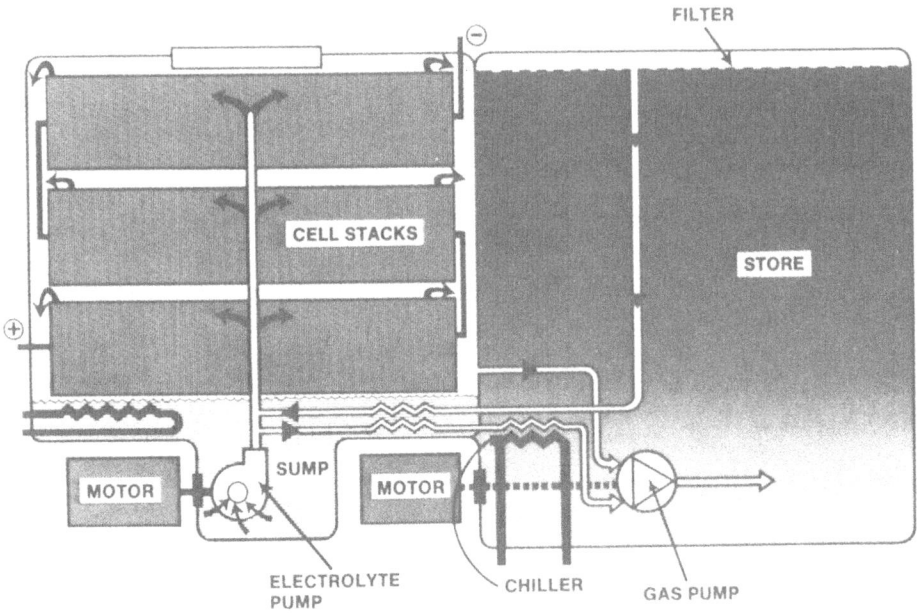

Figure 6.   Diagram of the zinc/chlorine system.

impurities in the electrolyte can cause spontaneous hydrogen evolution at the zinc electrode, resulting in a safety hazard, and a loss of efficiency.

Continuing work on the $Zn/Cl_2 \cdot 8H_2O$ system has included provision for recombination of hydrogen with chlorine, and additives to control the morphology of the zinc deposit. The zinc deposit morphology is important because zinc protrusions or dendrites can cause cell shorting. This is avoided by completely discharging all of the zinc every several cycles, and then carefully redepositing it in the presence of morphology-control additives.

A number of 50 kWh modules of $Zn/Cl_2 \cdot 8H_2O$ cells will be assembled to form at least 1 MWh of energy storage capability for test on a utility network in the BEST (Battery Energy Storage Test) facility in the near future. A summary of the status of this system is presented in Table V.

None of the systems with aqueous electrolyte discussed above shows promise of yielding a specific energy above 100 Wh/kg. A

Table V

$$Zn + Cl_2 \cdot 8H_2O \rightarrow ZnCl_2 + 8H_2O$$
$$E = 2.12 \text{ V}; \ 405 \text{ W} \cdot \text{h/kg Theoretical}$$

Status
> Specific Energy   66+ W·h/kg @ 3-4 W/kg
> Specific Power   70 W/kg for seconds
> Cycle Life   1400*
> Cost   >$100/kW·h

Recent Work
> Additives for Zn deposition
> Recombination of H_2 and Cl_2
> 35-50 kWh systems
> Systems components

Problems
> Complete discharge required
> Bulky
> Complex
> Low specific power
> Very sensitive to impurities
> Low efficiency
> Gaskets

---

*1 kWh system only, with electrolyte maintenance

number of high-temperature systems with nonaqueous electrolytes do
offer this possibility. The list of possibilities includes: (5)

$$LiA\ell/LiC\ell-KC\ell/FeS$$
$$Li_4Si/LiC\ell-KC\ell/FeS_2$$
$$Ca_2Si/LiC\ell-NaC\ell-CaC\ell_2-BaC\ell_2/FeS_2$$
$$Na/Na_2O\cdot xA\ell_2O_3/S$$
$$Na/Na^+ glass/S$$
$$Na/Na_2O\cdot xA\ell_2O_3/SC\ell_3A\ell C\ell_4 in A\ell C\ell_3-NaC\ell$$

Space limitations do not permit a discussion of all of these, so
only the more well-developed cells will be presented below.

The LiAℓ/FeS cell operates at 450°C, and makes use of a
LiCℓ-KCℓ molten-salt electrolyte. Work on this system is an outgrowth of earlier work on the Li/LiCℓ-KCℓ/S cell, which offered a
theoretical specific energy of 2600 Wh/kg. Difficulties with the
containment and control of liquid Li and liquid S led to the use
of solid LiAℓ and solid FeS as the reactants, but at a significant
sacrifice of cell voltage (2.3 vs. 1.3 V) and theoretical specific
energy (2600 vs. 458 Wh/kg). An essential increase in cell stability and lifetime was obtained, however.

A diagram of a typical LiAℓ/FeS cell is shown in Figure 7.
The electrodes are prepared by pressing mixtures of powdered
reactant and powdered electrolyte together to form plaques, which
are assembled in contact with current collectors, and enclosed by
fine-porosity particle retainer sheets. The separator between
the electrodes is boron nitride felt. The pores in the felt and
in the electrodes are filled with molten salt electrolyte, and
the cell is sealed. These cells typically operate at 450°C, and
have a voltage plateau of 1.1-1.2 V under current drain.

The status of the LiAℓ/FeS cell is presented in Table VI.
Specific energy values for 320 Ah multi-electrode cells are near
100 Wh/kg; cycle lives for cells with lower specific energies are
in excess of 300 deep cycles, with 500-800 cycles being not uncommon, corresponding to lifetimes of over 5000 hr.

Recent work on the LiAℓ/FeS cell has included the evaluation
of MgO powder separators as a replacement for the more expensive
BN felt separators, thermal cycling of individual cells as many
as 30 times with no loss of capacity, improved, lower resistance
current collectors (for higher specific power), and the testing
of 10-cell batteries of 320 Ah cells (3.8 kWh). The issues continuing to receive attention include cell shorting caused by
swelling and extrusion of the active materials from the electrodes,
agglomeration of LiAℓ during cycling, feedthrough development,
corrosion-resistant materials, especially for the current collectors, and thermal control of batteries. Thermal control is

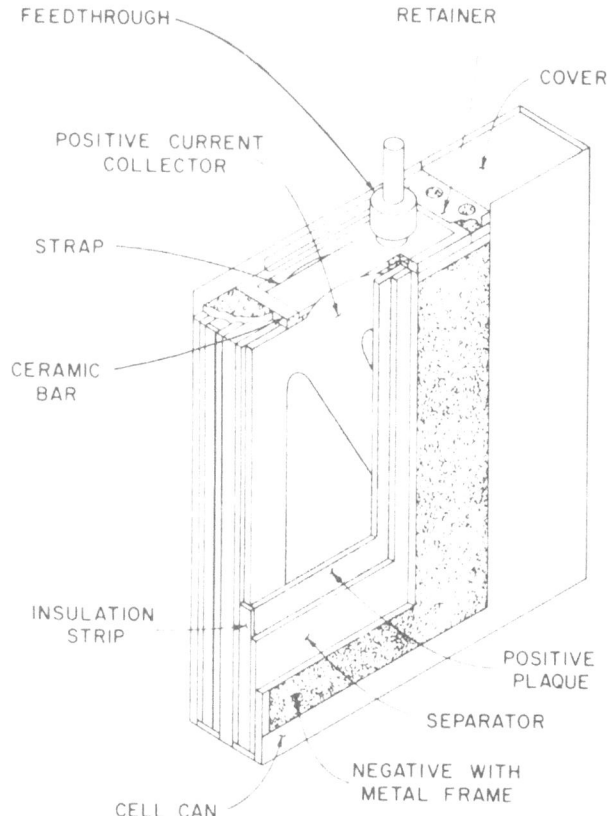

Figure 7.   Cutaway drawing of a LiAℓ/FeS$_2$
            cell.

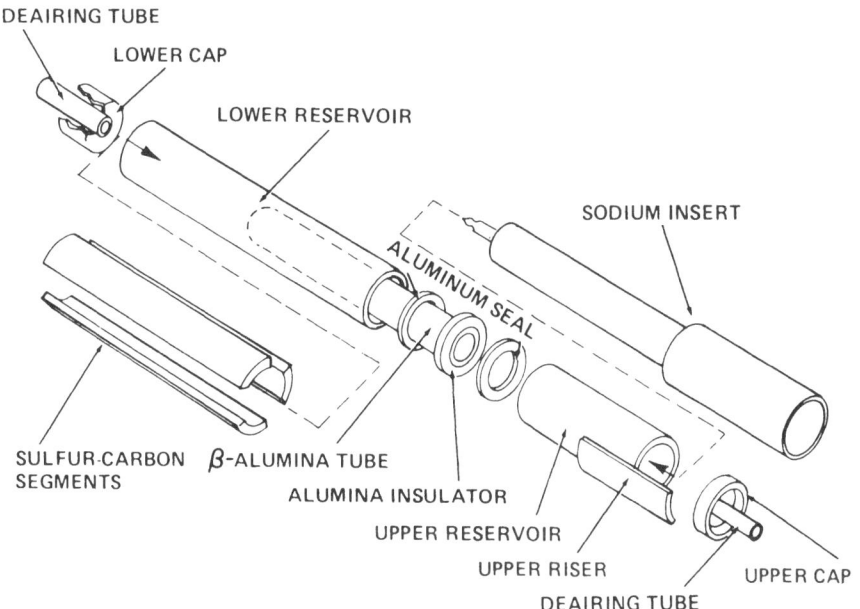

Figure 8. Exploded view of a Na/S cell.

Table VI

$LiAl/LiCl-KCl/FeS$
$$2LiAl + FeS \rightarrow Li_2S + Fe + 2Al$$
E = 1.33 V; 458 Wh/kg Theoretical
T = 450°C

Status
    Specific Energy   60-100 Wh/kg @ 30 W/kg
    Specific Power   60-100 W/kg, peak
    Cycle Life   300+ @ 100% DOD
    Lifetime   5000+ h
    Cost   >$100/kWh

Recent Work
    LiX-rich electrolyte
    Wetting agent for BN separators
    Powder separators-MgO
    Batteries of 320 Ah cells
    Improved current collectors
    Thermal cycling

Problems
    Low specific energy
    Low voltage per cell
    Cell shorting major failure mode
    Electrode swelling and extrusion
    Agglomeration of Li-Al with cycling
    Capacity loss
    High separator cost
    Leak-free feedthroughs
    Thermal control

important in vehicle batteries because the thermal losses through the insulation must be balanced against the heat generated by the use of the battery, in order to maximize the battery efficiency. High-performance, thin insulation materials are under development for this application.

A higher specific energy can be obtained by use of lower equivalent-weight reactants. The $Li_4Si/LiCl-KCl/FeS_2$ cell has a theoretical specific energy of 944 Wh/kg, offering the opportunity for a practical specific energy of about 200 Wh/kg, which should give an urban range of up to 400 km. Laboratory cells of this type have already demonstrated specific energies of 180 Wh/kg at low specific powers, cycle lives of 700 cycles, and lifetimes of about 2 years. (6) These cells are still in the laboratory stage of development, and have not yet reached a practical configuration for electric vehicle use. Some effort has been devoted to the

operation of small bipolar cell stacks. The problems needing additional work include corrosion-resistant current collectors for the
FeS_2 electrode (graphite and molybdenum are used now), low-cost
separators (BN felt is now used), solubility of sulfur-bearing
species in the electrolyte, leak-free feedthroughs, and thermal
control. Table VII shows the status of this system.

An alternative to the approach of using solid reactants and a
liquid electrolyte, is to use a solid electrolyte with liquid
reactants. The most well-known system of this type is the sodium/
sulfur cell with a solid electrolyte of $Na_2O \cdot 11Al_2O_3$ (beta alumina).
This cell makes use of the electrolyte in the form of a closed-end
tube, with one reactant on the outside, and the other on the inside. One version of this cell is shown in Figure 8, where the
sulfur is held around the outside of the electrolyte tube in a
graphite fiber matrix which serves as a current collector, and the
sodium is on the inside of the electrolyte tube. Thermocompression
bonded seals are used to close the cell. An operating temperature
of 350°C is used, where the electrolyte has an acceptable resistivity of about 5 ohm-cm (for the β'' alumina modification, containing Li_2O and MgO additives), and the reactants and some of the
products are molten.

Table VII

$$Li_4Si/LiCl-KCl/FeS_2$$
$$Li_4Si + FeS_2 \rightarrow 2Li_2S + Fe + Si$$
$$E = 1.8, 1.3 \text{ V}; 944 \text{ Wh/kg Theoretical}$$

Status
Specific Energy	120 Wh/kg @ 30 W/kg
	180 Wh/kg @ 7.5 W/kg
Specific Power	100 W/kg, peak
Cycle Life	700 @ 100% DOD
Lifetime	∿15,000 h
Cost	>$100/kWh

Recent Work
    Bipolar cells
    Li-Si electrodes
    BN felt separators
    70 Ah cells

Problems
    Materials for FeS_2 current collector
    Leak-free feedthroughs
    High internal resistance
    Low-cost separators needed
    Thermal control

Various versions of the sodium/sulfur cell have been investigated in laboratories located in several countries. The performances and lifetimes of the cells have been highly variable. The ceramic electrolyte has been difficult to produce with sufficiently well controlled properties that cell lifetimes are reproducible. Various electrolyte fabrication methods have been investigated, as have alternative solid electrolytes such as Nasicon ($Na_{1+x}Zr_2Si_xP_{3-x}O_{12}$), (5) and sodium borate glasses (5) (in the form of thin, hollow fibers).

As might be expected, the corrosiveness of the sulfur has been a major issue, and many materials have been investigated as possible current collector and container materials, or as coatings. The most corrosion-resistant materials have been graphite and molybdenum; some stainless steels and coated metals have also been used. Mass transport problems in the sulfur electrode have been ameliorated by the use of specially shaped graphite fiber current collectors, tailored-resistance graphite current collectors, and additives to the sulfur. At present, it is possible to achieve close to 150 Wh/kg, a cycle life of a few hundred to almost 2000 cycles, and a lifetime of up to two years with single Na/S cells. Several batteries of about 10 kWh have been tested, but these have had short times between failure of cells (hundreds of hours). Thermal cycling remains an important issue: freezing of cells usually causes failure.

Development of the Na/S cell continues with stationary energy storage being a main goal, and vehicle propulsion batteries being a secondary goal. Utility energy storage modules of about 1 MWh capability may be tested within the next few years. The status of the Na/S cell is summarized in Table VIII.

Other high-temperature cells are under investigation, but are at earlier stages of development than those presented here. For a discussion of some of the other systems, see Reference 5.

Some perspective can be gained with regard to the relative projected capabilities of the various batteries discussed above by examining Figure 9. The specific energy values above 100 Wh/kg are limited to the high-temperature systems, and these high specific energies are accompanied by the necessity of having a thermal control system for the battery. There are other candidate electrochemical systems which might be of interest for electric vehicles, including metal/air cells, and fuel cells. These are reviewed in Reference 2.

Table VIII

Na/Na+ Solid/S
$$2Na + 3S \rightarrow Na_2S_3$$
$$\bar{E} = 2.0 \text{ V}; \quad 758 \text{ Wh/kg Theoretical}$$

Status
    Specific Energy   85-140 Wh/kg @ 30 W/kg
    Specific Power   60-130 W/kg peak
    Cycle Life   200-1500
    Lifetime   3000-15,000 h
    Cost   >$100/kWh

Recent Work
    Batteries, ~10 kWh
    C_6N_4 additive to S
    Ceramic (TiO_2) electronic conductors
    Shaped current collectors
    Tailored resistance current collectors
    Sulfur-core cells
    $Na_{1+x}Zr_2Si_xP_{3-x}O_{12}$
    Thermocompression bonded seals

Problems
    Corrosion-resistant material for contact with S
    Low cost seals
    Low cost electrolyte
    Specific power is low
    Thermal cycling

IV. CONCLUSIONS AND FUTURE DIRECTIONS

Consideration of the material presented above, and its implications leads to the following conclusions and indications of appropriate future activities.

1. Electric vehicles will probably be produced in significantly larger numbers in the next few years, especially utility vehicles and vans with improved Pb/PbO_2 batteries.

2. The next generation of electric vehicle battery will probably be an alkaline system - Fe/KOH/NiOOH, or Zn/KOH/NiOOH. The zinc/nickel oxide system has the advantage of higher specific energy, and capability of being sealed. The zinc electrode re-

quires a longer cycle life before it can be commercially attractive for EV's (at least 300 cycles, 100% depth of discharge).

    3. The LiAℓ/FeS cell is very robust, and may offer some
attractive features for electric vehicles, but its specific energy

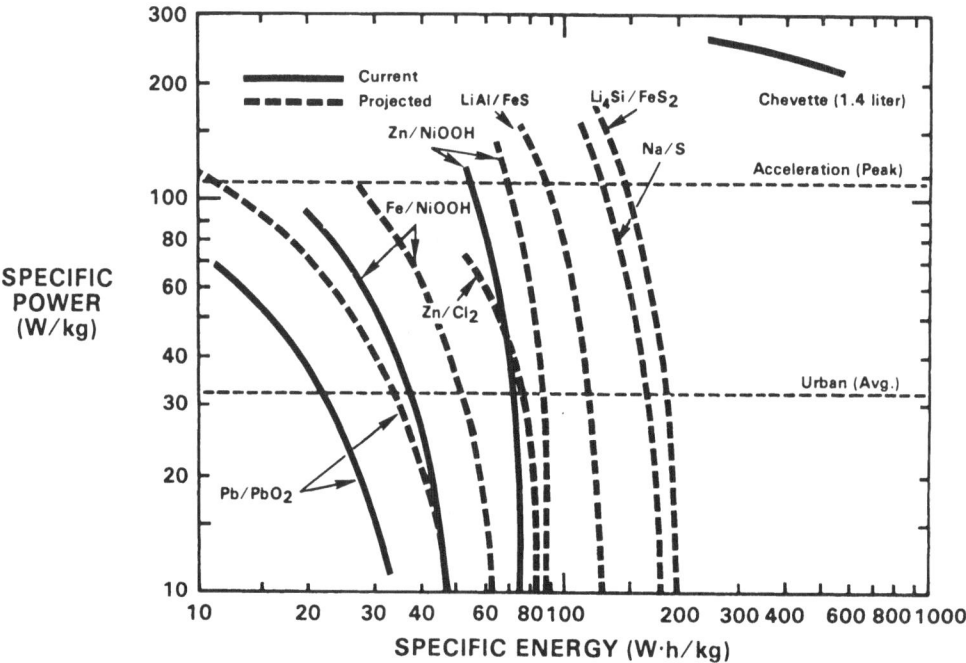

Figure 9.   Specific power vs. specific energy plot for various
            batteries.

of 100 Wh/kg is probably borderline for a high-temperature system.

    4. The Li$_4$Si/FeS$_2$ cell offers a very high performance, but
the materials problems need more work before a practical battery
can be produced.

    5. The Na/S cell family with solid electrolytes could prove
suitable for some mobile applications, but the thermal cycling
problem and materials problems need solutions.

    6. Systems involving flowing reactants (such as Zn/Cℓ$_2$,
Zn/Br$_2$, metal/air, fuel cells) are farther from practical applications in electric vehicles than the other systems discussed, because they tend to be bulky and complex. Fuel cells have not been
evaluated in realistic vehicle environments, so dynamic response
remains a question, along with cost and lifetime.

7. Future directions for batteries of even higher performance include a Li/S cell, which may require a solid, lithium-ion
conducting electrolyte, stable to lithium and sulfur. Failing a
Li/S cell, lower equivalent-weight reactants than Li_4Si and FeS_2
are attractive research topics.

8. Ambient-temperature, non-aqueous rechargeable cells with
lithium negative electrodes offer the possibility of high specific
energy without the inconvenience of high-temperature operation.

9. Rechargeable oxygen electrodes of low overvoltage, and low
cost could significantly change the outlook for metal/air cells and
fuel cells.

10. Low-cost, long-lived alkaline fuel cells may be attractive power plants for electric vehicles, using cracked ammonia as
fuel. This should be evaluated.

REFERENCES

1. A.L. Austin, B. Rubin, and G.C. Werth, Lawrence Livermore
    Laboratory Report UCRL 51221, May 30, 1972.
2. Elton J. Cairns and Earl H. Hietbrink, "Electrochemical Power
    for Transportation," in Comprehensive Treatise of
    Electrochemistry, Volume 3, J.O'M. Bockris, B.E. Conway, E.
    Yeager, and R.E. White, eds., Plenum, New York, 1981, p. 421.
3. M.L. Kyle, H. Shimotake, R.K. Steunenberg, F.J. Martino, R.
    Rubischko, and E.J. Cairns, in Proceedings of the 1971 IECEC,
    Society of Automotive Engineers, Inc., New York, 1971, p. 80.
4. E.J. Cairns and J. McBreen, in Industrial Research, June,
    1975, p. 56.
5. Elton J. Cairns, "Secondary Batteries--New Batteries: High
    Temperature," in Comprehensive Treatise of Electrochemistry,
    Volume 3, J.O'M. Bockris, B.E. Conway, E. Yeager, and R.E.
    White, eds., Plenum, New York, 1981, p. 341.
6. E.J. Zeitner and J.S. Dunning, in Proceedings of the 13th
    IECEC, Society of Automotive Engineers, Inc., Warrendale, PA,
    1978, p. 697.

This work was supported by the U.S. Department of Energy
under Contract W-7405-ENG-48.

HIGH ENERGY PRIMARY BATTERIES

Ralph J. Brodd

Inco ElectroEnergy Corp.
5 Penn Center Plaza
Philadelphia, Pennsylvania 19103

INTRODUCTION

Perhaps we should start by defining what we mean by high energy primary cells. High energy density means different things to different people. Energy density is expressed as the energy per unit volume or the energy per unit weight, and high is a relative term compared to low. For the purpose of this discussion, we will be talking about commercial applications, so that the energy density, watt hours per cu. in., is the most important term. The energy density per unit of weight is of lesser importance for commercial primary batteries. However, given two batteries with the same energy density per unit volume, the choice would be the lighter battery.

The discussion will compare the various battery systems in light of the criteria for successful primary battery systems given in Table 1. It is not possible to assign an order to the importance of each of these criteria, as their relative importance changes with the application.

In this discussion, the frame of reference for low energy primary batteries will be the regular Leclanche flashlight batteries. These have an energy density of about 2.7 watt hours per cu. in. on low drain. The mercury and silver batteries have an energy density of about 8 to 9 watt hours per cu. in. and fit my definition as high energy primary batteries.

I will limit my comments here to commercial battery systems. Only passing reference will be made to solid state systems. Thermal batteries and other high power military systems will not be discussed. Primary batteries have recently been reviewed.[1] When one mentions high energy density primary cells, almost everyone thinks of lithium batteries.

Although lithium does have the highest voltage of any of the metals, it has a relatively large equivalent volume. The equivalent volumes of aluminum and zinc and several other elements are smaller than lithium, due to the low density of lithium. This means that, on a volume basis (Ah/cc), lithium will be at a disadvantage compared to aluminum, magnesium, and zinc. On an energy basis (Wh/cc), lithium is slightly better than zinc, but aluminum and magnesium have greater Wh/cc than lithium. Batteries with aluminum and magnesium anodes would have higher energy density when coupled to the same cathode.

Table 1. Criteria for Primary Batteries

Mechanical and chemical stability
Self-discharge
Energy density
Power density
Leakage
Shape of discharge curve
Temperature range
Cost

In practice, several aqueous battery systems have energy densities equal or superior to many lithium systems. Tables 2, 3, and 4 summarize the properties of common high energy battery systems. It is noted that zinc-air has an energy density equal or superior to lithium systems. The only lithium system with equivalent energy density is the Li-SOCl$_2$ system.

AQUEOUS BATTERIES

The mercury cell still sets a standard for high energy density batteries. It was conceived and developed by Dr. Samuel Ruben in the 1940s. Since that time many new sizes and shapes have been introduced. The characteristics of the cell are noted in Table 2. The mercury cell discharges at a constant voltage of 1.35 volts. The discharge voltages are slightly higher if manganese dioxide is added to the cathode formulation. The principles of the cell design, cell balance, etc., were set down by Ruben and really have not changed over the years. Of course, new materials, e.g., separators, are used today. For many electronic applications, a higher voltage discharge is desirable. To satisfy this need, the silver cell system was developed and introduced in 1961. The cell has a design similar to that of the mercury cell except that monovalent silver oxide is used in the cathode instead of mercury oxide. Typical construction of these cells is shown in Figure 1. The silver cell discharges at a constant 1.6 volts. It can handle pulses upward to 50 milliamps per square centimeter. The overall cell capacity is about 10% lower than the equivalent size mercury cell.

Table 2.   High Energy Density Aqueous Systems

Designation	Cell Reaction	Nominal Voltage	Nominal Wh/in^3	Comments
Zn-Hg (mercury)	$Zn + HgO = ZnO + Hg$	1.35	9.5	Standard for comparison.
Zn-Ag (silver)	$Zn + Ag_2O = ZnO + 2Ag$	1.6	8.8	Higher unit cell voltage than mercury.
Zn-Ag(II) (divalent silver)	$Zn + AgO = ZnO + Ag$	1.6(1.8)	10	Increased capacity, initial voltage step.
Zn-air (air)	$Zn + 1/2O_2 = ZnO$	1.4	15	Teflon-bonded air electrode, limited activated life.
Zn-MnO$_2$	$Zn + 2MnO_2 = ZnO + 2MnOOH$	1.5	2.7	Low cost, listed for comparison.

Table 3.   Lithium Inorganic Electrolyte Systems

Designation	Cell Reaction	Nominal Voltage	Nominal Wh/in^3	Reference
Liquid Cathodes				
Li-SO$_2$	$Li + SO_2 = Li_2S_2O_4$	2.9	8	2-5
Li-SOCl$_2$	$4Li + 2SOCl_2 = 4LiCl + S + SO_2$	3.5	18	6-9
Li-SO$_2$Cl$_2$	$2Li + SO_2Cl_2 = 2LiCl + (SO_2, Li_2S_2O_4, Li_2S_2O_3)$	3.8	18	6-9
Solid Electrolytes				
Li-I$_2$	$Li + 1/2I_2 = LiI$	2.8	12	10
Li-Pbs/PGI$_2$	$4Li + PbS/PbI_2 = 2LiI + Li_2S + 2Pb$	1.9	8	11
Li-Br$_2$	$Li + 1/2Br_2 = LiBr$	3.5	20	12

Table 4.   Lithium Organic Electrolyte Systems

Designation	Cell Reaction	Nominal Voltage	Nominal Wh/in^3	Reference
Li-FeS	$2Li + FeS = Li_2S + Fe$	1.6	8	13
Li-FeS$_2$	$4Li + FeS_2 = 2Li_2S + Fe$	1.6	8	14
Li-CuO	$2Li + CuO = Li_2O + Cu$	1.5(2.4)	10	15-17
Li-CuS	$4Li + CuS = 2Li_2S + Cu$	1.8(2.27)	7	18
Li-MnO$_2$	$Li + MnO_2 = LiMn(III)O_2$	2.7	8	19,20
Li-CF$_x$	$Li + CF_x = LiF + C$	2.8	10	21,22
Li-V$_2$O$_5$	$2Li + V_2O_5 = Li_2V_2O_5$	3.5	11	23,24
Li-Ag$_2$CrO$_4$	$2Li + Ag_2CrO_4 = Li_2CrO_4 + 2Ag$	3.3	10	25

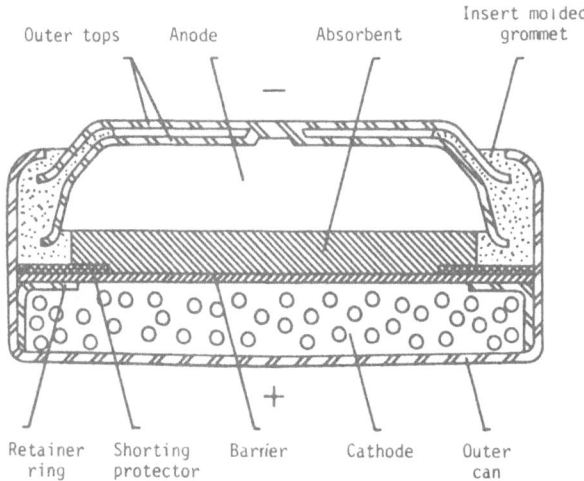

Fig. 1. Schematic cross-section of an AgO–Zn cell. [Reference 26]

Divalent silver oxide has higher energy density and has essentially twice the ampere hour capacity per gram of monovalent silver oxide. However, the divalent silver oxide is metastable in aqueous solution and thermodynamically will spontaneously decompose water with the evolution of oxygen. An examination of the pH potential diagram shows that silver oxide potential is above oxygen. That is, silver oxide should spontaneously liberate oxygen from aqueous solutions. With no kinetic hindrance, $2AgO + H_2O = 2AgOH + O_2$, oxygen evolution, should occur. This could produce a considerable pressure buildup inside the cell, leading to bulging and subsequent leakage, if not to cell rupture.

Certain electronic applications require substantially constant voltage discharges. The two-step discharge, one at 1.8 volts for divalent oxide and one at 1.6 volts for monovalent silver dioxide, often termed voltage instability or voltage-up, also has deterred the use of divalent silver oxide in commerce. Various means have been suggested to stabilize the divalent oxide by retarding or preventing the gas evolution. One approach is the use of solution additives such as gold, heavy metal ions, and sulfide, as well as other divalent ions.[12,27-29] The additives adsorb or react with the surface of AgO and change its reactivity toward water. While effective in reducing solution contact and oxygen evolution, this approach does not stop it completely unless a reduction of the surface occurs to form Ag_2O or silver. The goal of these processes or techniques is a silver cell which discharges at the monovalent voltage but with the increased capacity due to divalent silver oxide.

Soto-Krebs and others[30,31] have patented a technique for stabilization in which the monovalent oxide provides a contact to the current collector can and isolates the divalent oxide from the collector and electrolyte. During discharge, an exchange takes place to regenerate the monovalent material:

$$Ag_2O + H_2O + 2e = 2Ag + 2OH^-$$
$$Ag + AgO = Ag_2O$$

Thus the cell discharges at the monovalent voltage but the charge exchange permits the utilization of the coulombic capacity of the divalent silver.

Recently, Megahed and Davig[26] reported on a similar process in which the stability is controlled by coating the divalent silver oxide pellet with a thin layer of monovalent silver oxide. The divalent oxide pellets are treated first with a mild reducing agent, then with a strong reducing agent to form a silver layer on the surface of the cathode. Effectively, this treatment produces a cell with a discharge voltage of 1.6 volts. The performance of these cells is shown in Figures 2 and 3. It is noted that about 30% increase in cell capacity can be obtained using the divalent oxide cathodes. The common thread underlying all of these studies is the thought that stability results from a protective coating of monovalent oxide on the surface of the divalent silver oxide particle. The cathode construction is depicted in Figure 2.

The shelf life or self-discharge of the alkaline cells is controlled largely by the organic separator materials. The cathode materials are slightly soluble in the alkaline electrolyte. These dissolved positive active material species are very oxidizing. They can react directly to degrade the separator material. Separator layers function as a barrier to diffusion but lose their effectiveness by attack from the soluble cathode material. The chemical attack can result in destruction of the separator and formation of conductive filaments through the separator material, which form internal short circuits and discharge the cell. Self-discharge occurs when the soluble cathode species diffuse to the anode and react chemically with the anode material. Figure 4 shows that silver cells can be constructed with excellent shelf life.

Ruetschi[32] has discussed the principles which lead to longer life of silver and mercury cells. The organic membrane separators swell and absorb electrolyte to provide conductivity through the cell. The absorption of electrolyte is controlled by the concentration of the caustic and is less in sodium than in potassium hydroxide. The absorption of electrolyte also controls the silver (or mercury) migration through the membrane and thereby affects the rate of the chemical reaction of the soluble cathode material with the membrane. Membranes which absorb less electrolyte and are more chemically inert, such as the radiation-crosslinked polyethylene materials, show significantly longer storage life. However, since they absorb less electrolyte, they have an increased cell internal resistance. Reutschi predicts that the optimal design of a low rate cell would have extremely low self-discharge rates and could have more than 80% capacity remaining after 10 years of storage at room temperature.

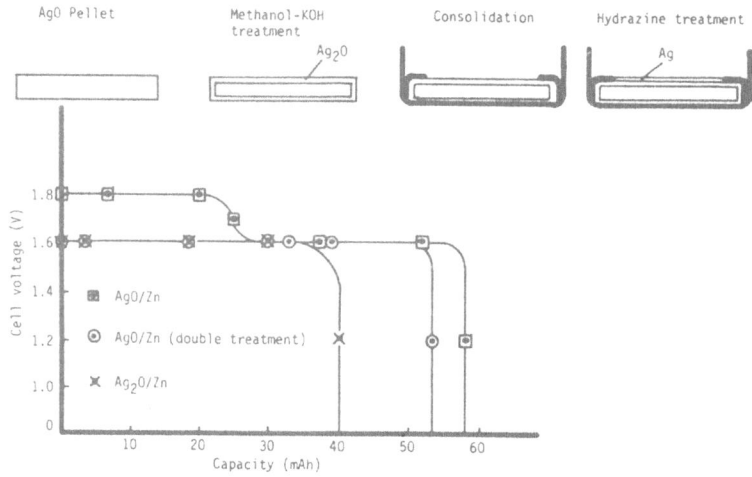

Fig. 2.   Discharge curves of SR 41 size cells with monovalent and
divalent cathodes (15 µA continuous at 21°C). [Reference 26]

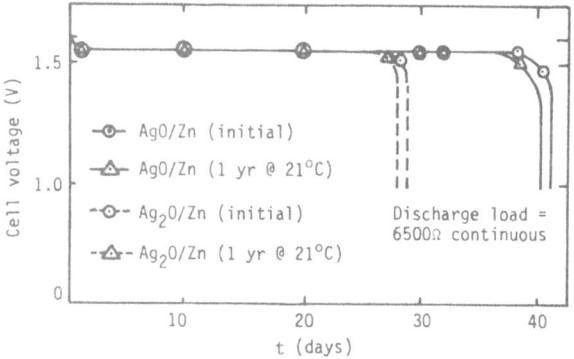

Fig. 3.   Discharge curves of SR 44 size cells.
[Reference 26]

A new microcalorimetric technique has been developed recently to
measure heat release in cells.[33-36]   The heat evolved can be used to
estimate the self-discharge reaction rate, provided certain assumptions
are made.   Table 5 lists several different reactions which can be
identified and which contribute to heat evolution.   The method has found
greater application in heart pacer battery studies.   For instance, if a heat
release from a 120 mAh mercury cell is 2 µW, in a year's time the cell

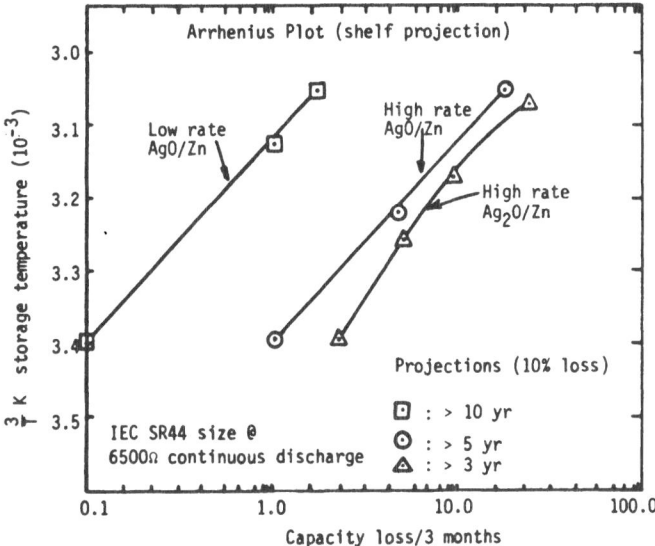

Fig. 4.   Arhennius plot of degradation vs. storage temperature
for monovalent and divalent silver cells. [Reference 26]

would lose 2/1.35 x 24 x 365 = 12 mAh or 10% of its original capacity.
This assumes the heat is due solely to a self-discharge reaction. Many of
the other reactions responsible for heat evolution in cells are noted in
Table 5.

The definition of leakage is the creepage of electrolyte or the
appearance of electrolyte outside of the cell. In commercial cells,
creepage of the electrolyte must be avoided. Many applications involve
sensitive electronic equipment which is damaged by contact with the
electrolyte.

The commercial cells of the high energy density systems are of
sealed construction. Seals are formed normally by compressing a plastic
grommet between two metal surfaces. When proper attention is paid to
the details, e.g., the compression of the grommet, the production techniques, and the proper design of the overall cell, leakage of the cell will
be minimized. It is observed normally that creepage occurs only on the
negative metal surface of the cell. The caustic concentration in this
creepage film is higher than the bulk electrolyte inside the cell. This
concentration difference gives rise to an electroosmotic force which can
cause solvent and ion transfer to occur between the two phases. Hull and
James[37] have proposed that the driving force for liquid transport is
electroosmosis associated with this ionic flow between the two electrolyte concentrations. The schematic of the phenomenon is given in Figure
5.

Table 5.   Processes Which Can Lead to Heat Release in Batteries

I.   Direct battery reaction —
     internal short:
     • Electronic

     a)   Metal spur penetrating the insulating gasket at seal.
     b)   Separator conductive, e.g., carbon particles sifting
          through conductor; ZnO pptn throughout separator.

     • Chemical

     a)   Diffusion of soluble active material to the other electrode,
          e.g., Ag^+ to the anode.
     b)   Redox shuttle, e.g., NO_3^- in Ni-Cd.

II. Chemical reaction, such as
     decomposition of active
     material with supposedly
     inactive component, or
     secondary chemical reactions

     a)   Oxidation of separator, CMC, or carbon by cathode or a
          soluble cathode species.
     b)   Reduction of solvent, i.e., H_2 on zinc, passive layers.
     c)   Disproportion, e.g., $Ag_2O = Ag + AgO$; $2Hg^+ = Hg^{+++} + Hg$.
     d) $2MnOOH + Zn^+ = Mn_2O_3 \cdot ZnO + 2H^+$.
     e)   Gas recombination, $H_2 + Ag_2O = 2Ag + H_2O$; oxygen ingress and
          reaction with anode.
     f)   Curing of plastics, e.g., epoxy, adhesives; $(PVP)_x I_2 =$
          $(PVP)_y + I_2$.

III. Physical processes
     (changes in physical state)
     of components)

     a)   Evaporation or absorption of solvent through seal.
     b)   Recrystallating of powders.
     c)   Stress relief.
     d)   Precipitation of insoluble products, e.g., ZnO from
          solution.

IV. Added effects during
     discharge

     a)   Entropy (unavailable energy).
     b)   Irreversibility, activation, resistance, diffusion.
     c)   Intermediate reactions.

The concentration difference for leakage appears to arise from oxygen reduction on the negative. Oxygen diffuses into the seal area through fissures, etc. When it reaches the electrolyte boundary, the oxygen reduction results in increased OH^- concentration locally. Corrosion of both the cell can wall and the zinc anode has the effect of increasing the OH^- concentration because of H_2 evolution. Thus the cell environment, O_2, and water vapor are important aspects of leakage. For instance, cells stored in dry N_2 have a much lower leakage. Also, the material of construction used for the anode cup, e.g., Ni, stainless steel, or gold, also plays a role in leakage.

On the other hand, Baugh et al.[38,39] proposed that the spreading force or the differing surface tension of the two concentrations leads to the liquid transport. Both surface tension and electroosmosis are important. The pressures developed for liquid transport may be quite large for both mechanisms. The vapor phase transport of water due to the differing alkaline concentrations can also contribute, especially at very high alkaline concentration.

In order to control electrolyte creepage, most manufacturers use a coating, on the plastic grommet, of a surface active material such as asphalt, polyamide, etc. These surface active materials smooth out surface imperfections in the cell parts and reduce oxygen ingress by removing air passages. They also affect the surface tension. The overall result is greatly reduced leakage when these compounds are used.

The stability of the seal materials must be considered, as they are subjected to severe mechanical stress and chemical attack for long storage periods. However, in contrast to other plastics, nylon does not undergo cold flow under mechanical stress within its elastic limit even after long periods of storage.

Recently, Kajita et al.[40] reported that removing surface imperfections by using chemically polished anode cups and a nylon gasket with high elasticity and low water content was very effective in eliminating leakage. They also reported that the degree of crystal energy in the nylon grommet was important in maintaining a good reliable long life seal.

In cells where leakage prevention is especially important, sodium hydroxide is generally used as the electrolyte. The cells will have a slightly lower rate capability because of the effect of sodium hydroxide on the reaction rate of the active material. This makes an interesting point. The mechanical aspects which led to leak-free cells, e.g., seals, manufacturing processes, etc., are of equal or greater importance than the electrochemical aspects in commercial cells. Attention to details associated with leakage, e.g., amount of seal compression, clean anode top surfaces, plastic stability, etc., can be the difference between success or failure in the marketplace.

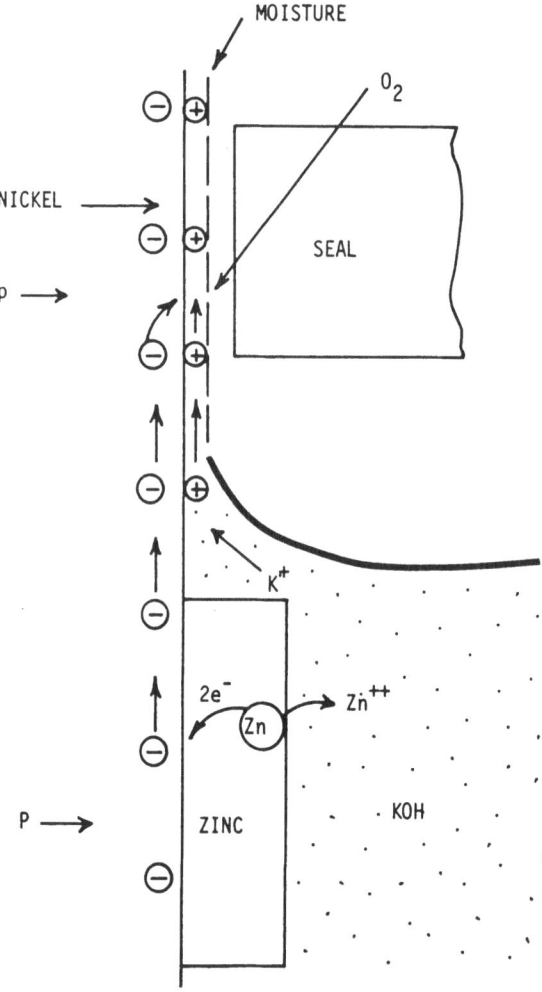

Fig. 5. Schematic of the spatial separation of the reactions at an didealized metal-seal interface. [Reference 37] Reprinted by permission of the publisher, The Electrochemical Society, Inc.

    The newest commercial high energy density aqueous battery is zinc-air.[41] It has about twice the capacity of existing mercury and silver batteries but discharges at a slightly lower voltage. Figure 6 presents a schematic drawing of a zinc-air cell. A tab is placed over the air access holes to prevent oxygen access to the electrode system. The difference between this cell and other button cells is the oxygen (air) electrode which replaces the solid cathode of mercury or silver cells.

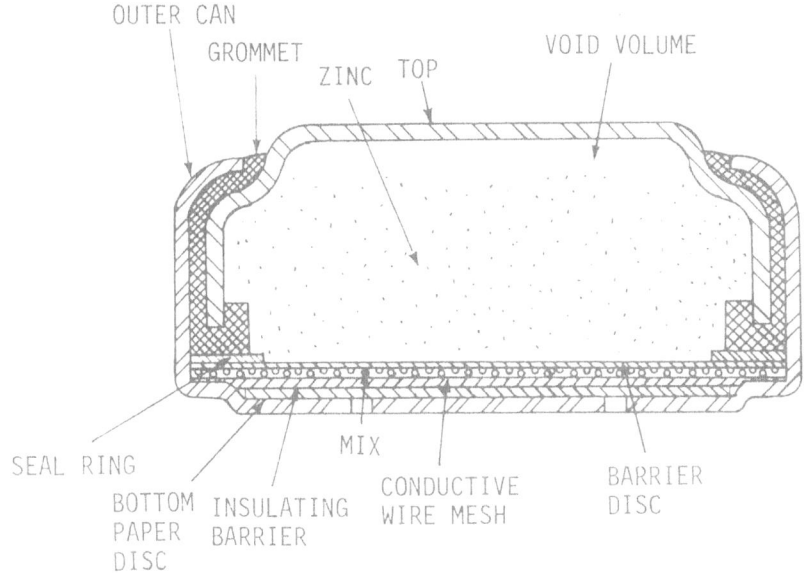

Fig. 6. Schematic cross-section of a zinc-air cell.

The oxygen electrode is a partially hydrophobic, porous plastic-bonded carbon structure. An activated carbon is used. Some manufacturers also use a catalyst, e.g., MnO_2 or other oxygen reduction catalyst. This extended surface area construction is responsible for the good high current performance, but limits activated operational lifetime to one month or so. The system has good shelf life in the unactivated state. A comparison of the performance of the aqueous battery system is shown in Figure 7. Each of the cells has a unique advantage over or distinction from the other battery systems.

The oxygen electrode has been the subject of extensive investigations. Oxygen reduction occurs via a peroxide intermediate mechanism, and the rate capability depends on the surface and catalyst. For low rate cells, no catalyst is required provided the electrode has reasonable surface area. New catalysts of the ferric-phthalocyanine type have high activity and offer the possibility of low cost electrodes.[42] Oxygen catalysis is discussed elsewhere in this Symposium. It may be possible to improve the discharge voltage through the use of more active catalysts.

Zinc is a common anode material for all high aqueous energy primary batteries. The anode is zinc powder suspended in an organic binder, e.g., carboxymethylcellulose. The electrode functions as a high surface area porous electrode and is capable of handling high current pulses. Mercury is added to inhibit hydrogen evolution. Other alloying agents, Pb and Cd, also reduce gassing. Solution additions of inhibitor of the glycol type also decrease corrosion and gassing. The electrochemistry of zinc has been discussed elsewhere.[43]

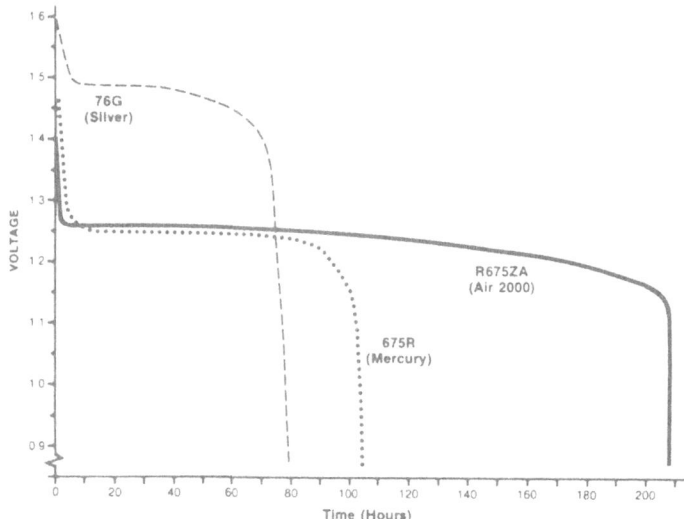

Fig. 7.   Discharge comparison of the same size zinc-air, mercury, and silver cells (62 ohms, 16 hours/day).

ORGANIC ELECTROLYTE LITHIUM BATTERIES

The first commercial organic electrolyte lithium battery was the Li-CF_x system used in Japan to replace mercury batteries in fishing floats. Since that time, two groups of cell systems have developed, those with 3.0 volt and those with 1.5 volt nominal open circuit voltage. The advantages and disadvantages, or problems, of lithium batteries are listed in Table 6.

Table 6.   Lithium Battery System Characteristics

Advantages	Disadvantages
Excellent shelf life	Limited production
Adequate energy density	Lower rate (pulse) capability
Choice of voltage	Need for hermetic seal (?)
Less tendency for leakage	Limited shipping approval
Superior low temperature	May not have higher energy
Low materials cost	density

The organic electrolyte cells are lower rate cells, compared to mercury and silver cells. This is due in part to the use of smooth rather than extended area anode construction. An added factor is the low exchange current (larger polarizations at high current) of lithium. Scarr[44] reported that the exchange current is about $10^{-4} A/cm^2$. The exchange current for zinc is about 10 to 100 times that for lithium. He also postulated a thin film formation on the lithium which can lower the exchange current even further. However, this thin film protects lithium from attack by the electrolyte and is responsible in large part for the good shelf life of organic electrolyte lithium batteries. Many organic electrolytes are unstable in contact with lithium since it can initiate polymerization, e.g., acetonitrile, and ring opening, as in propylene carbonate and tetrahydrofuran.[45-47]

The electrolyte characteristics have an important role in organic electrolyte batteries. In most cases the battery performance is not limited by the resistance of the electrolyte, although organic electrolytes have higher resistivities than aqueous electrolytes, as noted in Table 7. Cathode reactions tend to be slower, i.e., higher polarization in organic electrolytes than in aqueous media. Most organic electrolyte systems can handle pulses of 1-2 mA/cm^2. The interaction between the electrolyte and cathode determines the activity of the cathode. There seems to be a specific, but as yet unknown, relationship between cathode and electrolyte. A cathode will give good performance in one electrolyte but poor performance in other electrolytes. The nature of this specificity is not understood. It could be due to a variety of reasons, such as the availability of ions for reactions at the surface, the double layer structure at the cathode-electrolyte interface, and a higher activation energy for formation of the activated complex.

Table 7.   Resistivities of Several Common Electrolytes

Electrolyte	Approximate Resistivity (ohm-cm)
$KOH-H_2O$	1
$RbAgI_4$	5
$LiAlCl_4-SOCl_2$	50
$LiClO_4-PC$, DME	200
LiI, Al_2O_3	10^5
LiI	10^7

Blomgren[48] has pointed out that ion pairing and triple ion formation occur in these electrolytes. He estimated that about 80% form ion pairs and triple ions. Therefore, only a small fraction of the ions participate in the conduction process. This would account for the fact that the electrolytes are completely disassociated but have low conductivity, as noted in Table 7.   The electrolyte structure and ion pairing may also account for the low reaction rates of many cathodes in these electrolytes. Organic electrolytes generally have low freezing points and permit cell operation over a wider temperature range than do aqueous electrolytes.

Lehmann et al.[15] have described a low cost Li-CuO cylindrical cell with non-spiral construction. The cell could replace alkaline and carbon-zinc cell systems. These cells give up to three times the performance of alkaline and Leclanche cells.   Although the cell has a thermodynamic voltage of 2.4 volts, the cell discharges at 1.4 to 1.0 volts, depending on current drain. The system has good storage properties and can be stored for 3 weeks at 70°C without measurable loss of capacity. The system has good performance in miniature cells, as noted in Figure 8. Not shown is the tendency of all 1.5 volt lithium systems to exhibit initial high voltages, i.e., a voltage-up phenomenon. These voltages are removed by a slight amount of predischarge, i.e., "burn in", or various chemical treatments of the cathode to remove electroactive species responsible for the high voltage, e.g., oxygen adsorbed on carbon. The construction is very similar to that of present silver and mercury cells. The separator for lithium is most often a glass matte rather than the organic barrier membranes used for aqueous systems. Normally, low surface area sheet lithium is used for the anode instead of high surface area powders. This accounts in part for the low rate characteristics of present lithium cells. Not only is the exchange current lower, but the surface area is smaller than that of the gel anode. The Li-FeS$_2$ and Li-FeS systems have also been developed as low cost replacements for silver cells.

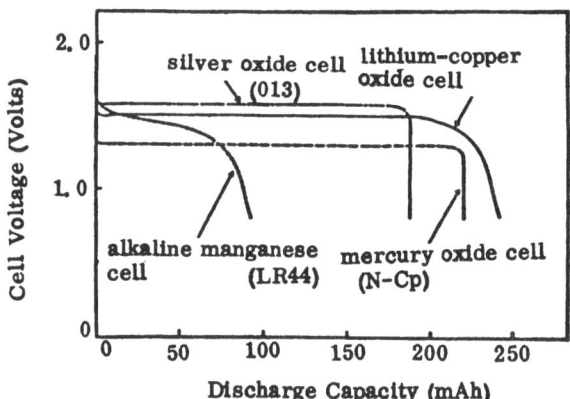

Fig. 8. Comparison of lithium-copper oxide cells with silver, mercury, and alkaline manganese dioxide cells. [Reference 17]

Figure 9 shows the construction of the fishing float cell for the Li-CF$_x$. The advantage of CF$_x$ is that the material does not dissolve in the electrolyte, and produces a conductive material, carbon, on discharge. The discharge voltage is in the same range as the Li-MnO$_2$ cell but remains relatively constant over the discharge period. The system also has better high rate performance compared to Li-MnO$_2$ in similar cell constructions. The system has good low temperature performance, as shown in Figure 10. The Li-CF$_x$ system has an energy density 3-4 times greater than Leclanche and alkaline cells and is essentially equal to mercury cells at lower drains.

The Li-MnO$_2$ cells have proved to be popular. They have good energy density and use low cost materials. The manganese dioxide used in the Li-MnO$_2$ cells is electrolytic MnO$_2$ (EMD) which is heat treated to modify the crystal structure and to remove water. Treatment at 350°C appears to give the best results. X-ray studies have shown that the MnO$_2$ reduction occurs with incorporation of lithium ions into the MnO$_2$ lattice, much as the proton is incorporated into the lattice during discharge in aqueous solutions. The sloping nature of the discharge curve noted in Figure 11 confirms the solid state homogeneous reduction reaction. A comparison of the discharge of lithium CF$_x$ and MnO$_2$ cells is given in Figure 12.

The Li-Ag$_2$CrO$_4$ and Li-CuS systems have found application as power supplies for heart pacers. These cells have excellent stability and can deliver pacemaker current drains for 6 to 8 years or more, depending on the pacer. Solid state Li-I$_2$ batteries account for over 90% of the pacer power supplies. The present state of solid electrolyte batteries is summarized in Table 8.

Table 8.   Characteristics of Solid State Batteries

Shelf life:	20$^+$ years
No leakage	
Wide temperature range:	
Operating	-20°C to +100°C
Storage	-60°C to +100°C
High energy density:	8-10 Wh/in^3
No separators and absorbers	
BUT	
Very low rate in present electrolytes	

Many cells used hermetic glass-to-metal seals and are virtually leakproof. The low cost techniques for sealing alkaline cells have been incorporated into many lithium cell designs. Organic electrolytes have less tendency to "creep" than do alkaline solutions and, as a result, the organic electrolyte cells have less tendency to leak. The electrolytes are less corrosive and less likely to damage contacts in devices.

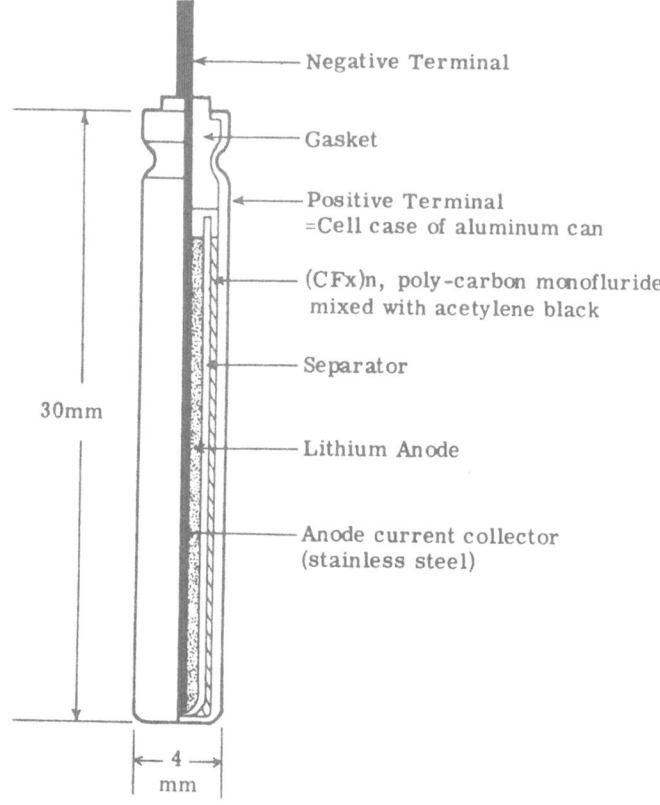

Fig. 9.   Schematic cross-section of a BR-435 Li-CF$_x$ cell.
[Reference 22]

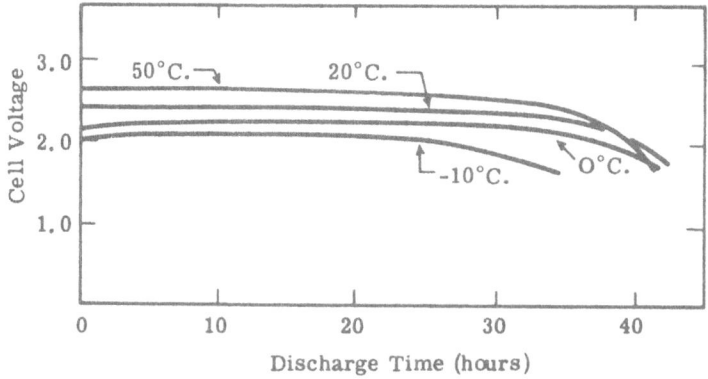

Fig. 10.   Discharge curve of BR-435 Li-CF$_x$ cells at various
temperatures. [Reference 22]

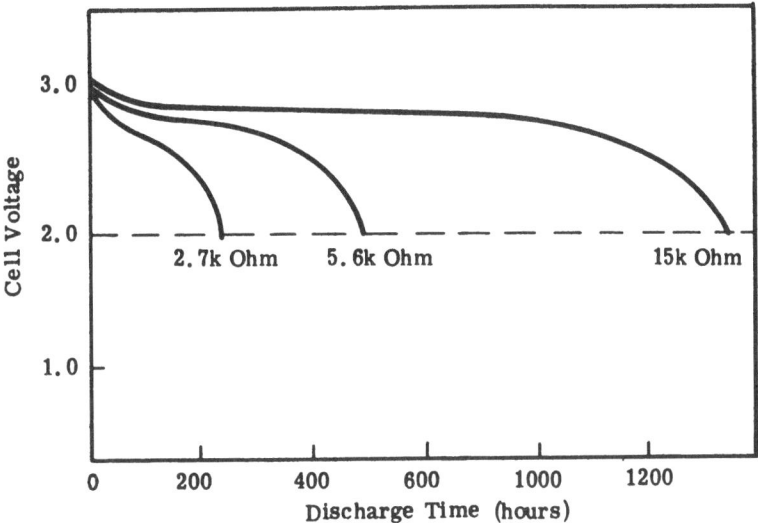

Fig. 11.   Discharge curves of pressed type cell (LF-1/2W) with
various loads. [Reference 20]

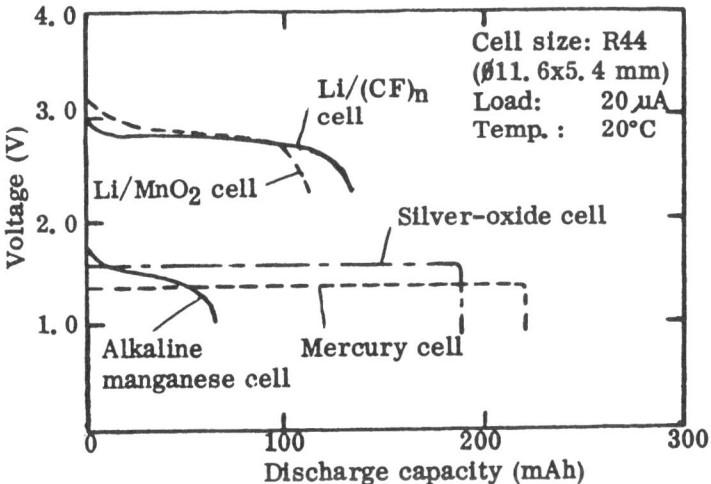

Fig. 12. Comparison of Li-CF$_x$ and Li-MnO$_2$ cells with mercury, silver,
and alkaline cells. [Reference 49]

In practice, the organic electrolyte lithium cells have been as safe as their alkaline counterparts except in the case of severe abuse conditions, e.g., incineration. Some solvents may be carcinogens. These must be avoided for commercial cells.

A common problem in battery design is the tendency for cathode swelling on discharge. This arises because the reactions produce insoluble lithium salts which precipitate in the cathode matrix. The volume of reaction products is often much larger than the reactants, in contrast to aqueous systems where the reaction products often have less volume than the reactants.

INORGANIC LITHIUM BATTERIES

The development of Li-SO$_2$ cells created a new concept in battery systems.[2,3] In these cells, the ionizing solvent, SO$_2$, also acts as the cathode reactant. Thus the cathode reactant does double duty and thereby improves the overall efficiency of materials utilization. Only one material is needed, whereas in the usual cell construction a different material is used for the solvent and the cathode reactant. As a result, the energy density is improved over that expected from an examination of the theoretical energy density. The Li-SO$_2$ cells have found wide use in military applications but have not achieved wide commercial acceptance.

Typical construction of Li-SO$_2$ cells is shown in Figure 13. Lithium foil serves as the anode. The cathode is a plastic-bonded carbon with a conductive aluminum mesh grid collector. The separator is usually a glass fiber matte. The cells are of wound construction, and some have a hermetic glass-to-metal seal. The electrolyte is LiBr in liquid SO$_2$ with a cosolvent, e.g., acetonitrile, added to lower the SO$_2$ vapor pressure and improve efficiency. Recently, Watson[50] described a process for SO$_2$ cell production. Typical discharge curves are shown in Figure 14. The cell gives excellent low temperature performance. Wilmar[51] reported that the Li-SO$_2$ system would give 4 to 5 times the performance of alkaline cells at low temperatures, e.g., -40°C.

One would expect SO$_2$ to react spontaneously and vigorously with lithium in the cell. The formation of a thin protective film of SO$_2$ reduction product (Li$_2$S$_2$O$_4$) on the surface of the lithium electrode effectively inhibits the SO$_2$ reduction. The breakdown of the film on circuit closure is responsible for the voltage delay phenomenon during discharge, but it also provides for good shelf life of the cell. The cells exhibit good performance, especially at low temperatures, due to the good transport properties of the SO$_2$ electrolyte and to the extended surface area electrode construction.

Although the cell reaction given in Table 3 is believed accurate, there are a large number of other reaction products possible from

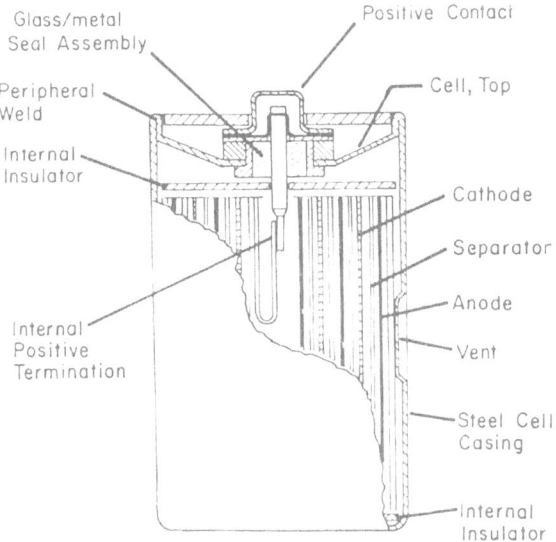

Fig. 13.   Schematic diagram of Li-SO$_2$ cell.
[Reference 52]

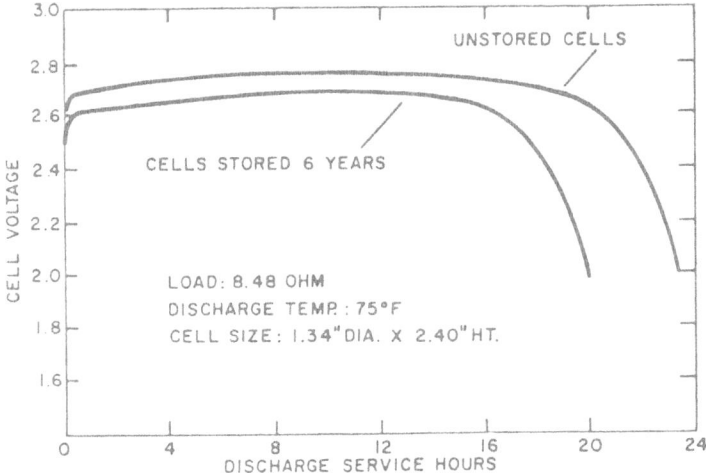

Fig. 14.   Discharge curve of Li-SO$_2$ cells fresh and after storage,
[Reference 52].   (These figures were originally pre-
sented at the Fall 1979 Meeting of The Electrochemical
Society, Inc., held in Los Angeles, California).

reduction of SO_2. As far as I can determine, there are no detailed reaction product analyses for these cells available in the literature.

Incidents of cell explosions have been reported for Li-SO_2 cells. The exact causes for this safety problem are not known, but several conditions for it have been noted. Shah[53] has reported spectacular results from drop tests, indicating a sensitivity to mechanical shock. Other postulated causes of explosion are inefficient discharge, long low temperature storage, cell reversal, exposure to the atmosphere, localized heating, and careless handling.

Dey[54] recently reported that the Li-to-SO_2 ratio was an important parameter affecting the safety. It may be possible that during very high current discharges severe cathode polarization can lead to lithium deposition in the cathode prior to exhaustion of the anode. The deposited lithium is very active and may lead to unsafe situations.

Another group of compounds has been identified which behave in a manner similar to SO_2 in lithium cells. These are the oxyhalides, $SOCl_2$, SO_2Cl_2, $POCl_3$, etc.[6-9] They are liquid at room temperature and are often referred to as liquid cathode cells. As with the Li-SO_2 cells, the liquid cathode construction results in increased efficiency in Li-$SOCl_2$ cell performance. The Li-$SOCl_2$ cells have been reported to deliver 18 Wh/in^3. This is significantly better than any other lithium battery. Only the zinc-air cell can approach this performance.

The construction of high rate Li-$SOCl_2$ cells is similar to that shown in Figure 13. The ionizing solute is $LiAlCl_4$, and the solvent is pure $SOCl_2$. The highest energy density cells have different construction. These may be characterized as a molded bobbin cell in which the plastic-bonded carbon forms a central cylindrical element. Cell sizes range from 12,000 Ah down to small coin-sized cells. The discharge voltage at various loads is shown in Figure 15.

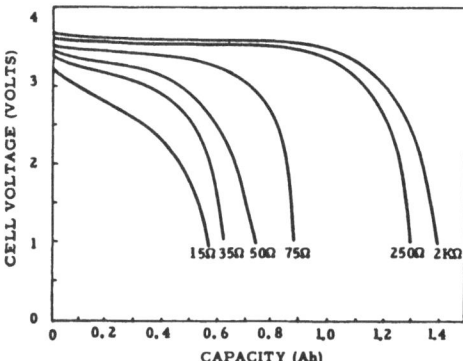

Fig. 15.   Discharge curve of Li-$SOCl_2$ 1.2 cm size cell at 25° C as a function of various loads, [Ref. 55]. (Originally presented at 29th Power Conference, June 1980, Atlantic City).

Film formation on the lithium during open circuit, mainly LiCl, effectively blocks the self-discharge reactions but results in a voltage delay phenomenon. Various techniques have been reported which reduce the voltage delay, including electrolyte purification and the application of protective organic membranes on lithium.[56,57] Recently, Gabano and Gelin[58] reported that electrolytes with Li_2O additions do not exhibit the delay phenomenon. This represents a considerable advance in cell technology.

The cell reaction and the mechanism of oxyhalide reduction have been the subject of numerous investigations. Bailey and Kohut[59] have analyzed the reaction products of cells discharged at various rates and at various states of discharge. They report that the overall reaction

$$4Li + 2SOCl_2 = 4LiCl + SO_2 + S$$

describes fully their findings. Blomgren et al.[60] suggest a mechanism which is consistent with this overall cell reaction and accounts for the observed sulfur chloride intermediate. The kinetics are not fully understood, but several intermediates, S_2Cl_2 and SCl_2, have been identified. Linear sweep studies have shown that the $SOCl_2$ reduction is affected by the electrode material. The reaction is hindered on nickel and platinum surfaces but occurs easily on carbon surfaces. The various reaction zones have been identified by intermediate seeding to confirm the identity of the reactants. A proposed mechanism for $SOCl_2$ reduction is:

$$SOCl_2 + e = SOCl\cdot + Cl^-$$
$$SO_2Cl\cdot = SO_2 + SCl_2$$
$$SCl_2 + e = SCl + Cl^-$$
$$2SCl = S_2Cl_2$$
$$S_2Cl_2 + 2e = 2S + 2Cl^-$$

There is little evidence based on reaction product analysis that SO_2 undergoes further reduction in thionyl chloride. Others have proposed an alternative reaction which involves formation of SO intermediate:

$$SOCl_2 + 2e = SO + 2Cl^-$$
$$SO = S + SO_2$$

This does not fit with the observation of SCl_2 intermediate.

Zupancic et al.[55] have described the performance of 1.25 Ah cylindrical cells using the lithium anode and stainless steel anode collector, a bonded carbon cathode current collector, and a felted fiberglass separator. The cell was sealed using a compression seal by crimping a fluorinated hydrocarbon polymer seal between the stainless steel can and cover. The performance of these cells is shown in Figures 15 and 16. The cells give good performance over the range of -40°C to 71°C. A voltage delay is observed which depends on storage temperature and storage time, as noted in Figure 17. The cells were reported to withstand electrical, mechanical and environmental abuse tests with no safety hazard except for skin, eye, or respiratory irritation caused by $SOCl_2$ vapors from open cells.

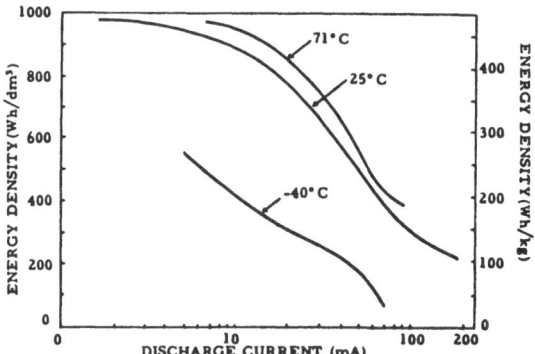

Fig. 16. Energy density of Li-$SOCl_2$ 1.2 cm size cell at various discharge conditions, [Ref. 55]. (Originally presented at 29th Power Sources Conference, June 1980, Atlantic City).

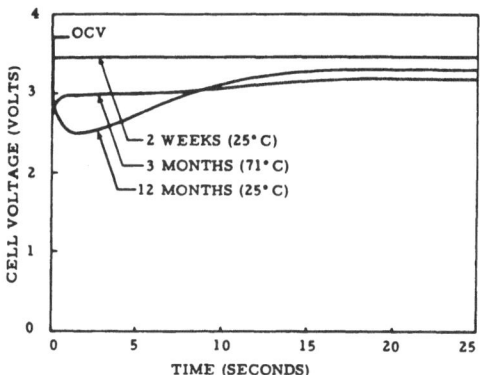

Fig. 17. Voltage delay of Li-$SOCL_2$ 1.2 cm size cell after various storage periods. Load 75 ohms, [Ref. 55]. (29th Power Sources Conference).

Abraham et al.[61] have discussed various safety problems of Li-$SOCl_2$ cells. They have been reported to explode (1) on short circuit, and (2) on reversal by forced overdischarge. On short circuit, a thermal runaway can occur which melts the lithium anode, destroying the adherent passivating film. Cathode-limited cells were safer than anode-limited cells. Many cells are thought to be anode-limited even though they are not lithium-limited. Poor lithium efficiency or poor electrical contact between lithium and the collector can make the cell anode limited. This type of anode-limited cell was found to be the most dangerous. This view is different from that of Dey,[54] who proposed that lithium-limited designs gave the safer construction.

COST

One of the most important criteria is the cost of a cell system. The relative material costs are probably the best way to compare the older and newer systems, given the vast difference in systems maturity. An estimate of these costs is presented in Table 9. The Li-FeS$_2$ system is significantly lower in cost than the silver cell and could replace it in the longer run. This is especially true if the high rate performance of the Li-FeS$_2$ system can be improved. In estimating the cost of Li-SOCl$_2$ cells, a ceramic hermetic seal was included. If a compression seal can be used, the costs decrease significantly. A value of $20/oz. was used to estimate silver cell costs.

Table 9. Comparison of Material Costs (cell size 11.6 x 4.2 mm)

System	Relative Cost
AgO/Zn	1.0
HgO/Zn	0.17
CuO/Li	0.26
FeS$_2$/Li	0.40
Cu$_2$S/Li	0.27
CF$_x$/Li	0.27
MnO$_2$/Li	0.26
SOCl$_2$/Li	0.74

[Reference 62]

SUMMARY

1. Mercury and silver cells will be the main systems for general purpose use.

2. Zinc-air will become the hearing aid power source.

3. The 1.5 volt lithium (Li-FeS$_2$) will replace mercury and silver for low rate applications based on cost. They do not have higher energy density.

4.   The 3.0 volt lithium (Li-CF$_x$ and Li-MnO$_2$) will become important
      for electronic applications. However, while they do not have
      significantly greater energy density, they do offer a lower cost than
      silver and mercury cells.

5.   Li-SOCl$_2$ will find wide use provided safety problems are resolved.
      It has a higher energy density than Li-SO$_2$, and the safety problems
      appear resolvable by construction techniques and electrolyte modifi-
      cations. It will find applications in miniature and cylindrical sizes.

                           *   *   *

ACKNOWLEDGMENT

      The author gratefully acknowledges the helpful discussions and
information supplied by N. Margalit, M. N. Hull, and B. C. Bergum.

REFERENCES

1.   R. J. Brodd, A. Kozawa, and K. V. Kordesch, J. Electrochem. Soc.,
      125:271C (1978).
2.   D. L. Maricle and A. K. Hoffman, U.S. Pat. 3,578,500 (1971).
3.   D. L. Maricle and J. P. Mohns, U.S. Pat. 3,567,515 (1971).
4.   P. Bro, R. Holmes, N. Marincic, and H. Taylor, in "Power Sources 5",
      D. H. Collins, ed., Academic Press, London (1975), p. 703.
5.   E. S. Brooks, Proc. 7th IECEC (1973), p. 71.
6.   G. E. Blomgren and M. L. Kronenberg, Ger. OLS 2,262,276 (1973).
7.   J. J. Auborn, K. W. French, S. I. Lieberman, V. K. Shah, and A.
      Heller, J. Electrochem. Soc. 120:1613 (1973).
8.   W. K. Behl, J. S. Christopulos, M. Ramirez, and S. Gilman, J.
      Electrochem. Soc. 120:1619 (1973).
9.   J. J. Auborn and N. Marincic, in "Power Sources 5", D. H. Collins,
      ed., Academic Press, London (1975), p. 683.
10.   B. B. Owens and K. R. Brennen, in "Progress in Batteries and Solar
      Cells", J.E.C. Press, Inc., Cleveland, OH (1980), Vol. 3, p. 38.
11.   C. C. Liang, J. Electrochem. Soc. 123:453 (1976).
12.   R. F. Amlie and P. Ruetschi, J. Electrochem. Soc. 108:813 (1961).
13.   Y. Uetani, K. Yokoyama, and T. Kawai, Proc. 28th Power Sources
      Conf. (1978), p. 219.
14.   M. L. Kronenberg, U.S. Pat. 4,163,829 (1979).
15.   G. Lehmann, G. Gerbier, A. Brych, and J. P. Gabano, in "Power
      Sources 5", D. H. Collins, ed., Academic Press, London (1975), p.
      695.
16.   Y. Jumel and M. Broussely, Proc. 28th Power Sources Conf. (1978),
      p. 222.
17.   H. Ogawa, T. Iijima, A. Morita, and J. Nishimura, in "Progress in
      Batteries and Solar Cells", J.E.C. Press, Inc., Cleveland, OH (1980),
      Vol. 3, p. 93.

18. B. C. Bergum, A. M. Bredland, and T. Messing, in "Progress in Batteries and Solar Cells", J.E.C. Press, Inc., Cleveland, OH (1980), Vol. 3, p. 90.
19. H. Ikeda, M. Hara, and S. Narukawa, Proc. Manganese Dioxide Symposium (1975), Vol. 1, p. 384; Proc. 28th Power Sources Conf. (1974), p. 210.
20. H. Ikeda, S. Ueno, T. Saito, S. Nakaido, and H. Tamura, Denki Kagaku 45:314,391 (1977).
21. M. Fukuda and T. Iijima, in "Power Sources 5", D. H. Collins, ed., Academic Press, London (1975), p. 713.
22. M. Fukuda and T. Iijima, Denki Kagaku 44:543 (1976).
23. M. Lang, J. R. Backlund, and E. C. Weidner, Proc. 26th Power Sources Conf. (1974), p. 37.
24. S. C. Levy, Proc. 27th Power Sources Conf. (1978), p. 52.
25. G. Lehmann, T. Rassinoux, G. Gerbier, and J. P. Gabano, in "Power Sources 4", D. H. Collins, ed., Academic Press, London (1973), p. 493.
26. E. A. Megahed and D. C. Davig, in "Power Sources 8", in press.
27. A. Tvarusko, J. Electrochem. Soc. 116:1070 (1969); U.S. Pat. 3,650,832 (1972).
28. B. Cahan, U.S. Pat. 3,017,448 (1962).
29. S. M. Davis, U.S. Pat. 3,853,623 (1974).
30. L. A. Soto-Krebs, U.S. Pat. 3,615,858 (1972).
31. R. J. Dawson, U.S. Pat. 3,484,295 (1969).
32. P. Ruetschi, in "Power Sources 7", J. Thompson, ed., Academic Press, London (1979), p. 533.
33. H. F. Gibbard, J. Electrochem. Soc. 125:353 (1978).
34. E. J. Prosen and J. C. Colbert, NBS Special Publication 400-42 (August 1977).
35. L. D. Hansen and R. Hart, J. Electrochem. Soc. 125:842 (1978).
36. D. F. Untereker, J. Electrochem. Soc. 125:1907 (1978).
37. M. N. Hull and H. I. James, J. Electrochem. Soc. 124:332 (1977).
38. L. M. Baugh, J. A. Cook, and F. L. Tye, in "Power Sources 7", J. Thompson, ed., Academic Press, London (1979), p. 519.
39. L. M. Baugh, J. A. Cook, and J. A. Lee, J. Appl. Electrochem. 8:253 (1978).
40. K. Kajita, A. Shimizer, and Y. Aetani, in "Progress in Batteries and Solar Cells", J.E.C. Press, Inc., Cleveland, OH (1980), Vol. 3, p. 99.
41. J. E. Oxley, in "Progress in Batteries and Solar Cells", J.E.C. Press, Inc., Cleveland, OH (1980), Vol. 3, p. 48.
42. R. J. Brodd, V. Z. Leger, R. F. Scarr, and A. Kozawa, NBS Special Publication 455 (1976), p. 253.
43. R. J. Brodd and V. Z. Leger, in "Electrochemistry of the Elements", Plenum Press, New York (1977), Vol. 5, ch. 5.
44. R. F. Scarr, J. Electrochem. Soc. 117:295 (1970).
45. J. C. Cessna, Corrosion 27:244 (1971).
46. V. R. Koch, J. Electrochem. Soc. 126:182 (1979).
47. V. R. Koch and J. H. Young, Science 204:499 (1979).
48. G. E. Blomgren, in press.

49. J. Watanabe, R. Okazaku, M. Nakai, and Y. Toyoguchi, in "Progress in Batteries and Solar Cells", J.E.C. Press, Inc., Cleveland, OH (1980), Vol. 3, p. 70.

50. T. M. Watson, Proc. 28th Power Sources Conf. (1978), p. 192.

51. E. D. Wilmar, in "Power Sources 6", D. H. Collins, ed., Academic Press, London (1977), p. 469.

52. B. Jagid, T. Watson, and S. M. Chodosh, Proceedings, Electrochemical Society, Inc. (1980), Vol. 80-4, p. 615.

53. P. U. Shah, Proc. 28th Power Sources Conf. (1978), p. 188.

54. A. M. Dey, J. Electrochem. Soc. 127:1886 (1980).

55. R. L. Zupancic, L. F. Urey, and V. S. Alberto, Proc. 29th Power Sources Conf., in press.

56. T. Kalnoli-Kis, U.S. Pat. 3,993,501 (1976).

57. K. French, P. Cuker, C. Persiani, and J. J. Auborn, J. Electrochem. Soc. 121:1045 (1974).

58. J. P. Gabano and G. Gelin, in "Power Sources 8", in press.

59. J. Bailey and J. Kohut, in "Power Sources 8", in press.

60. G. E. Blomgren, V. Z. Leger, T. Kalnoki-Kis, M. L. Kronenberg, and R. J. Brodd, in "Power Sources 7", J. Thompson, ed., Academic Press, London (1979), p. 583.

61. K. M. Abraham, P. G. Gudrais, G. L. Holleck, and S. B. Brummer, Proc. 28th Power Sources Conf. (1978), p. 225.

62. B. C. Bergum, "Miniature Lithium Cells, Now and Tomorrow", Wescon 80.

DISCUSSION

Dr. J. Kruger, National Bureau of Standards, Washington: Because of the stresses introduced during fabrication and because of the sulfur and chlorine species involved in the chemical reactions of the Li-SOCl$_2$ battery, could some of the safety problems of these cells be connected with stress corrosion cracking?

Dr. R. Brodd: Yes, it is possible that stress corrosion cracking may be the cause for some safety problems. The mechanical operations of cell fabrication can induce stresses especially around the seal area. Mechanical working is often used to affect a compression seal and welding techniques can induce thermal gradients leading to stresses. The cell interior is exposed to many different sulfur, oxygen and chlorine containing species. These are known to induce corrosion.

Prof. Karl Kordesch, Technical University, Graz, Austria: At the ISE Meeting in Venice researchers from SAFT, France, reported they had solved the delay and passivation of lithium in Li-SOCl$_2$ and Li-SO$_2$Cl$_2$ cells. Please comment.

Dr. R. Brodd: I presume you refer to the work of Gabano et al. in which they add Li$_2$O to form the electrolyte in thionyl chloride and sulfuryl chloride electrolytes. The discharge curves I saw indeed had little or no delay associated with lithium passivation. This represents a good step toward eliminating one of the problems facing the $Li-SOCl_2$ and $Li-SO_2Cl_2$ cells.

Dr. Elton Cairns, University of California, Berkeley: In your summary at the end of your lecture, you did not include the $Li-SO_2$ cell. Does that indicate that you imply the demise of the SO_2 cell?

Dr. R. Brodd: The $Li-SOCl_2$ has higher energy density, can operate over just as wide a temperature range, and it appears that its safety problems may be easier to solve. In the long run, therefore, I would expect the $Li-SOCl_2$ to replace present $Li-SO_2$ applications. This certainly will not occur overnight, as $Li-SO_2$ is a production item and $Li-SOCl_2$ is still in the development stage.

BATTERY DEVELOPMENT IN EUROPE

Wolf Vielstich

University of Bonn
Bonn, West Germany

INTRODUCTION

Battery work in Europe presently is focused on secondary systems for electrotraction and load levelling. An improved lead acid system (ca. 50 Wh/kg, 5 hr rate) is still considered as the number one candidate for both applications in the near future. For heavy traction (electric bus) a cycle life of more than 1000 cycles at the 2 hr operation has been demonstrated with present technology (up to 25 Wh/kg incl. container). Further intensive investigations are made in the field of Na/S, LiAl/FeS, Ni/Fe, alkaline and acid H_2/O_2 cells as well as in primary Al-batteries with oxygen or air as oxidant.

In Western Europe and especially in West Germany, the development of batteries and fuel cells is influenced strongly by impulses coming from government institutions. For this survey I therefore like to restrict my report mainly to the work supported by the German program on electrochemical energy conversion.

THE ENERGY SITUATION AS MOTIVATION OF THE BMFT PROGRAM

The motivation for the directions in research and development, supported by the department of energy and technology (BMFT), is provided by the general energy situation. As is shown in Fig. 1, in 1973 more than 50% of our primary energy is given as oil. But our own production covers only 5% of this amount. While the overall oil consumption had been somewhat decreased since 1973, the volume needed for vehicle propulsion was increasing to 57 energy units (10^6t SKE - lt SKE is ca l kWa) in 1978. The oil used is in the form of gasoline, diesel fuel and kerosene. In addition, 3.6

energy units spent for vehicle propulsion (coal or nuclear as primary energy) are converted into electricity mainly for the railway
system. Less than 20% (9.6 units) of the 60.6 units could be calculated as useful energy.

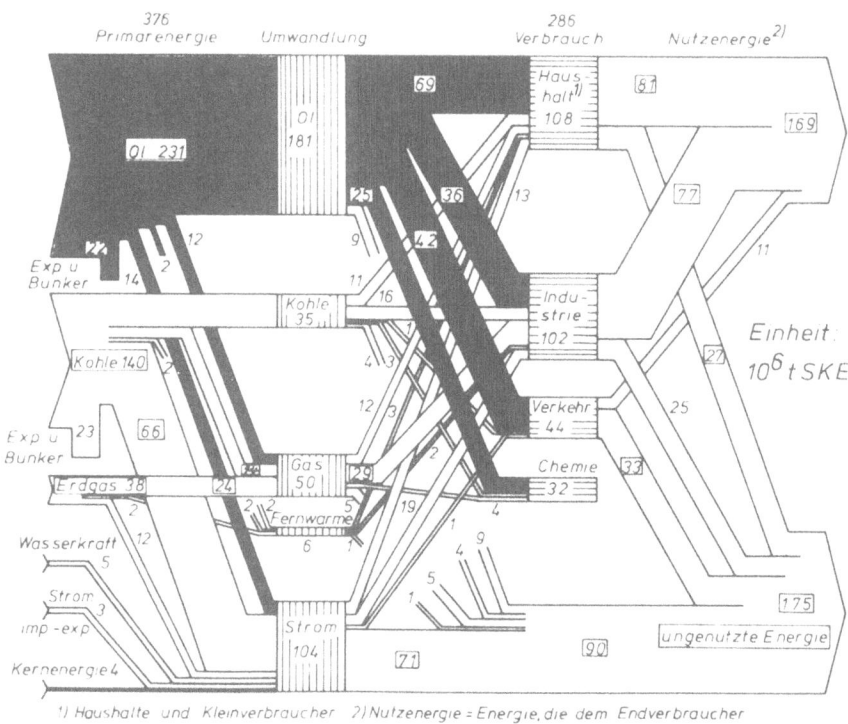

Fig. 1   Use of primary energy in West Germany in 1973, 10^6 kWa
          as energy unit, 1 kWa = 1 SKE ("Steinkohleneinheit")

     According to studies of the GES (1) (Gesellschaft für elek-
trischen Strassenverkehr) the oil needed for traffic can be partially replaced by coal or nuclear as primary energy via electrochemical energy conversion. The traffic statistics indicate that
in Germany and similarly in other regions of Western Europe most
passenger cars as well as buses and trucks travel less than 100 km
per day. So, already a short term development in batteries for
vehicles could help solve the problem of oil restriction. Assuming
that 75% of the passenger cars and 55% of buses and trucks in city
traffic could be electric, we could use an energy conversion as
suggested in Table 1. The electricity for recharging the electric
vehicle batteries could be obtained via nuclear energy. The rest
of ca 20 energy units as gasoline or diesel could be produced via
coal liquifaction [hydrogenation of coal or carbon monoxide (2)].
In the table an energy conversion factor of 50% is assumed.

Table 1. Electrotraction and the Use of Primary Energy
          in West Germany (10^6 kWa as energy unit)

Primary		Converted		Useful Energy
Oil	57	Fuel	53	9.6
Coal, etc.	3.6	Electric Power	1.2	
Coal	39^a	Fuel	19.1	
Nuclear,		Electric		10.0
etc.	28.2	Power	9.4	

Assumptions (1) City traffic 75% (55%) electric
             (2) Coal liquifaction η_{th} = 50%

[a]The 40 million tons of coal suggested for liquifaction are
the upper limit which can be made available for this purpose.

The situation discussed above suggests a strong effort in the
development of batteries for electrotraction. High energy densities
in the form of an economically reasonable secondary battery are
also desired for the application of load levelling. Therefore, in
the future more than 80% of the government support on electrochemical
energy conversion (during the last 5 years about 50 million DM)
will be concentrated on these two fields of application.

IMPROVED LEAD ACID BATTERIES FOR ELECTROTRACTION

The slow progress in the development of new secondary batteries
as NaS, NiZn or NiFe, suitable for mobile applications, has supported
the renewed interest in the lead acid cell for electric vehicles.
Furthermore, field testing of electric buses in three German cities
has demonstrated that present day technology is sufficient for this
application provided that the battery life can be extended over more
than 1000 cycles.

The Pb-battery tested until now in German buses did use 8 mm
tube-plates with nominal 360 V, 455 Ah and 6.100 kg. In the usual
2 hr service between charges the effective data have been 330V/350
Ah and a cycle life of 700-1000.

Obviously two factors have limited the time of operation (3) -
the oxidation of expander material at the PbO_2 electrode, and the
formation of acid layer of different density at the bottom of the
container.

Large organic molecules like Ca-ligninsulfonate adsorbed at
the lead surface are necessary to stabilize a sufficient porosity
of the negative plate. Fig. 2 shows the typical effect of the expander on a smooth lead electrode. With 0.2 g expander per liter

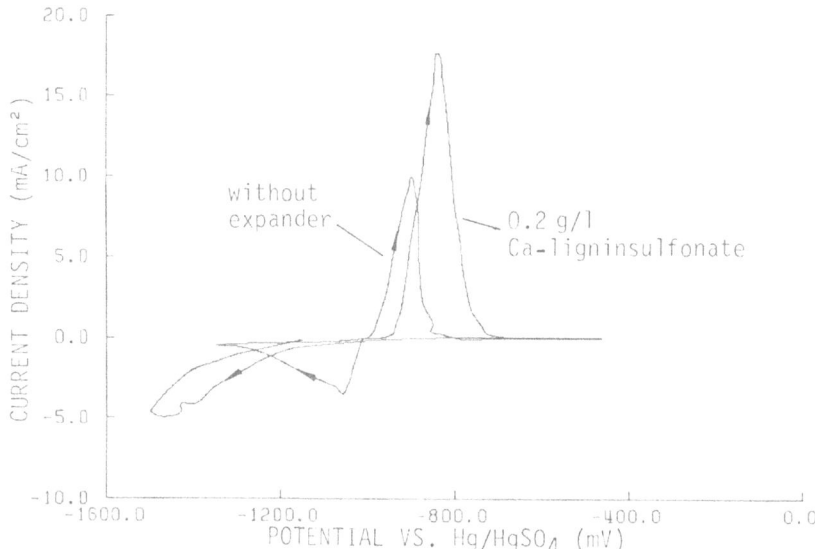

Fig. 2 Formation of PbSO$_4$ on smooth lead in 1N H$_2$SO$_4$ at 20°C,
        potential scan 10 mV/s, without expander and with addi-
        tion of 0.2 g/l Ca-ligninsulfonate (4)

the capacity is evidently enlarged but the reaction is somewhat
less reversible. Fig. 3 demonstrates that the effect is due to the
changed surface structure. In recent experiments (5) it has been
proved that via the addition of expander - discontinuous or in form
of a continuously dissolving "expander depot" - the number of cycles
is increased up to 1350 ± 50.

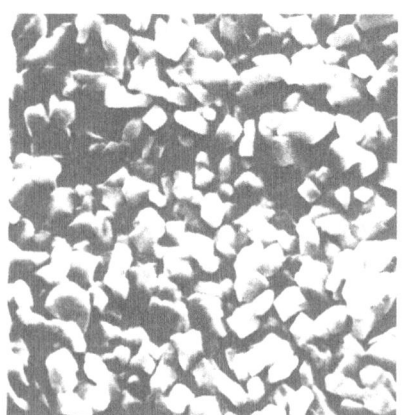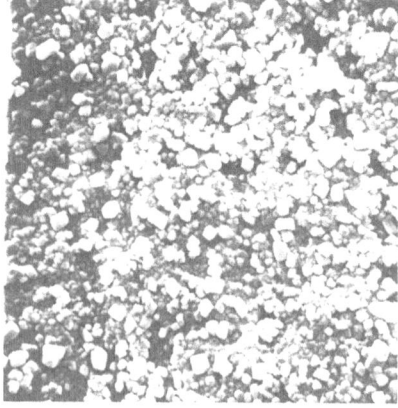

Fig. 3 Crystal structure during the formation of PbSO$_4$ correspon-
        ding to the experiment of Fig. 2, electron microscope mag-
        nification 5000. Left: without expander and right: with
        Ca-ligninsulfonate.

In order to avoid the acid layer formation VARTA has successfully constructed a low power electrolyte pump operating by air
circulation (5). The convection of the electrolyte during the
charge mode permits lowering the charge factor from 1.2 down to
1.03. The forced convection is more effective than the gassing
during overcharge.

For the passenger car the results of field testing obtained
recently by the GES with 6 mm tube-plates also look promising.
While using grid-plates only 200-400 cycles could be attained; the
replacement by tube plates extended the life of the battery to considerably more than 400 cycles. The Wh/kg data of the present 160
Ah passenger car battery are 34 for 5 hr rate and 25-28 for the 1-
2 hr rate. For the city traffic in West Germany, however, 38-42
Wh/kg for the 1-2 hr rate are required. There is experimental evidence that these can be obtained by newly designed cells. One
solution studied is the Eloflux battery according to Winsel (6).
In this mode of operation the electrolyte is forced vertically
towards the direction of the electric current through the plates.
Initial results on 4 mm plates indicate 140% increase in the effective capacity for the lead electrode (from 95 Ah/kg to 130 Ah/kg)
and 180% for the lead dioxide plate (from 60 Ah/kg to 110 Ah/kg).

The 40 Wh/kg (2 hr rate) battery will be sufficient for a car
of about 1000 kg. According to calculations of the GES, the price
of such a car will be only 15-20% higher than the gasoline version
--provided that the production figures have exceeded 100,000 to
200,000.

BATTERIES FOR LOAD LEVELLING

An extended application of secondary batteries in stationary
energy storage requires large amounts of active material. Therefore the components of the battery should be available for a relatively low price. Energy density is more important than power
density and the working conditions as temperature and mechanical
stability are not as important as in the case of a mobile application. Thus, in recent years Na/S, LiAl/FeS and Zn/Cl_2 are considered
as possible candidates. The use of these batteries in the electric
vehicle has also been discussed.

The somewhat exotic system Zn/Cl_2 is not being studied in
Europe. Investigations in the U.S. did result in the following
characteristic data: 33 mA/cm^2 at 35°C and 0.3 mm of electrode gap.
At this rather low current density the zinc corrosion by dissolved
Cl_2 corresponds to 10 mA/cm^2. With d>0.3 mm the voltage drop increases substantially and with d<0.3 mm the increase in natural
convection favors even more the rate of corrosion. On the other
hand, promising new results have been obtained with the two high
temperature systems.

The progress in the development of <u>NaS-cells</u> and batteries at BBC using tubes of β-Al$_2$O$_3$ as solid electrolyte (7), is shown in Fig. 4. In both figures, energy density and power density look promising even for mobile applications. The main problem is the limitation on the life of the cell because of the sensitivity of β-alumina to electrochemical cycling as well as to thermal cycling. In addition, the ceramic tool used to produce the electrolyte can be used 10-20 times only; this fact influences strongly the price of the battery.

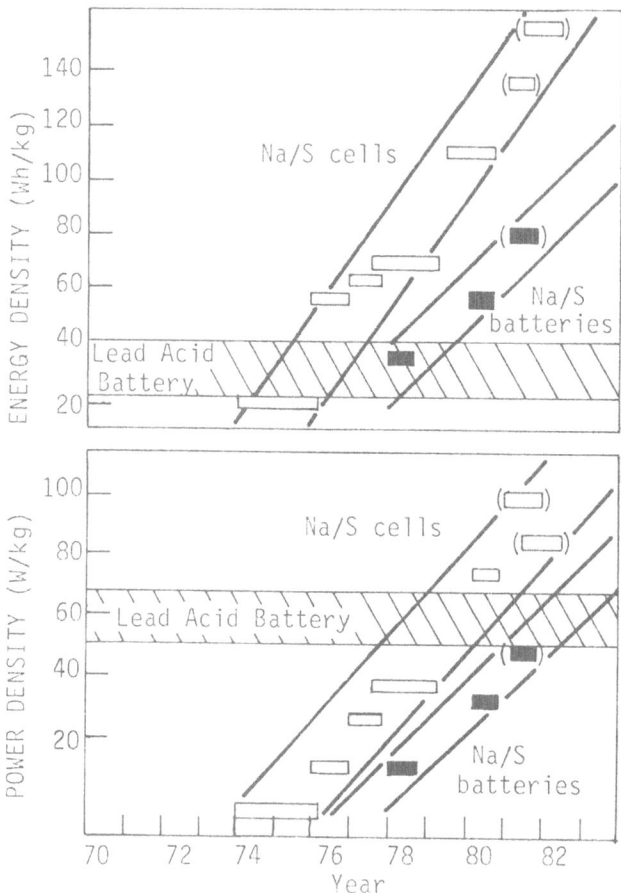

Fig. 4 Energy and power density of NaS-cells ▭ and batteries
        ▬ according to Brown Boveri Cie, Heidelberg (7); in
        dashed region: cells and batteries in production.

Experiments with NASICON as an alternative electrolyte indicate that this material is not stable in the Na$_2$S$_x$ melt.

A very important step forward could therefore be the recent development of a new electrolyte called "TITZICON" (8) with the

composition $Na_{3.1}Zr_{1.55}Si_{2.3}P_{0.7}O_{11}$. The ionic conductivity of this
sodium ion electrolyte is higher than that of β-alumina in the interesting temperature range ($2.2 \cdot 10^{-1}\Omega^{-1}cm^{-1}$ at 300°C, see Fig. 5)
– due to a three dimensional ion mobility compared with the two
dimensional mechanism in β-alumina. The stability against the
Na_2S_x has been proven to be excellent (one phase only).

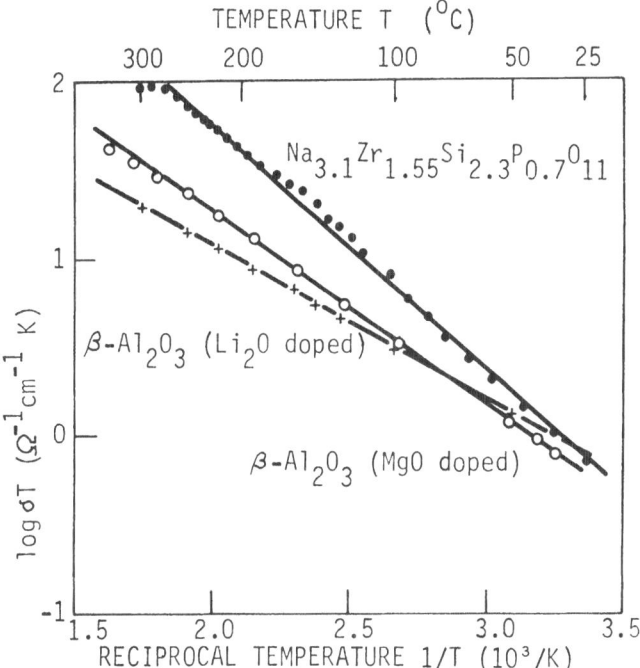

Fig. 5 New developed solid electrolyte Titzicon in comparison
        to MgO- and Li_2O-doped β-alumina according to Varta AG,
        Kelkheim (8); Arrhenius plot of the product conductivity
        times temperature as function of the reciprocal absolute
        tempreature l/T

    The second high temperature system LiAl/FeS shows a similar
state of the art and comparable electric data. One central problem
is the extremely high price of the BN-separator used. In this
respect the development of an inexpensive new powder separator
stabilized by two stainless steel grids is very interesting (9).
First experiments have shown more than 500 cycles at 90-80% FeS
utilization and more than 250 Wh/l at the 5 hr rate. The new electrolyte is cheap to prepare, the porosity and the grain size are
easily adjustable.

FUEL CELLS, H_2/O_2 AND REFORMER GAS/AIR

    In contrast to the situation in the U.S. the application of
fuel cells in power plants is not pursued in Europe. On the other

hand, intensive work during the last decade has resulted in the
development of alkaline and acid cells with possible applications
as emergency power and as small power sources in non-electrified
areas.

Alkaline H_2/O_2 modules working at 80°C and 1-2 bar overpressure
have been built at Siemens, Erlangen, for 7 kW of nominal power (10).
The successful introduction of a non-noble metal catalyst according
to Justi and Winsel is demonstrated in Fig. 6. A careful preparation of the catalyst, including the consideration of the influence
of the grain size has resulted in a remarkable success (11). The
respective voltage/current density plot is given in Fig. 7. Battery
units up to 20 kW have been built. A commercial production in
small scale has been started.

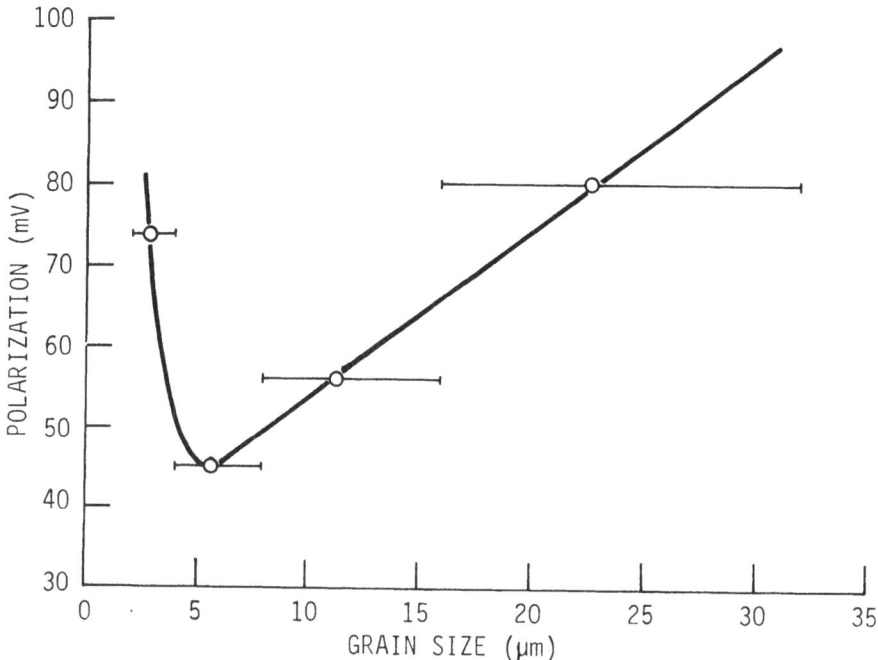

Fig. 6. Titanium doped Raney nickel hydrogen electrode at 500
        mA/cm^2 and 80°C (11), polarization as function of grain
        size of the "supported" catalyst layer

The acid reformer gas/air system, developed at AEG, Frankfurt,
uses H_3PO_4 as electrolyte. A special advantage is the introduction
of tungsten carbide as electrocatalyst at the anode (12). This
allows an operating temperature of 150°C only, due to the fact that
WC is not poisoned by the CO content of the reformer gas. The
recent characteristic data of this system are 50 mA/cm^2 at an average cell voltage of 500 mV. Feeding H_2 and O_2 into the cell, the

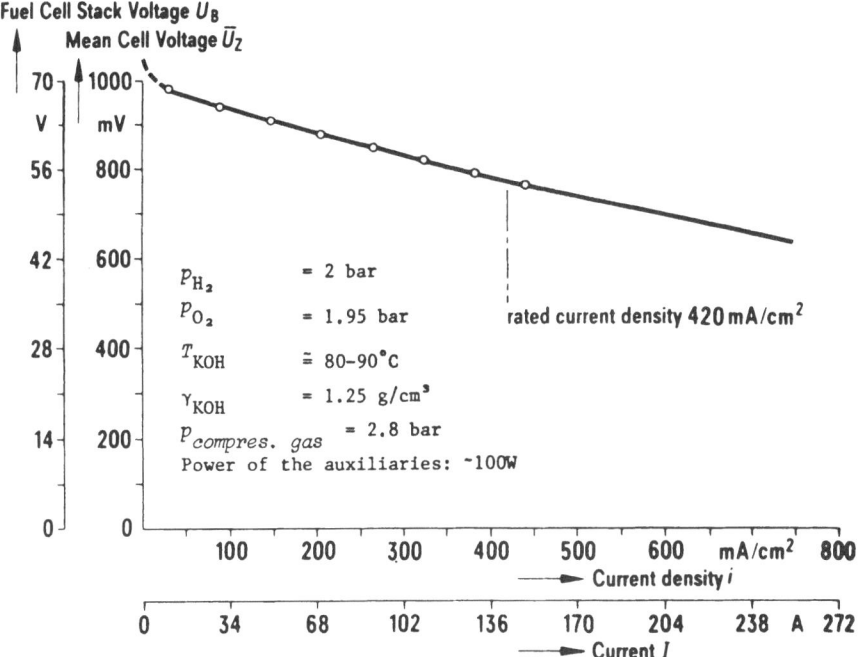

Fig. 7. Cell voltage as function of stationary current density
in a 70 cell $H_2/KOH/O_2$ module at 80°C according to
Strasser (10); note that the voltage deviation of single
cells at the rated current density of 420 mA/cm^2 is less
than ± 10 mV. [Reprinted by permission of the publisher,
The Electrochemical Society, Inc.]

current density increases to 100 mA/cm^2.

    At a fuel cell meeting in Fortaleza, Brazil, early this year
the application of this system in the power range of 1-5 kW was
discussed, considering alcohol from sugar plants as primary fuel.

ALUMINUM/AIR, A NEW APPROACH?

    Until recently aluminum has not been seriously considered as
active material in electrochemical power sources. This was mainly
due to its heavy corrosion in alkaline and acid media. A new interest (13) has been shown along two lines:

- high power cells with alkaline electrolyte, lowering the effect
  of corrosion via very high current densities (13,14,15), and

- high energy cells with "neutral" electrolyte.

    Due to its high power and energy density the alkaline system
is discussed for the application in electrotraction also (13,14).
But the recycling of the aluminum offers an overall efficiency of

7-8% only, referring to the ΔH value of the coal or to the nuclear
energy necessary for the aluminum refining process. In addition
there are unsolved problems in the overall chemistry of the complex
system.

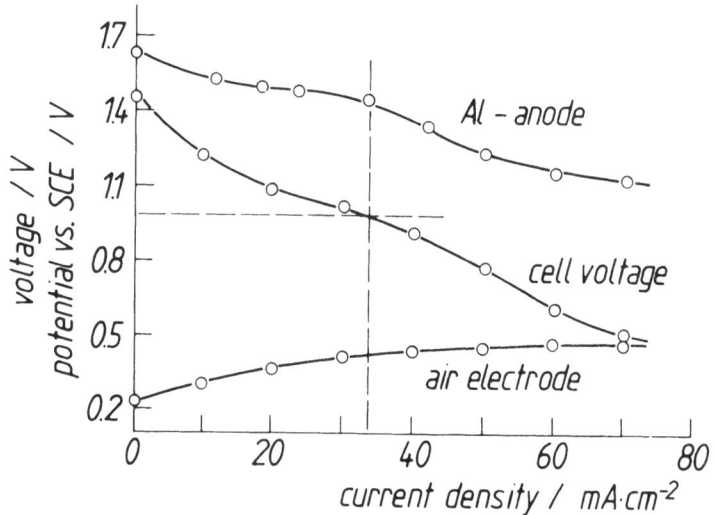

Fig. 8. Aluminum-air cell with 2M NaCl electrolyte according to
         Despic et al. (16)

Using 2 M NaCl as electrolyte in an aluminum/air cell offers
new possibilities in the application (16,17,18). With this electrolyte the loss of capacity via corrosion can be less than 10%.
The study of the electrode reactions shows that (a) a pH-gradient
is developed during the operation (18) and (b) the consumption of
water during the process has to be taken into account.

$$(-)\quad 2\ Al + 6\ H_2O \longrightarrow 2\ Al(OH)_3 + 6\ H^+ + 6\ e^-$$

$$(+)\quad 3/2\ O_2 + 3\ H_2O + 6\ e^- \longrightarrow 6\ OH^-$$

$$2\ Al + 3/2\ O_2 + 3\ H_2O \longrightarrow 2\ Al(OH)_3$$

The water necessary for the reaction limits the Ah-capacity of
closed systems to 300-500 Ah/l. The energy density depends very
much on the anode potential during discharge. It can be shifted
more than 400 mV by alloying the aluminum anode with ∿ 0.1% of

indium and/or gallium. Using an aluminum/indium alloy Despic et
al. (16) have obtained the results of Fig. 8. On the basis of this
data a mobile application is discussed. Fig. 9 shows the cell
design suggested to minimize corrosion. In order to control the
formation of the voluminous reaction product Al(OH)$_3$ the effect of
a pulsing movement of the electrolyte (amplitude $\sim$ 1.5 cm) is being
studied.

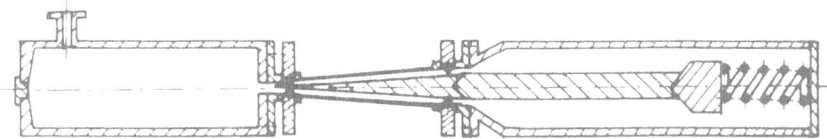

Fig. 9. Cell suggested by Despic and Milanović (17). Continuous
         shift of the Al-electrode minimizes corrosion.

     The adjustment of a suitable air electrode is of central impor-
tance in the development of "neutral" electrolyte Al/air cells.
The strong pH-gradient at the cathode suggests the use of a non-
noble metal electrode. Due to the alkaline environment active
carbon already is a suitable catalyst (Fig. 10).

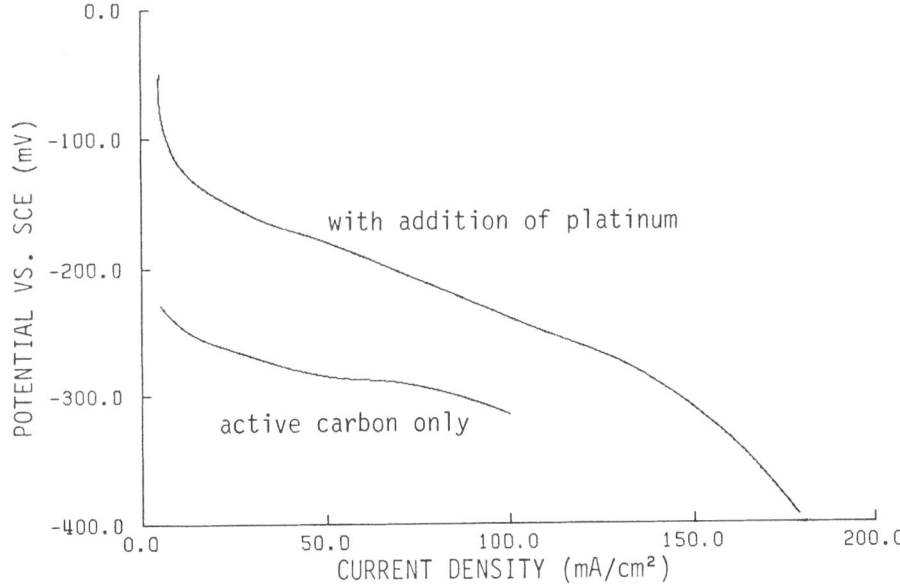

Fig. 10. Performance of "neutral" electrolyte air electrodes with-
          out noble metal catalyst according to Hoffmann, Holze
          and Vielstich (19), half cell test of 10 cm^2 electrodes
          in 200 cm^3 electrolyte, 2 $\underline{\text{M}}$ NaCl and 20°C.

The problems with the formation of the gel and the consumption of large amounts of water can be avoided using <u>cells open to sea water</u> as electrolyte (Fig. 11). In addition to the easy operation one has a very high energy to weight ratio available because only the weight of the aluminum has to be taken into account. Even at cell voltages as low as 1.0 volts and Ah efficiencies of only 60%-70%, more than 2.000 Wh/kg can be obtained.

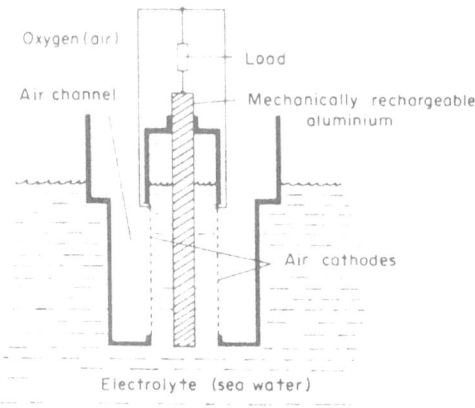

Fig. 11. Construction of an aluminum-air cell according to Ritschel and Vielstich (18), activated by excess (sea water) electrolyte.

REFERENCES

1. H.G. Müller, paper presented on May 9, 1980 at the Berlin meeting of the DGES.

2. W. Vielstich, Kohle als Basis, Nachr. Chem. Tech. Lab. <u>27</u>, 553 (1979).

3. H.G. Müller and J. Schulz, priv. communication.

4. G. Hoffmann and W. Vielstich, unpublished.

5. A. Kalbitz, H. Rabenstein and K. Kleber, DOS 28 55 313 (12.12. 78); A. Winsel and J. Schulz, in Elektrochemische Energietechnik, Der Bundesminister für Forschung und Technologie, Bonn 1980.

6. U. Hullmeine, W. Kappus, J. Schulz, E. Voss and A. Winsel, VARTA A.G., Kelkheim, unpublished.

7. W. Fischer, in Elektrochemische Energietechnik, Der Bundesminister für Forschung und Technologie, Bonn 1980.

8. U.V. Alpen, M.F. Bell and H.H. Höfer, J. Solid State Ionics, in press.

9. VARTA A.G., Kelkheim, priv. communication.

10. K. Strasser, H. Grüne and H.B. Gutbier, Power Sources $\underline{6}$, 569 (1977).

11. H. Grüne and K. Strasser, priv. communication.

12. R. Fleischmann, H. Böhm and J. Heffler, 14th Intersoc. Energy Convers., Vol. 1, p. 559, Boston 1979.

13. Proc. First Intern. Workshop on Reactive Metal-Air Batteries, Bonn, July 1979.

14. J.F. Cooper and E. Littauer, 13th IECEC, San Diego, 1978.

15. J. Ruch, Electrochem. Soc. Meeting, October 1977, Abstr. 42.

16. A.R. Despić, D.M. Drazić, M.M. Purenović and N. Ciković, J. Applied Electrochem. $\underline{6}$, 527 (1976).

17. A.R. Despić, priv. communication.

18. M. Ritschel and W. Vielstich, Electrochim. Acta $\underline{24}$, 885 (1979). W. Vielstich and L. Grambow, ETZ-A $\underline{99}$, 609 (1978).

19. G. Hoffmann, R. Holze and W. Vielstich, unpublished.

DISCUSSION

Professor Alvin Salkind, Rutgers University Medical School: Since the life of negative electrodes in many systems is largely controlled by expanders, what current research is being done on new expanders?

Professor W. Vielstich: New substances to be used as expanders are being studied, especially in connection with the "expander depot" service. In addition, fundamental investigations on the mechanism are in progress.

SOLID ELECTROLYTES

Robert A. Huggins

Department of Materials Science and Engineering
Stanford University
Stanford, California 94305

INTRODUCTION

Recent years have seen a rapidly accelerating interest in a group of materials which, although solid, exhibit many of the properties normally associated with liquid electrolytes. This results from the fact that electrical charge is carried through such materials primarily by the transport of ionic species, rather than by electrons or holes. In addition, a number of materials are now recognized as having very high values of ionic conductivity, and sometimes at quite low temperatures. This has led to a flurry of activity in the scientific community, as well as to a great deal of effort aimed at the practical utilization of this phenomenon for both scientific and technological purposes. Materials that exhibit rapid ionic motion are variously called fast ionic conductors, solid electrolytes, or even superionic conductors, and the term "solid state ionics" is now used to encompass both the scientific and technological aspects of their practical use.

Actually, the phenomenon of electrical charge transport by the motion of ions within solids has been the subject of a large number of research investigations over many years. This phenomenon was employed in a practical broad band light source known as the "Nernst Glower" (1-4), which conducts charge by the motion of oxide ion defects (5), almost 80 years ago.

This area has received greatly increased emphasis in recent years because of the discovery of a large number of new solid ionic conductors and recognition of the wide variety of their potential applications. Awareness of the potential utility of

195

these materials has arisen gradually, starting with a group of important papers by Wagner and his co-workers [6-9]. Somewhat later, this interest was accelerated by the discovery of two important families of materials which have unusually high values of ionic conductivity at surprisingly low temperatures. These were the silver-conducting ternary silver iodides, which were simultaneously discovered in England and the United States [10-13], and the alkali metal-conducting beta alumina family, which was discovered by workers at the Ford Scientific Laboratory [14-15]. The Ford group also showed [16] that solid electrolytes such as sodium beta alumina could be employed in a radical new design of secondary battery, employing liquid electrode constituents and a solid electrolyte, that could potentially have attractively large values of specific energy.

As a result of the attention drawn by such developments, activities in this area have increased rapidly in recent years. In addition to use in batteries and fuel cells, solid electrolytes are of interest for an increasing variety of applications. Oxide ion conductors are now widely used to monitor the exhaust gas of automobiles, and are being developed for use in a number of industrial applications for the measurement and control of the oxygen content of gases as well as of liquid metals. Fluoride ion-conducting solids and other ionically-conducting solids are now employed as specific ion sensors in various ambient temperature liquids. Solid electrolytes have been shown to be useful in heterogeneous catalytic systems, and are being investigated as possible components in new solid state electrochromic display devices. A number of other practical devices have also been demonstrated, such as switches, timers and coulometers, which depend upon the unique properties of solid electrolytes.

While each type of application has somewhat different requirements, they have a number of important common features. Some of these are as follows:

    High ionic conductivity
    Selectivity of ionic transport
    Negligible electronic transport
    Stability with respect to thermal decomposition
    Stability with respect to electrochemical decomposition
    Stability against reaction with species in the environment
    Ease of fabrication at satisfactory cost
    Suitable mechanical, other related properties (e.g., thermal
        expansion coefficient that matches other constituents)
    Ready availability and low-cost ingredients.
    Adequate safety (e.g., must not be poisonous).

With regard to their use in advanced batteries, solid electrolytes can be employed in any of several different functions:

First, they can be used as simple electrolytes, separating either liquid or solid electrode systems. An example of this is the well known sodium-polysulfide cell involving a solid beta alumina electrolyte.

Secondly, they can be used as ion-transparent physical separators in cells in which the primary electrolyte is a liquid.

Thirdly, solid electrolytes can be used as ion-transporting constituents within solid electrode structures. This was demonstrated several years ago in the case of ambient temperature cells in which modest quantities of an electrolyte were mixed with the electrode reactant materials in order to physically distribute the interfacial electrochemical reaction over a greater area to provide improved kinetics.

Fourth, solid electrolytes may be used as second electrolytes in systems in which the primary electrolyte is not stable in contact with one of the electrode materials. This approach, which has recently led to some new applications, will be discussed further below.

SPECIAL STRUCTURAL CHARACTERISTICS OF SOLID ELECTROLYTES

It is now quite apparent that materials which exhibit fast ionic conduction do so because of special characteristics related to their crystal structures. We do not yet fully understand the physical processes involved in the motion of ions in materials which exhibit fast ionic conduction. Nevertheless, the main features of the structural dependence of this phenomenon can be seen from a relatively simple structure-dependent model for the transport of ions through specific crystal structures which was presented a few years ago [17-20]. This approach involves the determination of the potential energy profiles along which the mobile species move within specific crystallographic arrangements of the other ions, with the basic assumption that the other constituents in the lattice are fixed in position on the time scale of the motion of the mobile ions.

This type of calculation was first performed on the body centered cubic alpha-AgI structure [17], and follows the general method initiated by Born and Mayer [21], in which the total energy is assumed to be the sum of two-body interaction energies between a mobile ion i and the surrounding lattice ions, j. This interaction is expressed as the sum of three types of terms, the electrostatic coulombic interaction, dipolar polarization (van der Waals) attractive interactions, and overlap repulsion between the closed shell ions. That is:

$$E_t = E_c + E_p + E_r$$

where

$$E_c = e^2 \sum_j q_i (q_j / r)$$

$$E_p = -(e^2/2) \sum_j \alpha_j (q_i / r^4)$$

and

$$E_r = b \sum_j \exp[(r_i + r_j - r_{ij})/\rho]$$

where q_i and q_j are the charges, α_j the polarization, r_i and r_j repulsion radii, b and ρ constants, and r_{ij} the distance between the mobile cation and the jth static lattice ion.

With the use of a computer, the total interaction energy between a single mobile ion, arbitrarily placed at any position within the crystal structure, and the other atoms in the lattice can be calculated. For simplification, it is assumed that all of the other atoms remain fixed in position, rather than relaxing to accommodate the position assumed for the mobile ion. It is also assumed that there is no interaction between nearby mobile species, but instead that all of the energy resides in the mobile ion-static lattice interaction.

By this method the variation of the total energy with the assumed position of the mobile ion within the tunnels which run through the body-centered cubic anion lattice was found for a series of cases, assuming different values of the cation repulsion radius. In addition to a series of mobile ion locations along the centerline of the interstitial tunnel, a three-dimensional array of off-center positions was also investigated.

One of the important results from this simple theoretical approach was the conclusion that the potential energy profile is such that the minimum energy path through the tunnel that runs between the anions typically does not follow the centerline. Instead, its location is strongly influenced by the value of the cation radius. In the case of small mobile cations there is a symetrical pair of minimum energy preferred paths which deviate toward the nearby anions along the tunnel wall due to the relatively large influence of the attractive polarization energy term compared to the repulsive term. In the case of larger cations the opposite is true, and two equivalent minimum energy paths were found. They deviate from the centerline in the opposite sense, away from the nearby anions, as a result of the predominance of the repulsion term. The positional variation of both these short-range interactions is greater than that related to the long

range coulombic term in this structure.   On the other hand, it is
the coulombic term that primarily determines the total energy of
the lattice.

As an ion progresses along the tunnel in this crystal structure
it passes through an array of anions which may be visualized as a
series of alternating north-south and east-west dumbbell pairs,
which define a series of apertures through which the mobile ion
moves.   This type of aperture can be seen to be quite favorable,
for it permits a wide variety of paths to accommodate the relative
magnitudes of attractive and repulsive forces as the mobile ion
translates through the tunnel.

After determining the potential energy profile within the tunnel, and thereby the minimum energy path, the variation of the
(minimum) energy with position along the tunnel can be calculated
for ions of any radius. The result is shown in Figure 1 for a few
different ionic sizes.   It is seen that the competing effects of
the polarization and repulsive terms, which are out of phase,
result in a flatter potential profile for ions of intermediate
size.

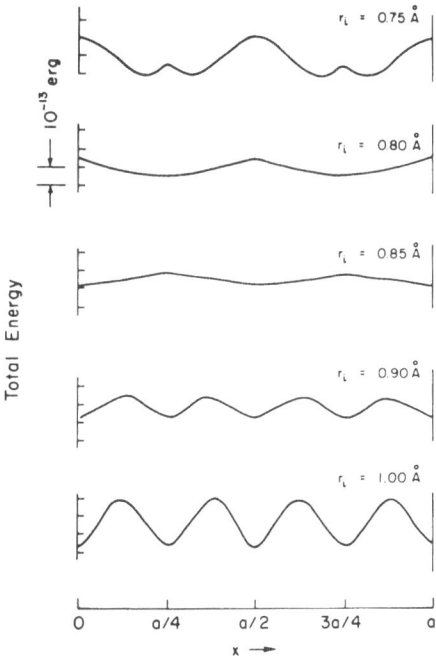

Figure 1.   Variation of energy with position along
minimum energy path for ions of different sizes. [Reprinted
from Hagenmuller and van Gool, eds. (20)]

If one interprets the peak-to-valley energy difference as the activation enthalpy for motion, the variation of that quantity with ionic size can be obtained, as shown in Figure 2. This result predicts that ions of intermediate size should be more mobile in this structure than either smaller or larger ones. Similar results have been obtained from calculations of this type in several other crystal structures [20].

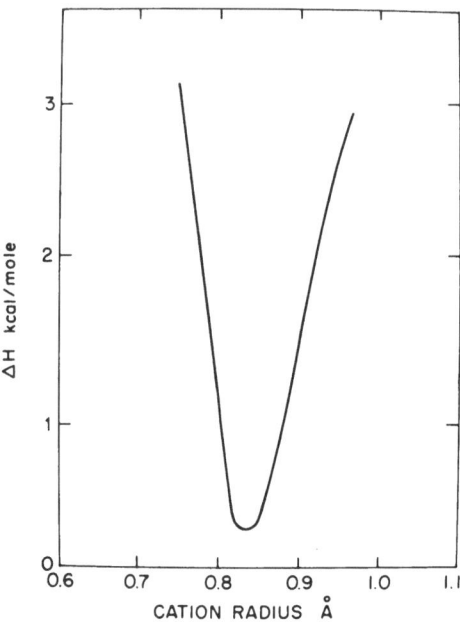

Figure 2.   Dependence of activation enthalpy for motion upon ionic size. [Hagenmuller and van Gool (20)]

These calculations also indicate that the mobile ions often occupy off-center positions at crystallographic lattice sites, and reside in shallow and assymetric potential energy wells. This should lead to large and anisotropic vibrational amplitudes and low jump attempt frequencies. Both of these have often been found experimentally. In addition, the minimum energy paths between lattice sites do not necessarily follow the direct site-to-site centerline. This has been verified recently in the case of fluoride ion motion in PbF_2 by the use of detailed x-ray diffraction techniques [22]. Since the shape of the potential profile along the minimum energy path is greatly influenced, and in some cases dominated, by short range forces related to the interaction between the mobile ion and the static lattice, which are extremely sensitive to both crystallographic features and the size of the mobile ion, any mobile ion-mobile ion interactions must be of only secondary importance.

THE ELECTROLYTE STABILITY WINDOW

While the potential utility of solid electrolytes surely depends to first order upon the magnitude and selectivity of the ionic conductivity, the practical utilization of such materials also requires that they meet the relevant stability requirements. As mentioned earlier, this means that they must be stable with respect to thermal decomposition, electrochemical decomposition, and reaction with other species in their environments. In addition, they must be utilized under conditions in which ionic, rather than electronic, species conduct most of the charge. Stated another way, such materials must be utilized within their appropriate thermodynamic and electrolytic stability windows.

Thus the practical utilization of materials as solid electrolytes is often limited to restricted ranges of temperature, pressure and chemical potentials. This matter has received relatively little attention to date, despite its obvious practical importance, as pointed out recently [23].

The electrolyte must be stable in the presence of the reducing conditions imposed by the negative electrode. A number of the systems being considered today involve the use of very aggressive materials, such as the alkali metals or their alloys, as negative electrode constituents. If the electrolyte is a binary phase containing the alkali metal, the question of stability is very simple if the phase diagram is known. The terminal (most alkali-rich) phase in the binary system will be thermodynamically stable in contact with the alkali metal, or its saturated solution. Any other will not.

Both LiI and Li_3N, for example, are good conductors of lithium ions, and are stable in contact with elemental lithium. However, some terminal binary phases dissolve some of the alkali metal at elevated temperatures. The excess alkali metal exists in the electrolyte as alkali metal ions and electrons. As a result, there can be an appreciable amount of electronic conductivity in such materials at high alkali metal activities. The alkali halides such as lithium and sodium chloride are examples of this, as they all dissolve their respective alkali metal components somewhat when molten, the lithium salts the least, those containing the heavier alkali metals more. This electronic conduction in the electrolyte leads to self-discharge in batteries containing such materials.

Electrolytes that are binary phases can also have stability problems at the positive electrode, where conditions are oxidizing. One can calculate the thermodynamic stability limit in such cases directly, if the Gibbs free energy of formation is known. For example, LiI will be stable up to 2.79 volts versus pure lithium at 25 C, whereas Li_3N is stable to only 0.44 volts versus

lithium at that temperature. Electrodes that impose potentials
more oxidizing than these values will react with the electrolyte,
causing it to be oxidized. Other reactions can also occur, of
course, due to chemical interaction with species in the electrode.
Displacement reactions are commonly found in this type of situation.

The stability of electrolytes that are composed of three or
more components is more complicated. In order to not get too
involved, we shall consider here only ternary (three component)
phases. Several such cases have been investigated in the last few
years [24-27], and the relevant principles are now clear.

When sufficient information is available and the stable phases
are known, one can use data on the Gibbs free energy of formation
of the various phases to construct the relevant ternary phase diagram and identify the stable tie lines. This then provides information about which phases are thermodynamically stable in the presence of others. For example, only electrolyte phases which share
two-phase tie lines or three-phase triangles with alkali metals
will be stable in contact with them. Likewise, one can readily
identify the phases that result from either reduction or oxidation
reactions.

In order to show the principles and methodology involved, a
hypothetical ternary system containing components Li, a second
metal M, and oxygen will be briefly discussed. For simplification,
let us assume that only two binary oxides of M, nominally MO and
MO_3, are stable at a temperature of 700 K (427 C), that there is a
single compound, LiM, formed between Li and M, and that there is
only one ternary phase, $LiMO_2$. We shall address our primary attention to the question of the stability window of the phase $LiMO_2$,
which might be a solid electrolyte of interest.

To help with the discussion, we shall assume that the values of
the Gibbs free energy of formation ΔG_f of all of the phases are
known. The values assumed are given in Table I. If they are not
all known, there are ways in which some of them may be deduced
from experimental data, as will become evident later.

The first task is to identify the stable two-phase tie lines in
the ternary phase diagram, so that areas of three-phase equilibrium can be identified. For this purpose we shall make the further
assumption that none of the phases has an appreciable range of
composition – that is, that they can all be represented by points
on a ternary isothermal Gibbs triangle; see Figure 3.

Some of the tie lines in Figure 3 are obvious, others have to
be determined from among two or more choices. An important case in
point is the question of whether the tie line between Li_2O and LiM

Table I.   DATA RELATED TO HYPOTHETICAL PHASE DIAGRAM

Data Assumed For Phases in Figure 3.

Phase	ΔG_f at 700 K (kcal/mole)
Li_2O	$- 121.5$
MO	$- 50.0$
MO_3	$- 80.0$
LiM	$- 20.0$
$LiMO_2$	$- 200.0$

Resultant Theoretical Values

3 Phase Triangle	Lithium Activity	emf (V) E vs. Li	Oxygen Press. (atm)
$Li_2O-LiM-Li$	1.0	0.0	5.3×10^{-76}
$LiMO_2-Li_2O-LiM$	1.2×10^{-5}	0.68	6.3×10^{-57}
$LiMO_2-LiM-M$	5.7×10^{-7}	0.87	6.3×10^{-57}
$LiMO_2-Li_2O-O$	1.2×10^{-19}	2.63	1.0
$LiMO_2-MO-M$	5.6×10^{-32}	4.34	6.0×10^{-32}
$LiMO_2-MO_3-O$	3.6×10^{-38}	5.20	1.0
$LiMO_2-MO-MO_3$	7.6×10^{-43}	5.85	4.3×10^{-10}

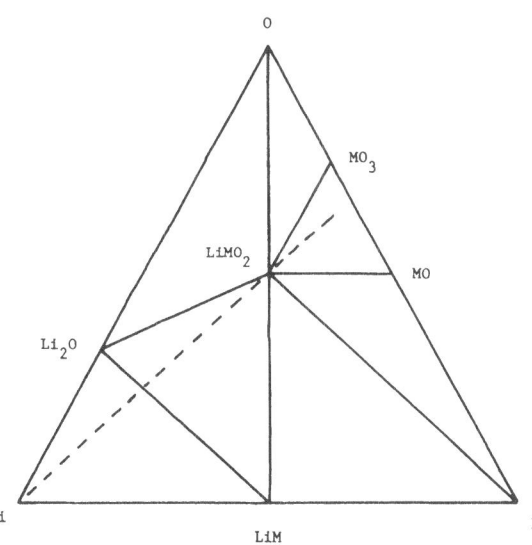

Figure 3.   Hypothetical isothermal ternary phase diagram.   Components are Li, M and oxygen.   Dotted line
indicates the path of the overall composition if Li is
either added to or deleted from the phase $LiMO_2$.

is stable, or whether one exists between Li and the phase $LiMO_2$
instead (tie lines cannot cross).If the $Li-LiMO_2$ tie line is present, the phase $LiMO_2$ will be stable in the presence of Li. On the
other hand, if the Li_2O-LiM line is stable, it will not be. The
answer to this question can be obtained by consideration of the
reaction

$$4 Li + LiMO_2 = 2 Li_2O + LiM$$

If this reaction would result in a negative Gibbs free energy
change, it will tend to go to the right, and the stable tie line
runs between Li_2O and LiM, rather than between Li and $LiMO_2$. It
can be seen from the data in Table I that this is the case for
this hypothetical system.

A similar question arises concerning the existence of a tie
line between Li_2O and M, or one between $LiMO_2$ and LiM. The relevant reaction to consider in that case is

$$4 M + 2 Li_2O = LiMO_2 + 3 LiM$$

Again, the sign of the value of the Gibbs free energy change
determines which way this reaction will tend to go, and therefore,
which pair of phases is stable. The phase diagram in Figure 3 is
drawn to correspond to the results of such a calculation, using
the data of Table I. There are also other possible tie lines in
the diagram, and they should be checked in this manner to be sure
that the proper phase diagram has been deduced.

These two-phase tie lines define a number of triangles. Inside
of each triangle three phases (those on the corners) are stable,
and the Gibbs phase rule teaches us that when three phases are in
equilibrium in a ternary system, all intensive variables are
fixed. What this means in this context is that the activities,
partial pressures or chemical potentials of all of the components
(ie. Li, M and oxygen) are independent of composition within any
given triangle. Their values can be calculated from the thermodynamic data. Likewise, the emf versus each of the components will
be constant for all compositions within each triangle, and can be
calculated (assuming an appropriate electrochemical cell).

If we are interested in the stability of the phase $LiMO_2$ with
respect to lithium, we should consider the thermodynamic parameters in the triangles $LiMO_2-Li_2O-LiM$ and $LiMO_2-MO_3-MO$, for the
overall composition will move into the former if Li is added to
the $LiMO_2$, and into the latter if Li is deleted from it.

The thermodynamic parameters relative to lithium within the
triangle $LiMO_2-Li_2O-LiM$ are obtained by considering the reaction

of Li with the $LiMO_2$ to form the other two phases. This same reaction was mentioned above in connection with the matter of tie line stability. The equilibrium constant K can be obtained from the Gibbs free energy change of this reaction from the equation

$$\Delta G_r = - RT \ln K$$

and since we assume that all of the phases are present at unit activity when the overall composition lies within this triangle, the value of the equilibrium constant is simply

$$K = a_{li}^{-4}$$

Thus the Li activity within this triangle is given by the relation

$$a_{li} = \exp(\Delta G_r / 4RT)$$

This value, given in Table I, is 1.2×10^{-5}. Thus we see that the phase $LiMO_2$ will not be stable in contact with pure Li. Indeed, not at Li activities above that value, for that is the limit above which the phases Li_2O and LiM will begin to form.

The limiting maximum value of Li activity at which $LiMO_2$ is stable can be converted into an emf (E) with respect to Li by use of the relation

$$E = - \Delta G_r / n F$$

where n is unity, and F is the Faraday constant. The resultant value of E is +0.68 V.

Now let us consider the question of the minimum Li activity at which $LiMO_2$ is stable. From the phase diagram we can see that deleting Li causes the overall composition to move into the triangle $LiMO_2$-MO_3-MO. The pertinent reaction to consider is

$$2 \ LiMO_2 = 2 \ Li + MO_3 + MO$$

In this case the equilibrium constant K is given by

$$K = a_{li}^2$$

so that the Li activity in this triangle is

$$a_{li} = \exp(- \Delta G_r / 2RT)$$

The emf versus Li can be found from the value of ΔG_r as before. In this case E is +5.85 V.

Thus we see that the phase $LiMO_2$ is stable over the range of Li activity equivalent to a potential between +0.68 and +5.85 V versus Li. However, it is not stable in the presence of elemental Li, reacting to form Li_2O and LiM.

In addition to the question of the Li stability window, the range of oxygen pressure $p(O_2)$ over which the potential solid electrolyte phase is stable can also be important, even though it may be used in a Li-transporting cell. This information can be obtained by finding the oxygen pressure in each of the same two relevant triangles.

The equilibrium oxygen pressure in the $LiMO_2$-Li_2O-LiM triangle is found from the Gibbs free energy change accompanying the reaction

$$LiM + O_2 = LiMO_2$$

The relation between the oxygen pressure and K for this reaction is

$$K = p(O_2)^{-1}$$

This gives an equilibrium oxygen pressure of 6.3×10^{-57} atm. in this three-phase triangle.

Similarly, the oxygen pressure in the $LiMO_2$-MO-MO_3 triangle is obtained from the relation

$$MO + O_2 = MO_3$$

Here, also,

$$K = p(O_2)^{-1}$$

Putting the proper numbers from Table I into these relations, one finds that the equilibrium oxygen pressure in this triangle is 4.3×10^{-10} atm.

These two oxygen pressure values define the oxygen pressure limits within which the $LiMO_2$ is stable. This requirement and the range of Li activity calculated earlier constitute the stability window of this phase. It must be used in an environment and with electrodes that lie within those limits along both the Li and oxygen coordinates.

Data for the Li activity and oxygen pressure in each of the triangles of this example are included in Table I. A discussion of another hypothetical example [23] and data on several actual systems [25-27] can be found elsewhere. It has also been pointed out

[23-25] that one can use these considerations to devise electrochemical cells with appropriate three-phase electrodes to experimentally measure thermodynamic quantities that may not already be
available. One can also readily obtain thermodynamic data from the
analysis of coulometric titration curves, and methods for doing
this have been discussed in the literature [28-31].

Data on the stability windows of a number of electrolytes are
included in Table II.

It should be emphasized that, despite the substantial amount of
effort being given to problems such as the recharging of elemental
Li electrodes in liquid electrolyte systems, it is not always necessary to have an electrolyte that is stable in the presence of
the alkali metal in order to have a useful alkali metal-based system. Indeed, the example discussed here points out that one can
obtain electrolyte phases with wide stability windows that may not
encompass the elemental alkali metal itself. The cell voltage
depends upon the difference in electrode potentials, not the absolute value of either one. Thus the critical quantity controlling
the maximum cell voltage is the width of the stability window, not
the placement of its limits on an absolute scale.

Therefore, it is worth considering the use of negative electrodes in alkali metal systems that are displaced from the alkali
metal potential somewhat, so that they lie within the stability
window of the electrolyte, rather than assuming that one must use
the alkali metal itself as the electrode. A number of binary and
ternary alkali metal systems are now known that have appreciable
capacities at fixed potentials, and thus might be used for this
purpose. Some of these are included in Table III. While their use
may appear to entail some sacrifice in cell voltage, it may be
advantageous in terms of stability in many cases.

SOLID ELECTROLYTE LAYERS FORMED IN SITU

Solid electrolyte layers can form as the result of reaction
between two materials: Oxidation and tarnishing are common examples of this phenomenon. The properties of this reaction product
can have a significant influence upon further progress of the
reaction. For example, if the layer is continuous, the reaction
between the phases upon its two sides can only proceed by the
transport of matter through it. It can also be shown that the
transport flux between the two reactant phases must be electrically balanced. This is a very important requirement, for in the
absence of an external current path, it means that two charged
species must be transported through the layer for it to grow
thicker.

Table II. SOME ELECTROLYTE STABILITY WINDOWS

Material	Li-Rich E vs. Li	Li-Poor E vs. Li	Temp.
LiI	0	2.79	25 C
Li_3N	0	0.44	25 C
LiCl	0	3.98	25 C
$LiAlCl_4$	1.68	4.36	25 C
$LiAlCl_4$	1.70	4.20	135 C
$Li_9N_2Cl_3$	0	2.5	100 C
$Li_{11}N_3Cl_2$	0	1.8	322 C
Li_6NBr_3	0	1.3	176 C
$Li_{13}N_4Br(LT)$	0	1.3	146 C
$Li_{13}N_4Br(HT)$	0	0.65	300 C
$Li_5NI_2(LT)$	0	1.9	98 C
$Li_5NI_2(HT)$	0	1.6	287 C
$Li_{667}N_{189}I$	0	0.9	313 C
$Li_{911}N_{27}I$	0	0.9	316 C
Li_2O	0	2.84	150 C
$LiNO_3$	2.5	4.2	150 C

Table III. LITHIUM RESERVOIR SYSTEMS AT VARIOUS POTENTIALS

System	Range of α	Voltage vs. Li	Temp.
$Li_\alpha Al$	0.08-0.9	0.300	400 C
$Li_\alpha Si$	0 -1.73	0.332	415 C
$Li_\alpha Si$	1.73-2.37	0.288	415 C
$Li_\alpha Si$	2.37-3.32	0.158	415 C
$Li_\alpha Si$	3.32-4.25	0.044	415 C
$Li_\alpha Ga$	0.15-0.82	0.565	400 C
$Li_\alpha Ga$	1.28-1.48	0.122	400 C
$Li_\alpha Ga$	1.53-1.93	0.091	400 C
$Li_\alpha Sn$	0.57-1.0	0.570	400 C
$Li_\alpha Sn$	1.0 -2.33	0.455	400 C
$Li_\alpha Sn$	2.33-2.5	0.430	400 C
$Li_\alpha Sn$	2.5 -2.6	0.387	400 C
$Li_\alpha Sn$	2.6 -3.5	0.283	400 C
$Li_\alpha Sn$	3.5 -4.4	0.170	400 C
$Li_\alpha Bi$	1.0 -2.82	0.750	400 C
$Li_\alpha Sb$	0 -2.0	0.910	400 C
$Li_\alpha Sb$	2.0 -3.0	0.875	400 C
$Li_\alpha Ti_3O_7$	1.0 -1.9	1.480	400 C
$Li_\alpha MnO$	0 -2.0	0.915	400 C
$Li_\alpha FeAsS$	0 -1.6	1.75	400 C
$Li_\alpha FeAsS$	1.7 -2.9	0.95	400 C
$Li_\alpha FeAsS$	3.1 -4.9	0.75	400 C

Thus the rate of growth of the reaction product layer is often controlled by the transport of a second or minority species, not the most prevalent or mobile species. As an example, zirconium oxide is very protective in oxidizing atmospheres, even though oxide ions are very mobile within it. The reason is, of course, that both other ionic species and electronic species have very low partial conductivities in this material, except under very reducing conditions.

In addition to the growth of the reaction product layer itself, the transport of various species through it can also be important. If it acts as a solid electrolyte, ions, but not electrons or holes, can move across the layer between the two adjacent phases. Under this circumstance, this layer can prevent a chemical reaction (between neutral species) or convert it to an electrochemical one (involving the transport of charged species).

There are a number of situations in which it would be desirable to have a reaction product layer form which acts as a protective barrier against further reaction, yet is ionically-conducting. One of these involves a battery system utilizing an electrode and an electrolyte which are not stable in contact with each other. If the reaction product layer conducts appropriate ions, it acts as a second, or auxiliary, electrolyte in series with the primary electrolyte in the cell.

The impedance to the transport of the ionic species through the solid electrolyte barrier layer acts in series with the other impedances in the system, and is thus preferably as small as possible. Its magnitude depends upon two factors, the ionic conductivity of the layer material itself, and the layer thickness. The conductivity is determined by the structure and composition of the layer, whereas the thickness is determined by the interaction of two phenomena, layer growth, and layer dissolution. The layer gets thicker with time as a result of the transport of neutral combinations of species across it, as mentioned earlier. Except when it is very thin, this generally has parabolic kinetics, ie. the thickness varies with the square root of time. At the same time, the thickness is being reduced if the layer material sublimes, or dissolves in one of the adjacent phases. The dissolution rate depends upon a number of factors, one of the more important of which is the fractional saturation of the adjacent liquid. The interplay of the layer growth and layer dissolution processes determines the steady state value of the layer thickness, and thus of its impedance.

Another important aspect of this situation is the fact that the dissolution of the protective layer also amounts to effective corrosion of the electroactive material in the electrode. This corrosion process reduces the efficiency of the cell, for it involves a reduction in the chemical energy stored in the system by a chemical, rather than an electrochemical, reaction.

There are a number of present battery systems in which a reaction product layer forms between a solid electrode and a liquid electrolyte with which it is not thermodynamically stable. In most cases, however, the layer has a low value of ionic conductivity and low solubility and gradually grows to an appreciable thickness. As a result, it contributes a large and deleterious impedance. The origin of the time delay phenomenon in some of the ambient temperature lithium-based cells is the growth of such a reaction product layer with a low ionic conductivity to an appreciable thickness.

Recent work has lead to the discovery of a new electrode/electrolyte system, involving the combination of a lithium or lithium-alloy negative electrode and an intermediate temperature molten oxide salt [32,33]. Lithium oxide is very stable, and in the presence of an oxidizing salt readily forms upon a lithium electrode. Earlier work [34,35] showed that lithium oxide is a reasonably good, but not outstanding, solid electrolyte, conducting charge by the transport of lithium ions. Thermodynamic data also shows that lithium oxide is stable over a wide range of lithium and oxygen activities. Thus a layer of lithium oxide can act as a protective solid electrolyte reaction product in certain environments. If it can be kept thin enough, it will not contribute a substantial additional impedance.

The lithium-lithium nitrate system is one of those that has these characteristics. Thermodynamic studies [32] have shown that the reduction limit of lithium nitrate occurs when it is equilibrium with lithium oxide and that lithium oxide is one of the products formed when lithium nitrate is reduced. As a result, those materials are stable in the presence of each other when the nitrate is saturated with the oxide. Preliminary measurements in the lithium nitrate-potassium nitrate eutectic, have shown that the solubility of lithium oxide in it is quite small [36]. Furthermore, the rate of growth of the lithium oxide layer is quite slow, so that it forms only a very thin layer under dynamic equilibrium conditions.

Experiments have shown that this thin solid electrolyte layer readily reforms if disturbed, and that it allows the reversible operation of lithium electrodes in these nitrate salts below the melting point of the electrode material. The ability to replate lithium during recharge is especially attractive, for it has been a stumbling block to the development of a number of other battery systems. It appears that this approach may lead to the solution of this type of problem in other systems as well.

In addition to establishing and demonstrating these principles, work on the lithium-nitrate system may well lead to new and potentially important battery systems. Therefore, some of the attributes of this system will be pointed out.

The eutectic composition in the $LiNO_3$-KNO_3 system (41 m/o $LiNO_3$) melts at about 135 C. In the presence of the Li_2O layer it can thus be used in lithium-based cells at temperatures well below those necessary for cells employing the common chloride salts (400-450 C). Its stability window lies between + 2.5 and + 4.2 V with respect to pure lithium. At higher potentials these salts oxidize to form NO_2 and O_2. At lower potentials they reduce, forming Li_2O and nitrite.

Primary cells can thus be constructed using an inert positive electrode material and a lithium anode, with an open circuit emf of 2.5 volts. Their maximum theoretical specific energy is about 1365 wh/kg at that voltage, and 1050 wh/kg if operated at a discharge potential of 2.0 volts and appreciable currents.

Secondary cells are also possible, providing that a reversible positive electrode operates within the electrolyte stability window, ie. between + 2.5 and 4.2 volts versus lithium. This provides the possibility of high voltage cells. However, most of the reversible positive electrode materials that have been employed with lithium cells to date operate at lower potentials and are thus not suitable for use with the nitrate salt. As an example, although it is probably not the optimum choice, V_2O_5 has been shown to reversibly accommodate lithium up to a concentration of at least x = 1 in $Li_xV_2O_5$, with a potential between 3.5 and 2.9 volts at 150 C.

It has also been shown that some binary lithium alloys can be used as negative electrode reactants without reducing the nitrate melt, indicating the formation of the lithium oxide solid electrolyte layer in those cases as well. This means that this concept can also be used at lithium activities less than unity, and at temperatures above the melting point of lithium.

ACKNOWLEDGEMENT

The author gladly acknowledges the numerous important theoretical and experimental contributions in this area by his students and colleagues at Stanford. The major portion of the financial support of work in this area at Stanford University currently comes from the United States Department of Energy under Subcontract LBL 4503110.

REFERENCES

[1]   W. Nernst, German Patent 104872 (1897)

[2]   W. Nernst, Z. Elektrochem. 6, 41 (1900)

[3]   W. Nernst and W. Wild, Z. Elektrochem. 7, 373 (1901)

[4]   H. Reynolds, Ph.D. Dissertation, Univ. Goettingen (1902)

[5]   C. Wagner, Naturwissenschaften 31, 265 (1943)

[6]   C. Wagner, J. Chem. Phys. 21, 1819 (1953)

[7]   K. Kiukkola and C. Wagner, J. Electrochem. Soc. 104, 308
        (1957)

[8]   K. Kiukkola and C. Wagner, J. Electrochem. Soc. 104, 379
        (1957)

[9]   C. Wagner, Proc. Int. Comm. Electrochem. Thermo. Kinetics
        (CITCE) 7, 361 (1957)

[10]   J. N. Bradley and P. D. Greene, Trans. Faraday Soc. 62,
        2069 (1966)

[11]   J. N. Bradley and P. D. Greene, Trans. Faraday Soc. 63,
        424 (1967)

[12]   B. B. Owens and G. R. Argue, Science 157, 308 (1967)

[13]   B. B. Owens and G. R. Argue, J. Electrochem. Soc. 117,
        898 (1970)

[14]   Y. F. Y. Yao and J. T. Kummer, J. Inorg. Nucl. Chem. 29,
        2453 (1967)

[15]   R. H. Radzilowski, Y. F. Yao and J. T. Kummer, J. Appl.
        Phys. 40, 4716 (1969)

[16]   N. Weber and J. T. Kummer, Proc. Ann. Power Sources Conf.
        21, 37 (1967)

[17]   W. F. Flygare and R. A. Huggins, J. Phys. Chem. Solids 34,
        1199 (1973)

[18]   O. B. Ajayi, Ph.D. Dissertation, Stanford Univ. (1975)

[19]   O. B. Ajayi, L. E. Nagel, I. D. Raistrick and R. A.
        Huggins, J. Phys. Chem. Solids 37, 167 (1976)

[20]   R. A. Huggins, "Crystal Structures and Fast Ionic Conduc-
        tion", in Solid Electrolytes, ed. by P. Hagenmuller and W.
        van Gool, Academic Press (1978), p. 27

[21]   M. Born and J. E. Mayer, Z. Phys. 75, 1 (1932)

[22]   K. Koto, H. Schulz and R. A. Huggins, to be published.
        in Solid State Ionics

[23]   R. A. Huggins, in Fast Ion Transport in Solids, ed. by
        P. Vashista, J. N. Mundy and G. K. Shenoy, North-Holland
        (1979), p.53

[24]   W. Weppner and R. A. Huggins, in Fast Ion Transport in
        Solids, ed. by P. Vashishta, J. N. Mundy and G. K. Shenoy,
        North-Holland (1979), p. 475

[25]   W. Weppner and R. A. Huggins, Solid State Ionics 1, 3
        (1980)

[26]   N.A. Godshall, I.D. Raistrick and R. A. Huggins, Mat. Res.
        Bull. 15, 561 (1980)

[27]   N. A. Godshall, Ph.D. Dissertation, Stanford Univ. (1980)

[28]   W. Weppner and R. A. Huggins, J. Electrochem. Soc. 125,
        7 (1978)

[29]   L-c. Chen and W.Weppner, Naturwiss. 65, 595 (1978)

[30]   W. Piekarczyk, W. Weppner and A. Rabenau, Z. Naturforsch.
        34a, 430 (1979)

[31]   W. Weppner, L-c. Chen and W.Piekarczyk, Z. Naturforsch.
        35a, 381 (1980)

[32]   I. D. Raistrick, J. Poris and R. A. Huggins, to be publ.
        in Proc. Lithium Battery Symp., Hollywood, FL, Oct. 1980,
        Electrochem. Soc., Pennington, NJ.

[33]   J. Poris, I. D. Raistrick and R. A. Huggins, to be publ.
        in Proc. Lithium Battery Symp., Hollywood, FL, Oct. 1980,
        Electrochem. Soc., Pennington, NJ.

[34]   R. M. Biefeld and R. T. Johnson, Jr., J. Electrochem. Soc.
        126, 1 (1979)

[35]   C. Ho, I. D. Raistrick and R. A. Huggins, unpublished work

[36]   J. Poris, I. D. Raistrick and R. A. Huggins, to be publ.

PLATING - NEW PROSPECTS FOR AN OLD ART

Uziel Landau
Chemical Engineering Department
Case Institute of Technology
Case Western Reserve University
Cleveland, Ohio 44106

In 1850, merely fifty years after Volta described his electrochemical pile thus providing the first power source, electroplating was already an established practice in Europe. Today,
electroplating is among the most widely spread chemical based
technologies, applied in a wide range of scales in thousands of
facilities. Surprisingly, many aspects of the field have not
significantly changed over the years, and many current plating
processes and facilities still closely resemble their older ori-
gins. The main reasons for this relatively slow technological
evolution have been the secrecy traditionally associated with
plating, which led to proprietary formulations and information
exchange mainly through patent disclosures, and the fact that relatively unsophisticated and easy to adjust plating processes
provided surface finishes which were quite adequate for early,
non demanding, applications. Consequently, much of the progress
in the field has been through "state of the art" concepts, rather
than based on systematic scientific investigations and application of engineering principles.

Recent requirements, for new processes and tighter specifications from the electronic and other high technology industries,
for conservation of critical metals and energy, and due to severe restrictions imposed by environmental considerations, are
all rapidly changing the field, bringing about a new scientific
and technological rejuvenation.

Scope and Applications

A wide variety of cathodic deposition processes such as
electroforming, electrowinning and electrorefining, and processes

215

at rechargeable battery electrodes, can all be classified under the heading of electroplating. Due to its limited scope, the present discussion mainly addresses those processes in which a relatively thin metallic layer is cathodically plated, in order to impart to the substrate different, usually superior, surface properties. A partial list of application includes:

Decorative and reflective coatings (mostly Au, Ag, Cr, Ni, Cu, Rh).

Corrosion protection (Zn, Cd - Sacrificial; Cr, Ni, Au, Sn - Protective)

Mechanical properties (Cr, Ni - wear resistance; In - solid lubrication)

Electrical conductivity and contact properties (Au, Ag, Cu, Ni, Rh, Pd)

Magnetic films (Fe, Ni, Co)

Soldering, bonding, and metallization (Pb, Sn, Au, Ni, Cu)

Etch masking (Pb, Au, Cu)

Catalytic and special surface properties (Au, Pt, Ag, Ru, Ni, Pd)

Often, in order to obtain unique properties, two or more metals are codeposited as an alloy. Plating on non-conductors is possible using electroless deposition processes in which the metal is plated through an autocatalytic redox reaction.

Plating of electronic components imposes perhaps the most demanding requirements on plating processes. Consequently, this area has been on the forefront of technological progress, often leading new developments in plating technology. The present discussion will therefore somewhat focus on this particular area.

Plating is an extremely important technology. Covering inexpensive and widely available base materials with thin (typically 0.1 - 10 μm) plated layers of more precious metals with superior properties, extends their use to applications which otherwise would have been prohibitively expensive.

Pending shortages and escalating prices of many critical metals (Table 1) are rendering plating not only economically important, but essential for maintaining in the future our present day technology.

Table 1

Reserves of Critical Metals[+]

Metal	Identified Reserves	Years to Depletion At Present (1979) Production Rate[*]	Projected Consumption Growth Rate (%/year)	Years to Depletion At Projected Growth Rate[‡]	Price (1979-81 est.avg.)
Aluminum	1.2×10^9 tons	75	6.4	27	$ .60/lb
Chromium	4.9×10^9 tons	466	2.6	99	$ .40-90/lb
Cobalt	9.9×10^9 lbs	157	1.5	81	$ 20/lb
Copper	3.4×10^8 tons	45	4.6	24	$ .70-1.00/lb
Gold	3.5×10^8 tr.oz.	9	4.1	8	$ 400-800/tr.oz.
Iron	7.1×10^{11} tons	1414	1.8	182	$ 75/ton
Lead	1.4×10^8 tons	40	2.0	29	$ .40/lb
Manganese	1.4×10^{10} tons	518	2.9	96	$ .50/lb
Molybdenum	6.3×10^{10} lbs	277	4.5	58	$ 9/lb
Nickel	9.0×10^7 tons	116	3.4	47	$ 2-3.50/lb
Platinum[**]	4.2×10^8 tr.oz.	63	3.8	32	$ 400-800/tr.oz.
Silver	5.4×10^9 tr.oz.	16	2.7	13	$ 8-30/tr.oz.
Tin	2.3×10^7 tons	90	1.1	62	$ 7.00/lb
Tungsten	2.9×10^9 lbs	29	2.1	23	$ 120/ton
Zinc	1.7×10^9 tons	283	2.5	84	$ .50/lb

+ After Meadows (1), unless otherwise indicated.

* Based on data from Mineral yearbook, U.S. Bureau of Mines, 1978-79 (2).

‡ Based on the formula: years to depletion $= \dfrac{\ln[(RXS)+1]}{R}$

where R = projected growth rate (column 4)

S = years to depletion at constant growth (column 3)

** The Platinum group includes platinum, palladium, irridium, osmium, rhodium and ruthenium.

Many critical metals are imported to the U.S. and to Western
Europe from third world countries. Imports to the U.S. according
to a recent report (3) include beryllium (100%), titanium (100%)
manganese (98%), cobalt (81%), platinum group metals (92%), tantalum (90%), chromium (89%), aluminum (86%), tin (86%), nickel
(70%), zinc (58%), tungsten (39%), iron (29%), copper (17%), lead
(14%). Political unrest in exporting countries has led in the
past to short supplies and rapid price fluctuations.

Although only a score of metals (Zn, Cu, Ni, Cr, Sn, Pb, Fe,
Cd, Au, Ag, Pt, Rh, Ru, Pd, Ir, In, Co) play a major role in
plating technology, all metals can, in principle, be cathodically
discharged. Many are likely to gain technological significance
in the future when new requirements and processes evolve. Metals
with reduction potentials more cathodic (negative) than manganese
cannot be plated out of aqueous solutions. Processes involving
solvents such as propylene carbonate, various ethers, amines, and
other solvents providing sufficient solubility and dielectric
constants have been developed but not yet commercially implemented
on a large scale for the deposition of aluminum (4,5,6)   and
other alkali and alkali earth metals (7,8). Deposition from molten
salts is also possible.

Sources of Information

Despite the traditional secrecy associated with plating,
significant published information is available.

Recent monographs include two books by Lowenheim: "Electroplating" is a practitioner oriented text (9), whereas "Modern
Electroplating" (10) provides a more comprehensive review of many
plating processes. "Electroplating Engineering Handbook" (11)
is hardware and process oriented. "Metal Finishing Guidebook and
Directory" (12) published annually, is shop oriented guide containing useful practical information and formulations.

Fundamental monographs include books by Fischer (13), Vetter
(14), and Bockris and Reddy (15). Engineering texts containing
material applicable to plating are books by Newman (16), and
Pickett (17).

Important series publication relevant to plating are
"Advances in Electrochemistry and Electrochemical Engineering"
(18), "Modern Aspects of Electrochemistry" (19), "Techniques of
Electrochemistry" (20), and Electroanalytical Chemistry" (21).

The list of scientific journals dealing with subjects
associated with plating includes the "Journal of the Electrochemical Society" (The Electrochemical Society, U.S.), "Electrochimica

Acta" (the International Society of Electrochemistry), "Journal
of Applied Electrochemistry" (Britain), "Journal of Electroanaly-
tical Chemistry and Interfacial Electrochemistry" (U.S.) and
"Elektrokhimiya". Trade oriented journals are "Plating and
Surface Finishing" (American Electroplaters Society).
"Transactions of the Institute of Metal Finishing" (Britain), "Sur-
face Technology" (Swiss), "Metal Finishing" (U.S.) and "Product
Finishing" (U.S.).

The Electrodeposition Process

The deposition process involves on the microscopic scale a
sequence of quite intricate events which are conveniently broken
into the following steps:

In the electrolyte:

1. Transport of ions towards the cathode by diffusion,
   electric migration, and convection. Near the electrode the convective contribution is negligible.
   Also, most plating solutions contain excess "supporting" electrolyte which does not participate in
   the electrode reaction, but suppresses the electric
   field and the migration of the plated ion. Hence,
   the plated ion is transported to the electrode
   mainly through diffusion down its concentration
   gradient.

2. The release of co-ions or complexing agents and
   their back-diffusion into the bulk.

At the electrode:

3. Adsorption of the discharging ions in the double
   layer.

4. Charge transfer, possibly through quantum mechanical electron tunneling.

5. Surface diffusion of the ad-atom (or ad-ion) to the
   growth site, typically a step, kink or screw disc-
   location, where less energy is required for the
   growth process.

6. Incorporation into the lattice.

Electrocrystallization is similar to other crystallization
processes, such as from the melt or from super-saturated

solutions, except that here the driving force is the applied
electric potential rather than thermal or concentration gradients.

Each of the listed steps involves kinetic limitations and a
corresponding energy dissipation. These are expressed in terms
of overpotentials, i.e., the voltage in excess of the thermodynamic equilibrium potential, required to drive the electrode
reaction at the specified rate. For convenience, the overpotential is divided into three main components:

$$\eta_t = \eta_o + \eta_c + \eta_a$$

Here, η_o is the ohmic overpotential associated with ohmic losses
in the electrolyte; η_c is the concentration overpotential required to overcome mass transport limitations; and η_a is the activation overpotential, dissipated in overcoming kinetic limitations
associated with all the surface processes lumped together, including the electrode reaction step.

These overpotentials, and particularly their relative magnitude, have a major effect on the properties of the deposit and
its distribution. Their exact role will be discussed later.

Modern instrumentation, recently developed analytical techniques, and the availability of high power computers are now providing new effective tools for unravelling the detailed mechanism of various deposition processes and their associated overpotential components.

The Interdisciplinary Nature of Plating

Plating technology involves surface phenomena, solid state
processes, and processes occuring in the liquid phase. As such,
it gains input from a number of scientific disciplines.

Studying the electrode processes has traditionally been
the main object of classical electrochemistry. This discipline
is currently undergoing a renaissance, and rather than concentrating on the traditional subject of electrochemical thermodynamics
- modern electrochemistry is focusing on kinetics, catalysis,
electrode reaction mechanisms, adsorption states of ions and
molecules, and crystallization.

Study of the transport phenomena which are associated with
the deposition process falls within the domain of electrochemical
engineering - a relatively new, but rapidly expanding engineering
discipline within chemical engineering. It is dedicated to the
study of the engineering principles underlying the operation of
electrochemical systems and in particular their design,

operating parameters, and control. It focuses on questions of
current distribution and voltage dissipation.

The interdisciplinary field of surface science, now in a
state of rapid progress, contributes to the understanding of the
electrodeposition process, mainly by providing modern analytical
tools capable of studying surfaces with higher resolution and
specificity than has so far been possible. Surface science has
in the past ten years particularly contributed to the understanding of electrocrystallization processes and adsorption on
electrodes. By providing an improved picture of surface structures, reaction mechanisms and electrocatalysis can be better
deduced.

Solid-State physics is yet another field which has lately
been contributing to the study of electrode processes. Modern
theories of charge transfer are based on quantum-mechanical
electron tunelling mechanism. The structure of oxide layers, and
the crystalization process can be modeled and explained in terms
of solid-state concepts. New types of electrodes which have recently become the focus of scientific interest and which hold much
technological promise, such as photovoltaic electrodes and ruthenium oxide catalytic electrodes, can be best explained by quantum-mechanical solid-state concepts.

Properties of plated films, in particular those for corrosion protection, are being studied in metallurgy and material
science departments. This discipline also focuses on interaction
and compatability between plated films and substrates.

Lastly, the rapidly expanding field of electronics is not
only a major consumer of plated products but it also provides
essential means, in terms of modern instrumentation, for developing improved and better controlled processes.

A serious problem facing plating technology is the traditional communication barrier existing between the scientific
community and the plating practitioners, making it difficult to
translate scientific progress into plating practice. Rather
than diminishing, this gap has widened over the years, partly due
to the reluctance of scientists to deal with practical problems
in systems which often are not well defined. Another obstacle
is the proprietary nature of many of the plating processes.
There are however indications that due to recent trends in
academic research funding, more fundamental research is shifting
towards practical problems. Tighter specifications on plating
processes also lead practitioners to seek more fundamental
answers to their problems, and to be more receptive to scientific
ideas.

Properties of Electrodeposits

    Rather than indicating a particular plating process as was
common in the past, the tendency nowadays is towards specifying
end point requirements for the plated films (22). These specifications vary with applications, however the most common properties include (23):

    Adherence - The fundamental question why certain metals
adhere to a given substrate whereas others do not, is not yet
completely understood. Adhesion to oily, unwetted, and oxidized
surfaces is often quite poor. Insufficient adhesion requires
plating an intermediate compatible layer between the substrate
and the final coating, or special activation procedures. A
universally accepted quantitative test for adhesion is still un-
available. The crude adhesive peel test is most commonly applied.

    Appearance - A somewhat subjective property relating to
color, reflectivity, smoothness, structure, and uniformity. Appearance is often not critical for non decorative applications,
however, an irregular appearance, such as an off-color deposit
is often indicative of a plating process which is operating outside its recommended range.

    Typically, when plating is carried at too high current
densities, close to the diffusion limiting current i_L, ($i/i_L \gtrsim 0.6$),
a dark "burnt" deposit is obtained. This appearance is indicative of powdery, dendritic, or rough deposit. When current
densities are too low ($\lesssim 1$ mA/cm^2), a dull, coarse deposit is obtained, possibly due to irregular nucleation and crystal growth.

    Roughness and texture - Smooth deposits are most often
preferred since they are shiny and reflective, less susceptible
to abrasion and wear, and are more compact and therefore less
porous. Due to the random shape and distribution of roughness
elements, it is difficult to specify and measure surface roughness. Mechanical, stylus type, surface profile followers can be
used to obtain highly magnified traces of surface profiles.
Their sensitivity can be quite high and elements as small as 10 Å
can be detected. The probe radius (~ 2.5 μm) sets the lower limit
on the roughness wavelength resolution. Surface roughness is
often expressed in terms of root-mean-square (RMS) values, however, maximum height of roughness elements is often specified.
Other means of measuring roughness is through reflectivity,
light scattering, or surface area measurements. Optical microscopy with polarized Nomarsky optics provides means of ob-
serving extremely small elements. Scanning electron microscopes
(now more abundant because of price reductions) provide three
dimensional view of rough surfaces over wide magnification range.

Roughness elements can vary in shape: powder, crystalline, dendritic, spongy, or needle-like. The texture depends mostly on the metal, the plating parameters (overpotential, current density, agitation and composition), and the presence or absence of additives. Particularly bothersome are dispersed, needle-like, protrusions occasionally appearing on otherwise smooth surfaces, and which are often referred to as "plating asperities." The reason for their appearance is yet unclear, however they are often associated with substrate defects or contamination.

When large surface area is indicated, rough deposits are preferable to smooth ones. Typical example is the "burnt" gold which has been shown to be superior to shiny gold in thermal compression bonding applications. Other exceptions include the dull nickel plated from the Watts bath which often is a preferable undercoat since it provides better adhesion, and the matt chromium plate. Another application which is likely to become more important in the future, is the production of high area surfaces. This includes the plating of catalytic electrodes, where the deposit can be chemically active (DSA or underdeposited Pt), and of solar panels.

Porosity - Films plated for corrosion protection should be pore-free. This is particularly crucial if the coated layer is more noble than the substrate (e.g., when gold is plated on copper) where enhanced localized pore corrosion rapidly progresses. Typically, pores in plated films are straight, cone shaped rather than convoluted. Thicker deposits lead to fewer exposed pores. Due to the small diameter of the exposed tips, pores cannot be effectively detected through an optical microscope. Porosity tests are based on the detection of the plume formed by corrosion products egressing through the pores, when the plated part is subject to enhanced corrosion by exposure to acid (HNO_3), corrosive vapor (H_2S), or anodic dissolution (electrographic porosity testing). No direct correlation is available between the observed spots and the actual pore sizes. Porosity specifications are given in terms of maximum size of tolerated corrosion spots, their density, and distribution (24).

Pores evolve most often due to substrate defects. An inclusion, dirt, oxidized spot, nonwetted region, or a gas bubble adhering to the surface, are often spotted at the base of a pore. Porosity can be minimized by effective substrate cleaning. This may include degreasing, electrocleaning, electropolishing, preplating with a smooth underlayer (often Ni), or preplating in a "strike" bath.

It can be anticipated that in the future by improving substrate preparation and maintaining correct plating conditions,

thinner layer of precious metal will produce nearly pore free
deposits. Recent applications of gold plating in the electronic
industry, where typically 100 micro-inches of hard gold are plated
on a nickel plated copper substrate, essentially yield pore free
deposits (25).

Thickness - The specified thickness of plated films depends mainly on their application. Due to cost increases of
many plated metals, the trend today is to specify thinner deposits
of higher quality. Except for sacrifical coatings (Zn, Cd),
there is no established correlation between the thickness of protective coatings and their performance. It is generally accepted
that the quality of plated protective films affects their performance more than their thickness. Thickness of plated conductive
or magnetic patterns, and of solder plating is naturally detri-
mental to the application.

The most critical parameter relevant to plate thickness is
the thickness uniformity. If thickness variations over the
plated part were minimal, most thickness specifications could be
substantially lowered. Thickness variations depend on the
"throwing power" of the system, a parameter which is a function
of the solution composition, the cell geometry, and the plating
conditions. Availability of more sophisticated calculations for
current distribution will lead in the future to improved cell designs and solution formulations providing more uniform plating.
This, combined with improved substrate preparation will probably
lead to thinner plate specifications, perhaps by a factor of 2.

Undercoating - Plating of intermediate layers is common
to many applications. Often, lack of adhesion between the substrate and the final coat requires such a layer, or it is
applied to impart mechanical properties, or act as a corrosion or
diffusion barrier.

Hardness - Requirements for hardness of plated layers
vary. Pure plated metals are often softer than metals which are
codeposited with small amounts of other constituents. Typical
example is gold which has a measured hardness of 50-80 knoop when
plated pure (>99.99), but when codeposited with traces (ppm range)
of cobalt or nickel can reach a hardness of 200 knoop. Hardness
is increased through grain refinement, or by "locking in" of the
grain structures by the codeposited metal. Hardness is also a
function of the plating parameters: typically, higher current density and lower bath temperature lead to increased hardness.

Composition - As far as end point requirements are concerned, composition of plated films is only important when it
affects performance. Examples include composition of deposited
alloys in magnetic films and the trace amount of cobalt or nickel

in hard gold films. Occasionally, codeposited material may diffuse to the surface, most likely through the grain boundaries,
accumulating and often causing harmful effects. Codeposited
cobalt has been shown to increase the contact resistance of
plated gold quite substantially by diffusing and accumulating
as an oxide on the outer surface. Few plated films are pure.
Components of the electrolyte, anodic decomposition products and
organic contaminants of the bath are often included in the deposit. Hard golds for example, were shown to contain significant
amount of organic polymeric material (45). Atomic percentages of
hydrogen, nitrogen and carbon included in hard gold deposits are
quite high.

        Contact resistance - Particularly critical property in
plated separable connectors for electronic applications. In view
of the many interconnections, and the extremely low voltage drops
permitted in modern electronic applications, contact resistances
must be in the micro-ohms range, and must remain stable at that
level over extended periods of time (40 years are specified in
the Bell System). Corrosion products, or oxide formation by
diffusion of codeposited non noble metals to the surface may lead
to significant increase in the contact resistance. Although it
is important, there is no accepted universal test for measuring
contact resistance, mainly because of the difficulty in applying
the contact testing probes to the surface in a reproducible
manner.

        Ductility - Indicates the ability of the plated film to
undergo deformation without cracking. It depends on the type of
material plated, the plating conditions, and on the thickness.
It is often measured by bending a plated foil over a mandrel or
by hydraulic or mechanical bulge tests.

        Internal stress - Plated films may maintain residual
stresses which can either be compressive or tensile. These may cause
geometrical distortion of non rigid parts, or in rigid parts may
lead to cracks or to stress corrosion cracking. This problem is
significant in electroforming, in fabrication by plating of fine
grids and micron-scale details. It is particularly critical in
a potential future application - the generation by plating of
submicron gold patterns for x-ray photo lithography masks. Here
the plated pattern is deposited on a thin polymeric film which
can be distorted due to stresses. Typically, internal stesses
in plated films are considerably lower than stresses present in
films produced by vacuum deposition. Furthermore, since stresses
in plated films strongly depend on the plating conditions, these
can often be selected such that no stresses are produced. It has
been demonstrated that by modifying the plating conditions, stresses
in gold deposits could be shifted from compressive into tensile.

The Plating Line - Present Practice and Future Improvements

Unlike older plating processes which involved only a single plating step, plating of the final coating is but one of numerous process steps the plated part undergoes in a modern plating operation. The sequence, intended to prepare the substrate for the final coating, varies considerably depending on the process, the materials involved, and the application. Process times vary from seconds to hours and typically are in the order of minutes. A typical representative sequence for plating hard gold on copper based alloy for electronic connectors includes the following steps:

1. Degreasing - can be vapor, solvent or detergent based.

2. Electrocleaning -cathodic current for surface scrubbing with evolved hydrogen.

3. Nickel underplating - 1 or 2 μm, often from a sulfamate bath.

4. Soft gold strike - high current density, low efficiency plating from a dilute gold bath for about 10 seconds yielding a very thin gold coating.

5. Hard gold plating - typically from gold potassium cyanide (1 tr. Oz./gal), cobalt or nickel hardened (1 ppm range), citrate or phosphate (or a mixed) supporting electrolyte. Thickness depends on applications, often 2.54 μm (100 micro-inches).

6. Final rinse - often two series tanks connected to a gold recovery system.

7. Drying - typically, hot air blowing.

Additional rinse stations are introduced between process steps to minimize cross contamination by drag-out. Bath dilution by rinse water carry-over is minimized by air blowing.

Future processes will undoubtedly be more complex including additional steps in order to provide higher quality finishes while applying thinner coatings. This trend is already seen in some recent installations (26). Additional possible improvements include:

1. Ultrasonic cleaning - quite effective in removing surface contaminants and oxides by causing surface cavitation. Commercial units are now available.

2. Electropolishing – a majority of plating defects arise due to substrate imperfections, roughness and asperities. These can be essentially eliminated through electropolishing which yields smooth, defect-free substrates. Furthermore, electropolished surfaces, when correctly plated produce extremely smooth deposits with excellent wear properties. A number of pitfalls are associated with electropolishing: Electropolished surfaces occasionally exhibit adhesion problems, particularly, when oxidized surfaces are produced. Also, elimination of drag-out of the viscous phosphoric acid from the electropolishing bath into the subsequent nickel cell is difficult and often requires an intermediate acid rinse (0.5 M H_2SO_4). Finally, electropolishing processes are difficult to control because of the high currents involved (~1 A/cm^2), the quite viscous electrolytes and the copper sludge produced.

3. Thicker nickel underplate – of about 2.5 μm, provides smooth, defect free substrate for the gold plating thereby permitting thinner gold plating of lower porosity. Raising the nickel content of the sulfamate bath permits using higher current densities to offset the longer plating time required.

4. Soft gold underplate – soft pure gold plate is often less porous than alloy hardened gold. A soft gold layer (~1μm) plated between the nickel and the hard gold layers is an excellent porosity barrier and often allows to reduce the total amount of plated gold.

5. Sophisticated rinse systems – which conserve water usage, minimize effluent volume and ease environmental pollution problems are likely to be introduced. These may involve cascaded systems, incorporated with metal recovery units.

Major Problem Areas

Cost and Availability of Metals – Escalating costs and diminishing supplies of precious and "strategic" metals (Table 1) affect plating technology in two major ways: First, many of these metals, including gold, platinum,(and platinum group), silver, nickel, cobalt, cadmium, chromium, tin and manganese are extensively used in plating applications and must now be conserved. The second major impact of increased cost and limited availability of critical metals on the plating technology is a positive one: More plated components are likely to replace in the future parts which so far have been fabricated in their entirety from expensive or rare base metals.

Environmental restrictions - Severe environmental restrictions on plating effluents have recently been imposed. In
particular, discharge of metals such as chromium and cadmium, and
of cyanides has been essentially disallowed. Consequently, new
plating facilities will have to undergo radical design changes
so as not to discharge harmful effluents, and existing facilities
will have to be modified.

Current distribution - Non uniform current distribution,
referred to in the plating practice as "poor throwing power",
leads to variable deposit thickness and possibly to varying properties. Gross non uniformities may lead to non-plated regions
or to excessive thickness in complex shaped parts. A quantitative
procedure for calculating practical current distribution problems,
which is accessible to plating engineers and practitioners in
the field is very much in need.

Process control - Most plating cells are poorly controlled. No on-line or feed-back control is available, even in high
speed plating facilities.

Poor control over shape and texture of electrodeposits -
Electrodeposits are often smooth and shiny but occasionally
rough and granular surfaces are produced, which can evolve into
dendritic or powdery texture. The phenomena leading to such
morphological changes are not yet completely understood, and consequently cannot be well controlled.

Understanding the role of additives in plating processes -
Various additives are often introduced in small quantities into
plating solutions in order to improve the quality of the deposit,
and extend the process operating windows. Recent patent literature is replete with plating additive formulations. Although
partial understanding of their role is now available, the complete
picture is far from being unravelled.

Precious Metals Savings

Precious metals - silver, platinum, palladium, rhodium and
in particular gold, are used extensively in electronic and decorative applications. Of the 6×10^6 troy oz. of gold consumed
in the U.S. in 1972, about 25% are estimated to have been in
plating (27). Escalating costs have motivated numerous investigations into means for precious metals savings, turning it into
perhaps the most active area of new plating developments. Among
the options explored:

Replacement by low carat alloys - Gold alloys with silver,
nickel, copper, zinc, cadmium, and palladium have long been used
for decorative purposes in the jewlery industry (28), where the

main requirements are color and wear resistance. Alloying gold imparts to it various tints and improves its mechanical proper- ties thereby giving it commercial appeal. Low carat golds, con- taining between 60-80% gold, alloyed with silver, copper, nickel or cadmium were considered for electronic applications (29,30). Mixed results were obtained indicating that in non-critical ap- plications such finishes are, in general, acceptable. It is questionable however, if gold is altogether essential for these applications. Low carat gold alloys typically do not meet the long term low contact resistance requirements, and do not exhibit the chemical and corrosion resistance of pure (or nearly pure) gold deposits. Despite the significant cost savings, these pro- cesses have not been widely accepted.

Replacement by less costly metals - Finding less noble replacements for precious metals has been a major objective for quite some time, yet no adequate replacement has been found.

Palladium plating is recently attracting much attention. It is less costly than gold ($140 vs. $450 per troy oz) and its lower density further increases the cost advantage. A major problem associated with palladium plating is the codeposited hydrogen adsorbed in the plated layer, often causing cracks. This problem has been significantly alleviated by a recently developed high efficiency buffered plating solution (proprietary formulation by Engelhard) which essentially eliminates hydrogen inclusion. Other serious problems associated with palladium coatings remain however. They are its reactive oxide, and its interaction with chlorides and a number of acids. Palladium coatings for critical applications must probably be overplated with a thin gold layer. The excellent corrosion resistance of tin-nickel alloys (65% Sn/ 35% Ni) has made them likely contenders for gold replacement (31). Tin-nickel alloys overplated with a thin gold flash for solder- ability have been applied on printed circuit boards by major electronic manufacturers. The major problem of tin-nickel coatings is their high stresses and brittleness which leads to cracking. Their future application in separable contacts is therefore questionable, although in many other applications they may provide a viable gold replacement. Other coatings which have been considered for gold replacement in electronic applications are bright tin, and tin-lead alloys (32). Their short term per- formance is acceptable, however they degrade rapidly.

Reduction of the plate thickness - Another means of con- serving precious metals is by modifying the thickness require- ments. Often, the specified thickness is arbitrarily assigned, and is not based on product performance. There is no clear cor- relation between gold thickness, its porosity, and wear properties, although the thicker the gold, the lower its porosity.

Improving the current distribution uniformity is yet another means of reducing the average thickness of plated coatings. Typically, it is the minimum tolerable thickness which is specified, and consequently any non uniformities lead to excessive plate thickness. Improved substrate preparation and undercoating may also provide for thinner coatings. Typically, smoother substrates require thinner coatings than rough substrates do, for similar protection, since substrate roughness is reflected in the deposit. Rough deposits wear rapidly and exhibit excessive porosity. Electropolishing provides one means of producing extremely smooth, clean substrates. Underplating an intermediate layer which may also provide a diffusion barrier for active substrates or corrosion products, is another. Nickel is often used for such intermediate layer in gold or chromium plating over copper.

Reducing losses - The amount of precious metals lost through drag-out and in process effluents is minimal. Effective recovery systems, mostly ion exchange columns, make such losses negligible. Accidental spillouts which cannot be handled by the recovery system are the source of most spillout losses. A major source of precious metal losses is due to plating processes operating outside their designated windows. These may result in excessively thick plated layers, non uniform thickness distribution, or altogether defective product which can only be used as scrap. Improved monitoring and control systems will minimize this loss.

Design modifications - Often relatively minor design modifications of the plated parts may lead to significant savings in the cost of plated materials. Parts can be modified such that they do not require much, or any, precious metals (e.g., high contact force, gas tight connectors), or such that minimal amount of selectively plated metals suffices.

Selective plating - Restricting the costly plated metal only to the functional region of the part through selective plating is perhaps the most effective method of precious metal conservation. Undoubtedly, most precious metals plating in the future will be carried out selectively.

High Speed Plating

Plated cells operating at high current densities provide important advantages in terms of effective utilization of equipment, labor, chemical inventory, and space. High speed plating is particularly critical for selective plating applications where rack plating is no longer possible. The high capital investment often associated with sophisticated equipment must be justified in terms of higher product throughput. Furthermore,

metal inventories in the electrolyte are more effectively utilized in high speed systems.

High rate plating is achieved by increasing the current,
however once the current density exceeds the recommended range,
the deposit deteriorates rapidly. An off-color, rough, dendritic,
or powdery deposit is observed, particularly when the limiting
diffusion current is approached. Recently, quantitative engineering approaches have been applied to the analysis of high current density plating. Observations (33) indicate that it is not
the absolute value of the current density which determines the
quality of the deposit, but rather the ratio of the plating
current density, to the limiting current density, i/i_L. When
the current density exceeds about 60% of the limiting current,
rough deposit is produced. At this point the mass transport
overpotential, ηc, becomes appreciable.

Consequently, a plating system can be operated at
quite high current densities, yet produce smooth deposits,
providing that the limiting current density, i_L, is high.
This latter term depends on the convective velocity, the higher
the flow, the larger i_L is. The exact relationship between the
convective flow and the limiting current is a complex one and
depends primarily on the geometry of the system, on the flow
regime (laminar or turbulent), and on the reactant concentration.

For laminar flow in a gap between parallel plate electrodes, the local limiting current density varies with position:

$$i_L = 0.9783 \ nFDC_b \left(\frac{v}{hDx}\right)^{1/3}$$

here, n is the number of electrons transfered, F - the Faraday
constant, C_b - the bulk concentration of the reactant, v - the
bulk velocity,   h - the gap between the electrodes, and x -
the position along the electrode at which the current is meas-
sured. Clearly, the mass transport limiting current is high
near the leading edge and decreases with the inverse cube root
of the distance along the electrode. In the turbulent regime,
where Re > 2500, the limiting current is considerably higher,
and does not vary with position beyond a short entrance length.
It is given by:

$$i_L = 0.01 \ \left(\frac{nFDC_b}{2h}\right) \ \left(\frac{2hv\rho}{\mu}\right)^{0.92} \left(\frac{\mu}{\rho D}\right)^{0.33}$$

Correlations for limiting currents in geometries other than
parallel flat electrodes, including free convection systems,
were compiled in a recent review (34).

Consequently, high speed plating facilities should operate with concentrated electrolytes ($i_L \alpha C_b$) and should employ high convective flow, preferably in the turbulent regime. The reactant concentration is often limited by the solubility, and in the case of precious metals also by inventory costs and drag-out considerations. Conventional plating cells often provide ineffective agitation, relying on natural convection or air sparging for electrolyte circulation. When high flow rates are required, closed channels of various geometries, or jet plating (35) must be employed. Alternatively, the plated parts can be translated at a fast rate through a stationary electrolyte.

It has been demonsrated (26,36,37) that by applying convective flow in the turbulent regime, good quality deposits can be obtained at current densities hundreds and thousands fold the normal plating rates. In one production application (26) cobalt hardened gold is selectively plated over relatively large curved surfaces at current densities of about 500 mA/cm^2. The gold is shiny, exhibits equal or superior physical properties, and its thickness variations are less than 10% over the entire plated region.

High speed applications may require special formulation of the electrolyte, particularly in the case of alloy plating when some of the constituents may be transport dependent and therefore more sensitive to flow variations. Properly formulated and maintained solutions, used in conjunction with well designed cells providing adequate anode area, do not exhibit the fast deterioration often reported to be associated with high speed plating.

High speed plating is a relatively new development. Increasing costs of new facilities will undoubtedly lead to wider implementation of this practice. Recently achieved engineering understanding of the fundamentals involved, will lead in the future to improved designs.

Compact Enclosed Plating Cells

Older plating facilities typically consist of a series of open tanks, usually containing hot and fuming corrosive plating solutions, into which the plated parts are manually dipped. This open tank configuration is not compatible with high-speed plating requiring high rate convective flow, or with selective plating where the parts must be carefully registered. Open tanks pollute the environment, require large exhausts, and are subject to excessive comtamination. They involve large, often expensive, electrolyte inventories, and occupy much floor space.

Replacing the open tanks with compact, enclosed plating cells offers tremendous advantages. Well designed channel-type cells employing forced turbulent convection are capable of handling products at rates thousands times faster than comparable plating tanks. Significant savings are associated with their smaller size and reduced contamination and evaporation. Smaller, precision machined cells provide more accurate positioning of the electrodes and consequently tighter controls on the plating parameters. The fast circulating electrolyte can be effectively filtered, replenished and analyzed. High technology and electronic applications require precision plating of fine, often micron-scale, features. This requires clean room environment which is not only expensive to maintain, but also foreign to classical plating technology. Compact, self enclosed plating cells essentially provide localized clean room environment for the plating process at a fraction of the cost. The plated parts can be inserted into such cells through slots, or carried through them in a strip form. Electrolyte drag-in and drag-out can be contolled more effectively in closed cells, in part due to the smaller plating racks used, or their absence. Finally, unlike open tanks, smaller, flow type channels can be rationally designed using engineering concepts to provide uniform current distribution and consequently more uniform plate properties and thickness.

Selective Plating

An effective means of conserving expensive plated materials without compromising their function is to restrict their application, by selective plating, to functional areas only. Strict end point requirements which prevent gold replacement in many applications, are turning selective plating into an increasingly important practice. Attesting to its importance is the large number of patents and technical publications devoted to the subject, over 170 were cited in the six year period 1970-1976. Although selective plating processes are more sophisticated and involve expensive equipment, economics indicate that in the future, almost all precious metal plating for non jewelery applications will be carried selectively.

Selective electrodeposition is usually obtained using mechanical or chemical masking or by controlled immersion. It always requires precise registration of the plated parts with respect to the cell, the electrolyte or the mask, and involves sophisticated equipment and costly handling. In order for it to be practical and economical, selective plating must therefore be performed at high rates. The high current densities associated with selective electrodeposition are often restricted by mass transport limitations and consequently, convective flow, often in the turbulent regime, must be employed,

Controlled immersion, or level plating, though limited to certain geometries, is the simplest selective plating technique. Since the electrolyte level must be precisely maintained, only moderate agitation is permitted and consequently only low plating rates are possible. Jet plating combines high agitation rate and localized electrolyte contact with the plated part. Plate thickness and properties are not as uniform as in other techniques, however the plating rates are extremely high (100-5000 mA/cm^2 for gold). Brush plating is an old selective plating method which has recently been automated and incorporated into a continuous selective plating facility for plating electronic contact strips and contact fingers on printed circuit boards (38). The rotating brushes offer significant advantages in terms of the high current densities possible, the relatively simple means of obtaining selectivity, and the ease of adjustment to different product codes.

A new quite different approach to selective plating has been applied in the Bell System (39). Selectivity was obtained by shaping the anode configuration and designing the cell such that highly non-uniform current distribution was generated on the plated part. High current was directed to regions where plating was desired and essentially no current to sections which were not to be plated. This technique lends itself particularly to irregularly shaped fragile parts, where contact with a mask is difficult. Its main advantage over jet plating is that uniform deposit can be obtained over larger areas. Many other novel techniques for selective plating and pattern generation are currently being explored. Among the more intriguing ones are laser enhanced selective plating developed at IBM (40), and selective plating by acoustical focusing (41).

In our own laboratories, we have recently developed a new technique for selectively plating layers of variable thickness (42). The technique provides control over the thickness variations which can be sharp (step wise), or gradual. Continuously modulated three dimensional surface features can be plated with resolution in the few microns range. The technique is based on plating through a permeable gelatin layer, covering the substrate and designed to provide the dominant resistance in the system. The gelatin is exposed, prior to plating through a variable (grey) density mask, and after proper development echos in its thickness the mask density. The metal plated through the variable thickness, and therefore variable resistivity gelatin, is inversly proportional in its thickness to the gelatin layer.

On-Line Process Monitoring and Control

Existing plating facilities are poorly controlled. Typically, the only direct process monitoring available are total
current measurements. Temperature and pH are often measured
only once a shift, composition - once a day. Monitoring of
plating processes is often done indirectly through quality control of product samples. When end point requirements are no
longer met, the process variables are adjusted, usually empirically, heavily relying on the experience of the operator.

The faster plating rates which rapidly produce large product volumes, and the stringent requirements on thickness,
distribution and properties of plated materials imposed by modern technologies, require automatic on-line monitoring, and
tighter, real-time controls. A major immediate application
is in the area of plating waste control.

The first difficulty encountered when attempting to imple-
ment a control system is the absence of proper sensors which
are stable in the adverse plating cell environments and which
can provide continuous composition readings. Recently developed
solid state electrodes exhibit improved stability, but are not
immune to interference by other ions. Another approach involves
the introduction of an automatic sampling device which periodically draws electrolyte samples into an auxiliary cell for
automatic analysis, e.g., by polarography or colorimetry (43).
Another process parameter, which is easily accesible, yet often
not monitored, is the applied voltage. A shift in the potential is an early indication of a process perturbation. Useful
information could be gained from the resolution of the voltage
into its overpotential components. This could possibly be
achieved by on-line transient measurements of superimposed perturbations or of naturally accuring noise. Frequency response
analysis seems quite promising for this application.

Availability of microcomputers and the acceptance of automatic control by many other process industries, combined with
an existing need, is now shifting the trend, and future plating
facilities will undoubtedly be equipped with some control features. High rates facilities for plating contact tabs on
printed wiring boards and for continuous reel to reel selective
plating of electronic contact strips, incorporating extensive
microprocessor-based controls, have recently been introduced
(44).

Steady State Operation

Most plating cells are batch reactors operating in an un-
steady state mode within established process windows. Concen-

trations of the reactants, bath additives, supporting electrolyte, anodic reaction by-products, and contaminants, vary continuously. Consequently, all product is not plated under identical conditions, and it deteriorates as the bath ages. When plated
products no longer meet end point requirements, the plating bath
is treated, replenished, or discarded. The situation is particularly critical in electropolishing cells where large amounts
of sludge are rapidly generated, in high current density plating
cells, when improperly designed, and in compact systems where the
electrolyte volume is small.

Two approaches can be adapted to improve the situation. The
first is continuous replenishment and treatment of plating baths
by filtering, ion exchange, dialysis, etc. The second approach is
a more complicated one and involves the application of interactive
controls to rectify any deviations as they occur. When such procedures are implemented, the common phenomena of bath "aging",
no longer occurs and plating baths could, in principle, be used
indefinitely.

Anode Design Considerations

The anode reaction is very much part of the cell process,
and since cells without transference are not practical, its products reach the cathode and may interfere with the deposition.
Anodes can generally be divided into two classes. Soluble anodes
where the dissolution is the reverse of the cathodic reaction,
and insoluble anodes, which most often evolve oxygen. Cells employing soluble anodes typically operate at lower potentials and
the reactant concentration is more closely maintained. The main
disadvantage of soluble anodes is their shape change causing deviations in current distributions and requiring periodic anode
replacement. Insoluble anodes offer dimensional stability which
makes them indispensable for precision plating and selective
plating applications. They are most often used in precious metal
plating. The higher overpotential associated with oxygen evolution reaction may lead to bath deterioration through the formation of undesirable oxidation by-products. Insoluble anodes are
often made of lead, platinum, platinized or ruthenized titanium,
or stainless steel. Catalytic, dimensionally stable anodes with
low oxygen overpotental have recently been developed for gold and
nickel plating (46,47). A useful combination of the two anode
types are basket electrodes. Here, a dimensionally stable basket
is fabricated from expanded mesh of insoluble metal, often titanium, and filled with soluble metal chips.

Current Distribution and Thickness Uniformity

Plating systems are often characterized in terms of their
"throwing power", essentially an empirical parameter expressing

the capability of the plating system to produce uniform thickness
deposits. Macro-throwing power refers to uniformity on the macro-
scale, across the entire plated part, whereas micro-throwing
power relates to the capability of the plating process to penetrate and uniformally cover microscopic details on the surface.
The latter is often a less critical parameter, since most plating
processes are always characterized by quite a high micro-throwing
power. Electrochemical engineers express the throwing power in
terms of a dimensionless parameter, the Wagner number, expressing the ratio between the activation to the ohmic resistances of
the system:

$$Wa = \frac{\kappa}{L} \left(\frac{\partial \eta_a}{\partial i}\right)$$

Here κ is the conductivity, L - a characteristic dimension of the
system and $\partial \eta_a / \partial i$ is the slope of the activiation polarization
curve. A high Wa number corresponds to a good surface coverage.

A more quantitative approach to current distribution problems is through rigorous solutions. Although the mathematical
expressions quantitatively describing the current distribution in
electrochemical cells can be rigorously formulated, explicit
solutions cannot be obtained, except for a few simple cases (16).
It is, therefore, often useful to first consider the extreme situations where the current distribution is dominated by a single
mechanism: A controlling electrolyte resistance yields the
usually highly non-uniform "primary" distribution. Including
an electrode reaction resistance term leads to a more uniform
"secondary" distribution. At the other extreme is the mass
transport controlled deposition which is flow dependent. The prevailing current distribution in a practical cell normally assumes
some intermediate value bound by those extremes. Determination of
even the simpler, primary current distribution in a real cell is
a non trivial problem. It involves solving, usually numerically,
the Laplace equation for the potential subject to the appropriate
boundary conditions. Finite differences or finite elements
techniques have successfully been employed (48,49). Detailed
solutions may however require a fine grid and many iterations.
Recently, a technique using Green's identity to transform the
problem to a boundary integral has been implemented (50). Although the equations involved are more complicated, a non itera-
tive procedure for primary distribution is obtained. Secondary
distribution problems are more unique to electrochemical systems.
They are solved similarly to primary distribution problems however they involve non linear boundary conditions and consequently are more complicated. At higher current densities, approaching the diffusion limiting current, a mass transport dependent "tertiary" distribution prevails. Its determination involves detailed knowledge of the flow and is quite difficult.

Solutions for some relatively simple configurations have been
obtained (16). Orthogonal collocation techniques have also been
successfully applied to the parallel plate geometry (51). Intermediate current distributions when no particular overpotential is
dominant are extremely difficult to solve. Newman has outlined a
general numerical procedure and obtained detailed solutions for
two simple geometries (rotating disc and parallel plate electrodes).

Environmental Considerations

     The most serious problem currently facing the plating industry
is coping with the extremely tight regulations which have recently been imposed on plating effluents by the EPA. These regulations specify maximum permissible effluent concentrations, all
in the ppm range, on most metals including cadmium, lead, nickel,
copper, zinc, boron, and iron. The specifications are particularly strict on effluent concentrations of cyanides (both total,
and amenable to chlorination) and of hexavalent chromium. Their
discharge levels are essentially set below the level of practical
detection (30 days average below 0.09 ppm). Temperature and pH
must also be adjusted before discharge. This not only started
a heated debate between industry and government officials on
whether the "best available technology" is capable of complying
with these regulations, but also requires quite expensive plant
modifications. Hexavalent chromium is reduced to its permissible
level, below 0.05 ppm, by a two step process which first involves
reduction in acidic environment to its trivalent form by either
SO_2, $FeSO_4$, $Na_2S_2O_5$, $Na_2S_2O_4$ or $NaHSO_3$, followed by raising the
pH to 8.5 - 9.0, where chromium hydroxide precipitates as a
sludge. Cyanides are most commonly destroyed by chlorination
in alkaline medium. The products are nitrogen and carbon dioxide.

     In the future (1983 or therabouts) even tighter regulations
will be imposed leading eventually to zero discharge requirements
on plating facilities.

     This implies entirely new designs for future facilities
which must be planned as totally enclosed units, with zero discharge or with only minimal dewatered and compacted solid wastes
release which can be used in landfills. Rinse water management
will be tighter, with significantly lower volumes and perhaps
total recirculation. Volume will be reduced through the use of
efficient spray nozzles which are considerably more efficient
than regular dip rinsing, and through cascading and internal recirculation. Contaminants will be removed from the effluents
through a combination of methods. These may include: plating
out on porous beds, ion exchange, dialysis, reverse osmosis, precipitation processes and flash evaporation.

Environmental considerations are also indicating some process modifications. Replacements are sought for cadmium plating
and for cyanide based electrolytes. Unfortunately, the metal
complexing properties of cyanides and their high solubility
renders them essential for gold and silver plating, and very desirable for zinc, copper and cadmium plating. Cyanide based
electrolytes have a good throwing power and the deposits are fine
grained and shiny. Nonetheless, acidic zinc plating processes
which do not involve cyanides are rapidly gaining popularity and
relacing the cyanide based alkaline process.

Morphological Changes In Electrodeposition - Roughness Evolution and Dendritic Growth

Depending on the plating conditions, electrodeposits can be
smooth and shiny or rough and granular. The latter often evolve
into dendrites, whiskers, powder or sponge type growth. Such
phenomena can be highly undesirable in plating applications, although can be used for the generation of metal powders. A recent
monograph (52) extensively reviews the problem· Although many empirical studies have been made, fundamental understanding of the
phenomena involved is lacking and the relationship between the
process parameters and the surface morphology have not been
quantitatively defined.

Recent investigations (53,54) indicate that roughness evolves
because of morphological instability of the plated surface,
brought about by extremely low interfacial reactant concentration,
C_o. In any given system, with fixed agitation (constant diffusion boundary layer thickness) and a constant bulk concentration,
high current density corresponds to a low C_o, and also to a large
concentration overpotential. It can be shown that the normalized
interfacial concentration C_o/C_b corresponds to the ratio i/i_L,
which was empirically established as a roughness criterion.
When the current density approaches the limiting diffusion current, C_o/C_b becomes vanishingly small. The model indicates that
any electrochemical system is inherently unstable and if plating
proceeds for long enough time, minute surface imperfections
will eventually evolve into roughness elements. A critical overpotential, below which roughness will take an extremely long
time to develop has been analytically derived and correlated to
the system's operating parameters, and to the properties of the
plated metal. An operating window corresponding to dendrite free
plating has also been established.

Plating Additives

One method which is extensively used for improving the
quality of deposits, and in particular make them shiny and smooth,
is the addition to the plating bath of small amounts (in the ppm

range) of organic or metallic additives. The term "plating
additives" is generic and encompasses a wide variety of chemicals
which affect the deposit in a number of ways. Their function and
mechanism of interaction is yet unclear, and their investigation
so far has mostly been empirical. Nonetheless, plating additives
are extremely important and establishing (nearly always by Edisonian
approach) the proper agents, most often determines the success
or failure of a given plating process.   Indicative of their
importance, is the plating patent literature which is mostly dedicated to plating additives disclosures. Often quite similar compounds are claimed, and the ensuing legal disputes highlight the
lack of fundamental understanding. Plating additives are highly
proprietary and are the main ingredient differentiating between
plating solutions from different suppliers.

Plating additives can be organic or metallic, ionic or non
ionic in nature. They are characteristically surface active
agents which adsorb on the plated surface and are often incorporated in the deposit. Broadly speaking, additives can be classified into four major groups: (a) Grain refiners, e.g., cobalt
or nickel in gold plating, which are codeposited in trace amounts
and generally impart to the metal hardness and brightness. (b)
Dendrite and roughness inhibitors are perhaps the most important
additives. They adsorb on the surface, covering it with a thin
layer and inhibit the growth of dendrite precursors. Organic
or inorganic additives belong in this class. The inorganic additives are preferable because of their stability. Typically
one or few monolayers of heavy metals, e.g., Pb, Bi, Sn, In, adsorbed in the underpotential region from electrolyte concentrations in the few ppm range   have been shown to be effective
dendrite inhibitors in gold or zinc plating. (c) A third class
of additives is the leveling agents which improve the throwing
power of the electrolyte, mainly by increasing the slope of
activation overpotential curve. (d) Wetting agents or sur-
factants are yet another group of additives whose main function
is to prevent pits and pores in the deposit.

Due to their minute concentration, control of plating additives is extremely difficult. However, maintaining their
concentration within the specified range is often critical to
proper operation of the bath. Lead additions in the 5-50 ppm
range inhibits dendritic growth in zinc plating from acid
electrolytes; however, when the lead concentration exceeds a few
hundred ppm, zinc whiskers often appear. Similarly in gold
plating, lead in the 1 ppm range enhances the brightness of the
deposit; however, when its concentration increases, dull deposits
are produced.

Because additives are typically present in extremely small

concentrations, their transport towards the electrode is nearly
always under diffusion control. Additives are therefore quite
sensitive to flow variations. In general, because of the
difficulties associated with proper control of additive containing
baths, non-additive formulations are preferable whenever they
produce acceptable deposits.

Future Directions

     The electroplating field has been recently subject to an
extremely rapid technological progress. Economic and
technological considerations indicate that this trend will
continue in the future.

     Perhaps the most important contributions are yet to come
from the scientific community. Improved understanding of the
electrocrystallization process, including the effects of additives
is very much in need. Modern surface techniques are now capable
of providing detailed information about chemical composition,
structure, and oxidation states on surfaces smaller than 1 mm^2,
with detection sensitivity of less than one thousandth of a
monolayer. These can supplement classical electrochemical
techniques, which are now capable of extremely fast response
and routinely provide sensitivity in the microamperes and micro-
volts range, but lack the specificity and resolution of the
vacuum based techniques. A major disadvantage of the latter is
their limitation to solid-vacuum or solid-gas interfaces, whereas
electrochemistry involves mostly solid-liquid interactions. In
the future, light scattering techniques with UV, laser or x-ray
may provide means for directly exploring gas-liquid interfaces.
Laser-Raman spectroscopy is particularly exciting since it may
provide information on the adsorption states and orientation of
species on electrodes.

     Improved understanding and modelling of the interaction
between the system parameters, including the flow, and the
deposition process is required. In particular, understanding
and modelling of morphological changes during the deposition
process are useful. On the macroscopic scale, improved, more
universal,and particularly more accessible procedures for
calculating current distributions on electrodes are likely to be
developed. These will provide the necessary foundations for
quantitative engineering approach to design of electroplating
systems, and to rational scale-up procedures from laboratory
experiments to industrial practice.

     New plating processes for electrodeposition of aluminum,
magnesium, and other metals which so far have not been electroplated, are likely to be implemented. Plating of alloys and

composite materials will probably be expanded to produce plated
layers with improved and unique properties. The plating of
catalytic electrodes for various applications is likely to be
expanded. One possible intriguing application is the plating of
solar panels, photo-electrodes and other devices with semiconductor type materials. Improved control over shape and
properties is likely to bring about a deeper penetration of the
electronic industry, possibly replacing a number of vacuum
deposition processes. The recent progress in the modelling of
pulse plating and the improved understanding of its effects on
the deposit are likely to increase its commercial applications.

Electroplating processes will be carried out in the future
at high rates in automatically controlled facilities which will
be totally enclosed, and environmentally compatible. Electroplating will be a precision operation where extremely thin
coatings of uniform thickness and properties, and of low porosity,
will be applied, often selectively, to an increasing number of
products.

References

1. Meadows, D.L., et al., "Dynamics of Growth in a Finite World",
    Wright-Allen Press, Inc. Cambridge, Mass., 1974.

2. U.S. Bureau of Mines, "Minerals Yearbook 1978-79", U.S.
    Government Printing Office, Washington, D.C., 1980.

3. Reported in "Battelle Today", June, 1978. Battelle Memorial
    Laboratory, Columbus, Ohio.

4. Brenner, A., "Electrolysis of Nonaqueous Systems" in "Advances
    in Electrochemistry and Electrochemical Engineering", C.W.
    Tobias and P. Delahay (eds.), Wiley, New York, 1969.

5. Peled, E. and E. Gileadi, J. Electrochem. Soc. 123, 15-19,
    1976.

6. Daenen, T.E.G., Philips Research Laboratories, Eindhoven,
    Netherlands. Private communication.

7. Tobias, C.W. and J. Jorne, U.S. Patent No. 3,791,945 (1974).

8. Brenner, A. and J.L. Sligh, Trans. Inst. Met. Finishing, 49,
    71-78 (1971).

9. Lowenheim, F.A., "Electroplating", McGraw-Hill, New York,
    New York, 1978.

10. Lowenheim, F.A., (ed.), "Modern Electroplating", 3rd ed.,
    John Wiley, New York, New York, 1974.

11. Graham, A.K. (ed.), "Electroplating Engineering Handbook",
    3rd ed., Van Norstrand Reinhold, New York, New York, 1971.

12. Hall, N. (ed.), "Metal Finishing Guidebook and Directory", Metals and Plastics Publications, Hackensack, New Jersey (published annually).

13. Fischer, H., "Eleklrolytische Abscheidung und Elektrokristallisation von Metallen", Springer, Berlin, 1954.

14. Vetter, K.J., "Electrochemical Kinetics", Academic Press, New York, New York, 1967.

15. Bockris, J. O'M. and A. Reddy, "Modern Electrochemistry", Plenum Press, New York, New York, 1970.

16. Newman, J., "Electrochemical Systems", Prentice-Hall, Englewood Cliffs, New Jersey, 1973.

17. Pickett, D.J., "Electrochemical Reactor Design", Elsevier, New York, New York, 1977.

18. Delahay, P. and C. W. Tobias (eds.), "Advances in Electrochemistry and Electrochemical Engineering", John Wiley, New York, New York.

19. Bockris, J. O'M. (ed.), "Modern Aspects of Electrochemistry", Buttersworth, London.

20. Yeager, E., and A.J. Salkind (eds.), "Techniques of Electrochemistry", Wiley Interscience, New York, New York.

21. Bard, A.T. (ed.), "Electroanalytical Chemistry", Marcel Dekker, New York, New York.

22. Baker, R.G. and T.A. Palumbo, Plating, 58, 791-800, 1971.

23. Sard, R., H. Leidheiser, and F. Ogburn (eds.), "Properties of Electrodeposits - Their Measurements and Significance", The Electrochemical Society, Princeton, New Jersey, 1975.

24. Clarke, M., "Porosity and Porosity Tests", Ibid., Chapter 8, 122-141.

25. Bacon, D., U. Landau, and R.L. Meek, The Western Electric Engineer, 22, 13-19, 1978.

26. Koontz, D.E. and G. F. Helgesen, The Western Electric Engineer, 22, 26-31, 1978.

27. Steinmetz, R.W., Plating, 61, 443-446, 1974.

28. Foster, A., Electroplating and Metal Finishing, 26 (2), 19-22, 1973.

29. Nobel, F.I., D.W. Thomson and J.M. Leibel, Plating 60, 720-727, 1973.

30. Krishnamarthy, S., Electroplating Metal Finishing 28, 18-24, 1975.

31. Antler, M., IEEE Trans. on Parts, Hybrids and Packaging,
     PHP-11, 216 (1975).

32. Donaldson, J.G., Plating 61, 225 (1974).

33. Landau, U., LBL-2702, Ph.D. Thesis, University of California,
     Berkeley, January, 1976.

34. Selman, J.R. and C.W. Tobias, in "Advances in Chemical
     Engineering", Drew et al. (eds.), Vol. 10, Academic Press,
     New York (1978).

35. Sewell, B., Trans. Inst. Metal Finishing, 50, 121-124 (1972).

36. Safranek, W.H. and C.H. Layer, Plating 62, 121-125 (1975).

37. Reported in "Search", General Motors Research Laboratories,
     Warren, Michigan, Nov.-D-c. 1978.

38. Eidschun, C.D., Microplate, Inc., St. Petersburg, Florida.
     Private communication.

39. Landau, U. and D.E. Koontz, U.S. Patent No. 4,001,093 (1977).

40. von Gutfeld, R.J., E.E. Tynan, R.L. Melcher and S.E. Blum,
     Appl. Phys. Lett. 35, 651 (1979).

41. Drake, M.D., Extended Abstracts, The Electrochem. Soc.
     Meeting, Pittsburgh, Pa. 78(2), 434 (1978).

42. Tomaswick, K.M., M.S. Thesis, The Department of Chemical
     Engineering, Case Western Reserve University, Cleveland,
     Ohio, August 1980.

43. Okinaka, Y., D.W. Graham, C. Wolowodiuk and T.M. Putvinski,
     The Western Electric Engineer, 22, 72-81 (1978).

44. Turner, D.R., W.R. Evarts, R.W. Duston and E.L. Byars,
     The Western Electric Engineer, 22, 32-40 (1978).

45. Munier, G.B., Plating, 56, 1151 (1969).

46. Okinaka, Y. and C. Wolowodiuk, U.S. Patent 4,067,783 (1978).

47. Scarpellino, A.J. and G.L. Fisher, Extended Abstracts,
     Electrochem. Soc. Meeting, St. Louis, Mo. 80(2), 1029
     (1980).

48. Fleck, R.N., D.N. Hanson and C.W. Tobias, Lawrence
     Radiation Laboratory Report, UCRL-11612, University of
     California, Berkeley 1964.

49. Klingert, J.A., S. Lynn and C.W. Tobias, Electrochim. Acta
     9, 297 (1964).

50. Dordi, Y.N., M.S. Thesis, Department of Chemical Engineering,
     Case Western Reserve University, Cleveland, Ohio, May 1979.

51. Caban, R. and T.W. Chapman, J. Electrochem. Soc. $\underline{123}$, 1036 (1976).

52. Calusaru, A., "Electrodeposition of Metal Powders", Elsevier, New York, 1979.

53. Shyu, J.H. and U. Landau, Extended Abstracts, The Electro-chemical Society Meeting, Los Angeles, California, 79(2), 427 (1979).

54. Oren, Y. and U. Landau, Electrocrystallization Symposium, Extended Abstracts, The Electrochemical Society Meeting, Hollywood, Florida, 80(2), 1018 (1980).

ON THE PRODUCTION OF HYDROGEN AS AN ENERGY MEDIUM

J. O'M. Bockris

Department of Chemistry
Texas A&M University
College Station, TX 77843

INTRODUCTION

The supply of liquid and gaseous fossil fuels has now reached a maximum insofar as Western countries are concerned (1), including supplies from the Middle East. Even if there is no catastrophic interruption of supplies from Saudi Arabia, the net supply of fossil fuels available to the United States and European countries will turn down during the 1980's. The standard of life is proportional to the energy consumed. To avoid a massive retrenchment of economic welfare in the western world, with obvious geopolitical consequences, the failing fossil fuel supply must be replaced. The replacement sources are agreed upon. They are: coal; atomic breeder reactors; and solar energy, collected upon a massive scale in desert sites or upon the sea.

Although synthetic gasoline from coal is a possibility, it would be some twice times more expensive, as measured for a unit distance of car driving in urban transportation, than hydrogen. The latter is the only fuel which could be obtained from atomic and solar sources, except electricity, which supplies only 20% of our energy needs and cannot be economically sent over distances of greater than about 400 miles.

Were hydrogen to be obtained from coal, the reason why this source of energy is regarded as unacceptable (the greenhouse effect) would no longer be valid for the CO_2 could be injected into the sea from shore-side plants (see below).

247

Were hydrogen to be the fuel developed from coal and used with atomic and solar sources, the same fuel would be usable from the time synfuels are made from coal onwards without change as the sources become atomic or solar. A Hydrogen Economy is a consequence of a departure from consumption of the fossil fuel cache.

The preparation of hydrogen upon a massive scale from water is a vital matter. Here, the cutting edge in this field will be described.

FROM COAL AND STEAM

The production of hydrogen from coal is the reaction which can occur between coal and water.

$$C + H_2O \rightarrow CO + H_2$$

$$CO + H_2O \rightarrow CO_2 + H_2$$

Hydrogen could be produced massively from coal. The conversion efficiency would be 98%. When methane is formed from water and coal one needs to go to much higher temperatures (e.g., 900°C; - for hydrogen, 600°C) and high pressures (hydrogen, normal pressures).

Because the Carnot cycle is involved in the manufacture of the electricity, there can never be hydrogen from the electrolysis of water which would be as cheap as hydrogen from coal. Thus, the fact that hydrogen from water by means of electrolysis is 100.0% pure, whereas that from coal is 98% pure is not a significant disadvantage. Coal contains little hydrogen. Coals are essentially CH whereas methane is CH_4. The function of the coal in reacting with water is to release hydrogen from water.

Disadvantages pertain to the direct production of H_2 from coal.

A. In utilizing coal to produce heat, it is not necessary to crush the coal. However, in the reaction of coal with steam, the only part which reacts is that in contact with the steam, and therefore a powdering of the coal is necessary. This costs.

B. When coal is used as a raw material one impurity produced is SO_2. Much American coal contains 5% sulfur. For a 1,000 megawatt electricity plant, running on coal for 1 year, the sulfur would have the size of a city block some six floors high. The only practical way sulfur can be removed is to put it back into the mine. This costs.

The most dangerous pollutant from coal is the CO_2. The greenhouse effect has often been expounded (2). CO_2 will start to play

a role in the world's temperature by 2000. By 2050, the average
world temperature would be increased by 5°C. This increase would
not be uniform: minimal at the equator and maximal at the poles
where the temperature increase by 2050 would be 10°C. Polar ice
would melt. The rise throughout world sea levels has often been
calculated and figures of the order of 10 meters have been given.
The flooding of very large amounts of primary land area would hence
be a consequence of using coal to get natural gas or synthetic
gasoline. Were plants producing hydrogen to be built along the
shore, CO_2 exhausts could be led into the deep sea. This would be
a suitable place and would have little effect upon the atmospheric
CO_2 (3), even though the coal supply were burnt. This seems the
only way in which coal can be safely used.

    Billings (4) held a symposium from which a range of prices
for hydrogen from coal was $3.00 to $5.00 per MBTU. The corresponding value for predictions for liquid gasoline from coal would
be $6.00 - $12.00 per MBTU.

    It is logical to avoid coal and go to the building of reactors
and solar plants. Because of the inertia of vested capital it is
likely that coal will be used to supplement oil. Oil companies
may not wish to go down the hydrogen path. The short term (5-10
year) profit picture is better if they produce synthetic gasoline
because: (a) It would cost more than hydrogen, i.e., be more pro-
fit-bearing; (b) Synthetic gasoline would imply no change in the
distributive apparatus, storage included.

    Were they to accept the hydrogen solution of avoiding the
greenhouse effect, plants would have to be built on the shore,
pipelines laid, cars made with modified engines. The distributive
apparatus would have to be changed, and the burners etc. used at
present would have to be changed.

    However, apart from the hesitation of the oil companies les-
sening short term profit for long term gain, there is the difficulty
of digging enough mines before oil exhausts. Thus, it takes 10
years to start a mine, because of legal procedures. Were the entire energy supply to be obtained from coal, say, by 2025, then
there would be a manpower demand of several percent of the American
work force. Mines would have to be opened at a rate of around one
per day. There is thus no time now to build the mining operation
we need whilst we still have oil, within our present econo system.

    Would it be possible to open mines fast enough with a coalition
government in military operation with all labor under DOD control?
The alternatives seem to be this: along the coal path or economic
collapse, and widespread death from cold and starvation; or widespread flooding throughout the world.

FROM HEAT AND WATER

It is known that, by the use of lenses, several thousand degrees can be attained from solar heat. Solar collectors could be arranged so that the heat is concentrated and water directly decomposed to hydrogen (6).

The problems are less that of attaining the temperatures (3000°C) than of separating hydrogen and oxygen to avoid a recombination. It could perhaps be done in several ways:

A. An ultracentrifuge utilizing tungsten.*

B. An oxide membrane. But the chemical stability of any oxides in hydrogen is doubtful.

C. A magnetic separation.

There are general difficulties:

(a) Finding materials stable for sufficient time;
(b) Attaining the requisite temperature, because of re-radiation.

In view of these difficulties it is best to leave direct decomposition, and proceed indirectly. This is a well known theme. However, the principle must be stated. It is sought by means of closed cycles, to balance off the negative and positive entropy of the partial reactions. As ΔG^0 for the decomposition of water to hydrogen and oxygen is positive, it is desirable to have a positive value of the net $T\Delta S$. When a positive value of ΔS is present, the reaction is at a high temperature, when negative, at a low temperature. The mass of gas involved in the cycles has to have its temperature and pressure changed: work has to be done. The changes of P and T in cycling are analogous to those of a Carnot cycle. Consequently, there will be a use-up of some of the heat of reaction in achieving the cycle. The production of hydrogen by such cycles involves, therefore, a Carnot efficiency factor.

Such factors come to be 0.2-0.5, and thus not far from the 0.4 involved in the manufacture of electricity from heat. However, there are other factors which would seem to make less likely good net economics the chemical cycle method of obtaining hydrogen: it will have a larger cost factor than that of the electrolytic method, because of the high temperatures involved. Each part of the cycle will need a separate group of apparati, rather than the single one of the electrochemical method. There may also be a difficulty that some of the materials will be "left over" because there are side reactions. These side reactions will use up material which had

---

* However, tungstic oxide would probably evaporate.

been costed once and for all; if there is even 0.1% lack of cyclicity the economics will be made less good.

In spite of these gloomy indications, there is more research on the chemical cycle method for the production of hydrogen than upon any other. (7)

FROM ELECTRICITY AND WATER

This method separates its own products; is well known and simple; gives 100% Faradaic efficiency; gives at least 85% efficiency of conversion of electricity to hydrogen; and there are many people around the world who know something of the technology of building electrolyzers.*

The difficulty with the electrolytic method is the energy balance and relates to the source of energy rather than to the efficiency of the use of electricity. In making electricity by conventional methods, the efficiency is 39%. If one takes the efficiency of the electrolysis device to be 0.85, one has an efficiency of 0.3. It is conceivable that efficiencies of 0.5 could be attained by the chemical cycle method.

Two points favor heat and electricity compared with coal, and compensate for the possible increase in efficiency which might come in a thermolysis. (1) The chemical decomposition of coal has a scaling factor: if the production is greater than 10^6 SCF per day, the electrolysis method is cheaper than direct chemical conversion (8). (2) If the energy source for the electricity is atomic or solar, environmental degradation is lessened.

Electrolysis is economically advantageous at any scale with old hydroelectric plants. The cost of electricity therefore is the cost of maintenance and thus small, e.g., 1¢ per kilowatt hour. Such electricity is available, for example, from the Bonneville Dam. It could be excellent to use it for refueling a hydrogen bus system, etc.**

---

* However, the United States is poorly off in this technology, having no major company with an outstanding record in it. The two largest companies in the world which make electrolyzers are the Lurgi Company of Germany and the de Nora Company of Italy.

** However, there are economic difficulties in beginning such a scheme, because, although running buses on hydrogen obtained from this source would be cheaper than running them on gasoline there is a capital investment in the conversion costs. Such a situation seems a worthy case for government help, i.e., the socializing of the costs of such energy-helpful enterprises by the granting of loans at, e.g., 4%.

Old electrolyzers can be improved by the application of modern electrode kinetics, though none of this has reached practice at present. Electrocatalysis is the obvious way in which electrolyzers can be made to function at lower voltages than the 2.0 volts which is the value used in classical and existent (big) electrolyzers. The work of Srinivasan (9), and Appleby (10) in particular, has shown how much the potential can be cut, by new work in oxygen electrocatalysis, in which conducting oxides have been used rather than the noble metals. (Such work was begun by Damjanovic, Sepa and Bockris (11), working with the tungsten oxides, in 1967.) It is possible to reduce the voltage for electrolysis, at 500 ma cm^{-2} to 1.6 volts at 100°. There are plenty of indications that the voltage can be reduced further, e.g., to 1.5 volts at the same current density, by the use of cathodic catalysts, and finally, perhaps, even to 1.4 volts, by the use of oxygen electrocatalysts in which the Tafel slope is RT/2F or even less.

Another approach is to avoid the evolution of oxygen. Thus, the potential for the decomposition of hydrogen iodide is less than that for water. Hydrogen could be cheaply obtained from hydrogen iodide electrolysis but one has to arrange a path whereby iodine can be returned to iodide. This costs. Numerous examples of such cyclical processes have been worked out. One of them, which was first suggested by Moulton and Juda (12), is the use of waste SO_2, which should be available from stack gases. The voltage for the decomposition of water can then be halved in this way. The product is sulfuric acid: eventually the storage would be that of SO_3. Such a process involves paying for the heat.

In HCl electrolysis, the chlorine could be returned to O_2 by means of the Deacon process. The thermal energy needed is not the free energy calculated for G° data but that, plus the "over-free energy," an allowance for the inefficiency of the process [Appleby and Bockris, 1980 (13)].

Among newer methods in water splitting is the high temperature method. In this, steam is electrolyzed at 1000° (14). The entropy change in the dissociation of water is positive and therefore the standard free energy becomes less positive as the temperature increases. Hence, it uses less electrical energy to make electrolysis at high temperatures. At first, one has the impression that one is decomposing water with increasingly less energy as the temperature increases. This is not so without a corollary. The corollary is that the system cools. If the system is to be kept at, e.g., 1000°C, heat must be injected from outside. The cost of this heat must be added to the decreasing cost of the electricity.

The fact that one has to add heat to the cell to keep it warm does not mean that one has not gained by electrolyzing steam. There is the reduction towards negligible value of the activation over-

voltages of the reactions. The heat added to the cell contributes
to the synthesis of hydrogen and oxygen from water without passing
through the Carnot cycle so that it may be that it results in a
cheaper process than electrolysis at lower temperature. On the
other hand the method is carried out with refractories, e.g., which
act as electronic conductors as well as those (e.g., ZrO_2) which
act as ionic conductors [c.f. Badwal, Bevan and Bockris, 1980 (14)]
and it is not known what the economics of such refractories could be.

It seems likely that, were research to be pursued on the above
methods (and many other possibilities not stated) the result could
be a good one. At present, little research on electrolysis is
going on in the United States and one of the difficulties of stimu-
lating a massive research effort for such a process is the lack of
electrochemists trained to look at the solid state chemistry of
the devices.

FROM LIGHT AND WATER (CONTRIVED)

The observation by Fujishima in a Japanese laboratory in the
early 1970's that when he held TiO_2 in the presence of water and
light some bubbles were evolved, led to experiments by Fujishima
and Honda (15) in which the TiO_2 was combined in a cell with platinum and subjected to controlled radiation. Photoelectrolysis took
place, and after some confusion in which it was not clear whether
the cathodic reaction was the evolution of hydrogen or the reduction of oxygen, it was shown that under certain circumstances (different pH's in the two compartments) it was possible to produce
hydrogen and oxygen (16).

The work of Fujishima and Honda has stimulated a large amount
of work on photo-electrolysis and names which have become known in
this situation are those of Bard, Wrighton and Heller, respectively,
in the United States. Gerischer has contributed more generally to
the elucidation of the semiconductor-solution interface. The work
of Memming and Pleskov is also well known.

All these contributions to photoelectrolysis have been oriented
towards the concept that redox reactions at an anode could be coupled
with reactions at cathodes to give electricity. They have not concentrated upon the splitting of water. Indeed, those who have concentrated upon this problem have come from another side, the biological side, and among these the leading names are those of Calvin
and of Graetzel (18).

It has been often stated that water-splitting would be difficult
to achieve without corrosion. Such statements involve the limitations of the thermodynamic approach. When, in 1980, the same considerations were made kinetically, more positive conclusions were
reached (19).

It is necessary to recall at what potential the reactions occur in the splitting of water. Guruswamy and Bockris (20) developed results obtained by Ohashi, McCann and Bockris (21). The onset potential was about 0.4 volts away for the cathodic potential, and the anodic reaction about 1.4 volts. This gross shift from the reversible potential alters the considerations of the thermo-dynamics of corrosion and makes the situation less difficult.

It is important to outline the theoretical concepts on which this is based. Many considerations given to photoelectrolysis, particularly, for example, by Butler and Ginley (22) have been based upon considerations of the charge carriers after they have been stimulated from the valency band to the conduction band by incoming photons. Such considerations involve hole-pair recombination, diffusion to the surface, and transfer through the surface to the solution. Several workers have considered these processes without taking into account the potential difference across the interface and the quantum mechanical aspects of this passage. Thus, Wilson (23), Reiss (24) and Reichmann (25) have paid no detailed attention to the step at the interface, and Butler and Ginley (22) have neg-lected it completely. Such neglect is less incredible when it is recalled that most of the treatment of the semiconductor-solution interface - including that of Gerischer (26) - has taken the posi-tion that the surface state concentration is not important, so that, although there must be a potential difference in the Helmholtz layer between the semiconductor in solution, potential difference variations occur inside the semiconductor. However, it is unclear at present to what extent this view of a negligible role played surface state is correct; and to what extent the alternate view, originated by Green (17) in 1957 (the semiconductor-electrolyte interface has a high degree of surface states) is the more correct (as it seems to be on p type electrodes).

Among the desiderata which influence this situation is that of Bard et al. who have referred to the idea of Fermi level pinning. This phrase arose because, when the surface state concentration is high, and the change of electrode potential is largely by a change in potential across the Helmholtz layer, the Fermi level in the bulk of the semiconductor does not change as the electrode potential changes. Bard (28) found a number of p-type electrodes to which this situation applied.

The first formulation of kinetic theory equations for photo-electrolysis was that of Bockris in 1954 (29). The work of Bockris and Uosaki (30) in 1978 was the first to formulate the equations in molecular kinetic terms, taking into account the presence of surface states. The various considerations were:

1. The excitation of electrons in the valency band to the con-duction band, which was assumed to be an equilibrium process under

the so-called "normal" Fermi statistics.

2. Diffusion of the created electrons and holes. The net direction is determined by the Nernst-Planck equation. Bockris and Uosaki took into account in this process a lifetime of the electrons within the semiconductor, i.e., they allowed for collisional de-activation and the passage of the electron from the point of its creation from photons, at a distance which has an order of magnitude of 1 micron, and its diffusion to the electrode-solution interface. In such diffusion there can be motion against the potential gradient.

3. At the electrode-solution interface consideration must be given to recombination. No quantitative allowance was made for this by Bockris and Uosaki (30). The probability of passage across the boundary was allowed for by means of a Gamow term [cf. Khan, Wright and Bockris (31)].

The physical considerations involve:

A. The energy gap.
B. The electron affinity.
C. The energy states of relevant ions in the solution.
D. The factor β which controls the potential gradient between the semiconductor and the solution.

What has to be done, therefore, is to take the equations of the theory of Bockris and Uosaki - or some improvement thereon - and find out the maximization of the various parameters which are in this equation, thus giving the optimum semiconductor for anode and cathode which then has to be sought and perhaps synthesized.

The corrosion properties have to be acceptable. This is a difficult requirement and the Achilles heel of the situation at this time.

Another matter which should be raised here is that the theoretical model of electrode processes leads to a schism within electro-chemistry. Thus the theoretical models as used by Gerischer (33) and related to those used by Levich and the Russian group, are different from that which is used by Bockris and Uosaki. In the Bockris and Uosaki model, electrons in the semiconductor tunnel to solvated ions in the double layer. In parallel work by Gerischer, the states which are relevant in solution are fluctuations on the ground state. The states which are relevant in the model of Bockris et al. (34) are the vibration-rotation ions of the solvated ions, excited from the ground state. Fluctuations around the ground state are discounted because of the results of a calculation due to Bockris and Sen (35) which shows that the availability of states of 0.5 ev away from the ground state is too rare (due to fluctuations) to be of interest to the orders of magnitude involved in electrode kinetic velocities.

One of the differences between the two models is that used by
Gerischer et al. leads to a relation (when applied to thermal
kinetics) between the current and potential which is grossly incon-
sistent with experiment, i.e., the current should be parabolic
with potential, whereas, in fact, there is a strict Tafel linearity
even for redox reactions (36).

These discrepancies in basic concept are worth resolving because
of their importance in considerations of the theory of photon-
electrode kinetics.

In 1977, L. Handley (37) suggested that the d levels in transition metal oxides could be utilized as states for activation in
photo anodes. Such states are near together and more likely to
give the needed energy level differences of about 1 ev for solar
energy absorption than with energy levels in s and p bands. A
similar suggestion was made by Tributsch (38). The hoped for advantage of this concept lies not only in the availability of the
levels but in the strong lattice energies which the transition metal
oxides have. Such strength should give lessening corrosion.

One of the more interesting substances which has been tried out
on the basis of this concept is that of "lanthanum chromite". This
is formed by heating La_2O_3 and Cr_2O_3, mixed in stoichiometric pro-
portions, upon a titanium base to 1250°C in A. XPS analysis of the
surface shows that the substance contains lanthanum chromate, lanthanum chromite, and titanium chromite. It is active as a photoanode and gives efficiencies which are higher than those of titanium
dioxide or strontium titanate.

A series of other perovskites, for example, lanthanum vanadate,
lanthanum titanate, lanthanum platinate, lanthanum aurate, and
others, has been examined by Guruswamy, Keillor and Bockris (40)
as photoanodes and the activity, although good, is not better than
that of the so-called lanthanum chromite.

Many interesting effects of these compounds can be obtained.
For example, it is possible to evolve chlorine (41) from sea water
with reasonable efficiency by means of photo-aided electrolysis
which reduces the net potential needed to produce the chlorine by
about 1/2.

Temperature effects are of value in the evolution of hydrogen
from water by photoelectrolysis. Most workers reduce the temperature effect by filtering out the IR radiation. However, this gives
rise to a loss of up to 1/4 of the energy in the solar spectrum
and it seems reasonable to allow this heat to be absorbed by the
solution and the electrode, giving rise to a reduction of overpotential and some transfer of the heat to the electrical situation. A
net efficiency of some 6% in the photo-aided electrolysis of water

(this efficiency includes allowance for the energy used by the
battery).

In summary, an efficiency of 10% is a reasonable aim in the photo-
aided electrolysis of water. The aim in efficiency without photo-
assistance should be 5%.

The advantage of the wet electrolysis is that only one plant
will be necessary to obtain stored solar power.

The use of n and p silicon as anode and cathode, respectively,
is an aim of photoelectrolysis. There could be a possibility of
doing this, utilizing an appropriate protecting arrangement for the
anode.

LIGHT AND WATER (BIOLOGICAL SYSTEMS)

The attraction of this approach (e.g., blue green algae irradi-
ated in solutions from which the nitrogen has been removed but give
hydrogen) is that the substances, being natural ones, could be produced cheaply and in abundance by breeding processes.

However, it may be that absorption of light is high but the conductivity is low so that it is not likely that high rates for hydrogen and oxygen evolution will be obtained. Hydrogen and oxygen
come off mixed together; and the lifetime of the materials is around
one to ten days.

Basic studies showing the electrochemical nature of these
photosynthetic organisms was carried out in 1977 and 1978 by Bockris and Tunulli (44). They showed that one could extract photosystem I and photosystem II from spinach. Absorbing these separately on platinized platinum gave rise to a light sensitivity on
the surface of the system. With photosystem II irradiation gave
an anodic current and evolved oxygen. With photosystem I, a cathodic current arose from irradiation and this was probably due to the
formation of hydrogen. Efficiencies were of the order of 1-3%.

The practicality of biomass systems depends upon raising the
efficiency by one order of magnitude. Research in this area has
been particularly little. Some photosynthetic reactions are reported to occur at 10% efficiency. There is little possibility of
utilizing a biomass approach for massive production of hydrogen
fuel unless the efficiency is more than 5%.

DIRECTIONS OF RESEARCH

Among the goals of research in the massive production of hydrogen are the following:

1. The seeking of chemical cycles which involve temperatures of not more than 800°C (to couple with the heat available from atomic reactors) but involving simple organic substances, together with water, to reduce corrosion.

2. Research on waste products, e.g., SO_2 and NO, to avoid the anodic reaction in electrical water splitting, without having to put in heat to drive some chemical reaction.

3. The theory of photoelectrolysis.

4. The development of new materials in photoelectrolysis.

5. The development of coatings for silicon in photoelectrolysis.

6. Temperature effects in photo-electrolysis.

7. The breeding of more helpful algae (e.g., those able to withstand the presence of oxygen).

PRACTICAL IMPLEMENTATION: THE EXTREMELY DEMANDING TIME SCALE

The U.S. Government program in the production of hydrogen, apart from the program supported by the JPL in chemical cycles, is a small one. Upon being approached in matters connected with electrolysis, DOE officials state that they are convinced that the General Electric membrane process (proton-only passage between electrodes) is the superior process: they will not research other methods!

The speed of practical implementation is of extreme importance. Root and Attanasi are the latest in a series of analyses of world oil supplies which show that the 1980's will see the maximization of the supply and a continuous descent will then occur. The time which it takes to build a new system is such that building on a massive scale of convertors for hydrogen splitting during this decade is essential if the energy supply per head in the U.S. is not to fall. The present national policy, which aims at producing synthetic gasoline, is irrational because of the much greater cost per unit of work done on this fuel than on hydrogen.

SUMMARY

A major method of decomposing water is to react it chemically with coal which gives the cheapest synthetic fuel. However, the lastingness of coal supply is shorter than is supposed, at the most 40 years with some estimates going down to 15 years. The setting up of large-scale plants to convert coal to hydrogen may therefore be impractical. Electrolysis of water using atomic energy is a possibility. The light-oriented methods offer the best hope of cheaper methods. Synthetic fuels in great amounts will be needed even in the present decade. Absence thereof will cause a downturn in living standard.

REFERENCES

1. D. Root and E. Attanasi, The American Association of Petroleum Geologists Bulletin, (1978).
2. .F. Niehaus, Report of the Nuclear Center at Juelich, JUEL pp. 11-65, (1975).
3. G.N. Plass, Tellus, $\underline{8}$, 140 (1956).
4. Chemical and Engineering News, June 25, p. 7 (1979).
5. J.O'M. Bockris, "Energy: The Solar-Hydrogen Alternative," Halsted Press, New York (1975).
6. S. Ihara, Conference Proceedings, First World Hydrogen Energy Conference (March 1-3, 1976, Miami), Vol. II, p. 5B-55 (1976).
7. J. Funk, Internat. J. Hydrogen., $\underline{6}$ (1980).
8. S.P. Chakravarty and K.S. Varde, Conference Proceedings, First World Hydrogen Energy Conference (March 1-3, 1976, Miami), Vol. III, pp. 4C-5, (1976).
9. S. Srinivasan, G. Kissel, P.W.T. Lu, F. Kulesa, and C. Davidson, Brookhaven National Laboratory, Contribution No. 22163 (1977).
10. A.J. Appleby and G. Crépy, 151st Electrochemical Society Meeting (Philadelphia, May 9-12, 1977), Symposium Proceedings on Electrode Materials and Processes for Energy Conversion and Storage, p. 382, (1977).
11. D.B. Sepa, A. Damjanovic and J.O'M. Bockris, Electrochim. Acta, $\underline{12}$, 746 (1967).
12. W. Juda and D. McL. Moulton, Chem. Eng. Symp. Series, p. 59, (1972).
13. J.O'M. Bockris and A.J. Appleby, Int. J. for Hydrogen Energy, $\underline{6(1)}$, (1980).
14. J.O'M. Bockris, J. Bevan and S. Badwal, Electrochim. Acta, $\underline{25}$, 1115 (1980).
15. A. Fujishima and K. Honda, Nature, $\underline{238}$, 38, (1972).
16. A. Fujishima, K. Kohaya Kawa, and K. Honda, J. Electrochem. Soc., $\underline{112}$, 487 (1975).
17. M. Calvin, Faraday Discussion, $\underline{70}$, 70/22 (1980).
18. M. Graetzel, Faraday Discussion, $\underline{70}$, 70/20 (1980).
19. H. Gerischer, Faraday Discussion, $\underline{70}$, 70/6 (1980).
20. V. Guruswamy and J.O'M. Bockris, Solar Energy Materials, $\underline{1}$, 141 (1979).
21. K. Ohashi, J. McCann and J.O'M. Bockris, Nature, $\underline{266}$, 610 (1977).
22. M.A. Butler and D.S. Ginley, J. Electrochem. Soc., $\underline{125}$, 228 (1978).
23. R. Wilson, J. Appl. Phys., $\underline{48}$, 4297 (1977).
24. H. Reiss, J. Electrochem. Soc., $\underline{125}$, 937 (1978).
25. J. Reichmann, Appl. Phys. Lett., $\underline{36}$, 1, (1980).
26. H. Gerischer, "Physical Chemistry: An Advanced Treatise," (H. Eyring, D. Henderson and W. Jost, eds.), Vol. 9A, pp. 463-542, Academic Press, New York, (1970).
27. M. Green, J. Chem. Phys., $\underline{31}$, 200 (1959).
28. A.J. Bard, A.B. Bocarsly, F.F. Fan, E.G. Walton, and M.S. Wrighton, J. Am. Chem. Socl, $\underline{102}$, 3671 (1980).

29. J.O'M. Bockris, "Modern Aspects of Electrochemistry," (J.O'M. Bockris, ed.) Chapt. IV, Butterworths, London, (1954).

30. J.O'M. Bockris and K. Uosaki, J. Electrochem. Soc., $\underline{124}$, 98 (1977).

31. S.U.M. Khan, P. Wright, and J.O'M. Bockris, Electrokhymia, $\underline{13}$, 914 (1977). Republished in English as Electrochimia, $\underline{13}$, 914 (1977).

32. V. Levich, "Physical Chemistry: An Advanced Treatise," (H. Eyring, D. Henderson, and W. Jost, eds.), Vol. 9B, Chapt. 12, Academic Press, New York (1970).

33. H. Gerischer, "Special Topics in Electrochemistry," (P.A. Rock, ed.), p. 35, Elsevier Press, New York (1977).

34. J.O'M. Bockris and S.U.M. Khan, "Quantum Electrochemistry," Plenum Press, New York (1979).

35. J.O'M. Bockris and R.K. Sen, Mol. Phys., $\underline{29}$, 357, (1975).

36. F.C. Anson, N. Rathjen, and R.D. Frisbee, J. Electrochem. Soc., $\underline{117}$, 477 (1970).

37. L. Handley, Report on a Study in Photoelectrochemistry, Flinders University, School of Physical Science (1978).

38. H. Tributsch, Ber. Bunsen. Phys. Chem., $\underline{81}$, 361 (1977).

39. V. Young and V. Guruswamy (in course of publication).

40. V. Guruswamy, H. Keillor and J.O'M. Bockris, Solar Energy Materials, in press (1980).

41. V. Guruswamy and J.O'M. Bockris (in course of publication).

42. V. Guruswamy and J.O'M. Bockris (in course of publication).

43. G. Neil, D.J.D. Nicholas, J.O'M. Bockris, and J.F. McCann, Heliotechnique and Development, $\underline{1}$, 481 (1976).

44. J. O'M. Bockris and M.S. Tunulli, "Bioelectrochemistry" (H. Keyzer and F. Gutmann, eds.), Plenum Press, New York (1980).

45. Report of the Australian Academy of Sciences on Solar Energy Research in Australia, No. 17, p. 22 (1973).

PHOTOELECTROCHEMICAL CELLS

Adam Heller

Bell Laboratories
600 Mountain Avenue
Murray Hill, New Jersey 07974

ABSTRACT

Photoelectrochemical solar cells based on semi-conductor liquid junctions have been developed over in the past six years into efficient converters of sunlight into electrical energy or hydrogen. Solar conversion efficiency has increased from less than 1% to 12% and operational life from hours to months. The new cells retain about 70% of their single efficiency when made with chemically formed, polycrystalline materials, far more than other types of solar cells.

Gains in performance paralleled the understanding of the chemistry of surfaces and grain boundaries in semiconductors. Through such understanding it became possible to chemically manipulate the position of surface and grain boundary states and thus to reduce losses due to electron-hole recombination. Chemisorption of Ru^{3+} ions on n-GaAs photoanodes increased the efficiency of the single crystal n-GaAs$|K_2Se-K_2Se_2-KOH|$C cell to 12%, and diffusion of Ru^{3+} and Pb^{2+} into grain boundaries of polycrystalline n-GaAs films on graphite quadrupled the efficiency to 7.8%. Surface oxidation of p-InP led to the first efficient photocathode, which, in contrast to photoanodes, is not prone to oxidative photocorrosion at high light intensities. Light tends to cathodically protect the semiconductor surface. With the p-InP photocathode, 11.5% solar-to-electrical conversion efficiency has been achieved in the p-InP|VCl$_3$-VCl$_2$-HCl|C cell. Chemisorption of silver on grain boundaries of p-InP films on graphite increases the efficiency of the polycrystalline photocathode six-fold to 7%.

By photoassisted electrolysis of water 12% solar conversion efficiency has been reached in the p-InP (Rh)|4MHClO$_4$|IrO$_2$-Ir(TaO$_3$)$_4$-Ti cell, and a similar efficiency has been reached in the cell p-InP(Ru)|2M-HCl-2M KCl|Pt(RuO$_2$), in which hydrogen and chlorine are produced. These cells represent the most efficient photoelectrochemical systems for conversion of sunlight into energy stored in fuels or in useful chemicals.

INTRODUCTION

In 1839, Becquerel published the first report on the conversion of light into electrical energy at a junction between a solid and an electrolyte.[1] An all solid, selenium based photovoltaic cell was made 38 years later, in 1877, by Adams and Day.[2] There has been continuous activity in the field ever since, with studies on Cu$_2$O, Tl$_2$S and Ag$_2$S and Ag$_2$S dominating the work in the early part of this century. The era of efficient conversion of sunlight into electrical energy began, however, only later and coincided with the evolution of modern semiconductor science. Solar cells based on junctions in semiconductors became, and remain to this date, the most efficient converters of sunlight. Their efficiency exceeds that of photosynthesis in plants and of other photochemical systems, including recent systems based on organized structures in solution.

The silicon p-n junction cell was invented by R. S. Ohl, whose patent was filed on May 25, 1941.[3] In 1954, the efficiency of this cell was increased to 6% by Chapin, Fuller and Pearson.[4] A Cu$_2$S/CdS p-n junction cell, of comparable efficiency, was also reported in 1954 by Reynolds et al.[5] The pace of study of photovoltaics accelerated after 1954. In 1958, silicon cells were used to power the transmitter of Vanguard I, the first U.S. orbiting satellite. The satellite's radio operated for eight years before it failed because of radiation damage. Silicon cells powered all communication satellites since the launching of the first in 1962, when their efficiency already exceeded 10%.

While efficient solar cells were being developed,
W. H. Brattain, co-inventor or the transistor, laid
down in his classical paper with G. C. B. Garrett
the foundations of the physics and the electrochemistry
of illuminated semiconductor electrolyte junction.[6]
Their paper was followed by the introduction of many
concepts of modern photoelectrochemistry, such as the
flat-band potential,[7] the TiO_2 anode,[8] surface states
at semiconductor electrolyte interfaces[9] and photocorrosion of these interfaces.[10]   These concepts, like
those of Ohl, Garrett, Chapin, Fuller and Pearson
originated at Bell Laboratories in the 1941-1962 period
and are associated with the names of Boddy, Brattain,
Dewald and Turner.   Regenerative photovoltaic cells,
based on junctions between semiconductors and redox
couple solutions, were first described in 1960 by
R. Williams of RCA Laboratories.[11]

It is significant that the generation of
scientists who founded today's solid state science and
technology, though profoundly familiar with semiconductor electrolyte junctions, decided to introduce
only technologies based on semiconductor-semiconductor
(p-n) junctions and on metal-semiconductor (Schottky)
junctions.   One reason for this was the high capacitance
associated with the double layer at the interface, which
limited the frequency of potential devices.   The devices
were, however, not considered even for use in rectifiers
and other low frequency applications, because of poor
maintenance of their electrical performance.   Adsorption of impurities at the interface, passivation,
and corrosion changed the electrical behavior, making
these, in sharp contrast with other solid state devices,
irreproducible and prone to failure.

While substantial progress was being made in the
understanding of the science of semiconductor electrolyte interfaces in the 1960's, the interest in using
these in solar conversion grew following the publication of a 1972 paper by Fujishima and Honda,[12] who
used the TiO_2 anode[8] for photoassisted electrolysis of
water and of a 1975 paper by Gerischer on solar cells
based on junctions between semiconductors and redox
couple solutions.[13]

My interest in the field originated in discussions
with Gerischer in 1974. Gerischer saw that semiconductor liquid junction cells would be simpler to make than
p-n junction cells because their junction is

spontaneously formed when a semiconductor is immersed
in a redox couple solution. I recognized that the cells
might allow the use of simple, chemically formed,
films of polycrystalline semiconductors instead of
single crystals, without the severe penalty incurred in
p-n junction and other cells. It seemed to me that
liquid junction solar cells might bring the use of
chemically formed, polycrystalline thin films a step
closer to reality.

    Two factors have prevented simple, chemically
formed films of semiconductors from forming efficient
p-n junction solar cells. First, the cells were
shunted, by preferential transport of dopants in grain
boundaries, in the diffusion step forming the junction.
Second, the high recombination velocity of electrons
and holes in grain boundaries of covalent semiconductors introduced prohibitive losses in efficiency.
While the recombination velocities are acceptable in
ionic semiconductors such as TiO_2, $SrTiO_3$ and ZnO,
all of these have excessively large band gaps and are
useless for solar conversion. The bonds in potentially
useful semiconductors with appropriate (1.1-1.6eV)
band gaps such as Si, GaAs and InP are most covalent.
The breaking of these bonds introduces states closer
to the center of the band gap and these states are
effective electron hole recombination centers. We
shall discuss the problem of recombination of
electrons and holes in grain boundaries of covalent
semiconductors in a subsequent part of this paper.

    The shunting problem is described schematically
in Fig. 1. To produce a p-n junction solar cell[4]
the surface of a p or n type material is doped with
atoms that invert the type (p to n or n to p). These
atoms are then thermally diffused to the optimal junction depth. Accurate positioning of the junction is
essential for cell performance. Since the rate of
diffusion in grain boundaries greatly exceeds the
rate of diffusion in the bulk of a crystal, it has
not been possible to properly position the boundary
between the p and n regions in polycrystalline semiconductors. Rapid diffusion of the dopant in grain
boundaries shunts the solar cells, i.e., the dopant
traverses the entire depth of the semiconductor
reaching, at some points, the wrong electrical contact.
Because the semiconductor liquid junction cell avoids
the need for doping and diffusion, it seemed that it
might open the way to the use of simple polycrystalline

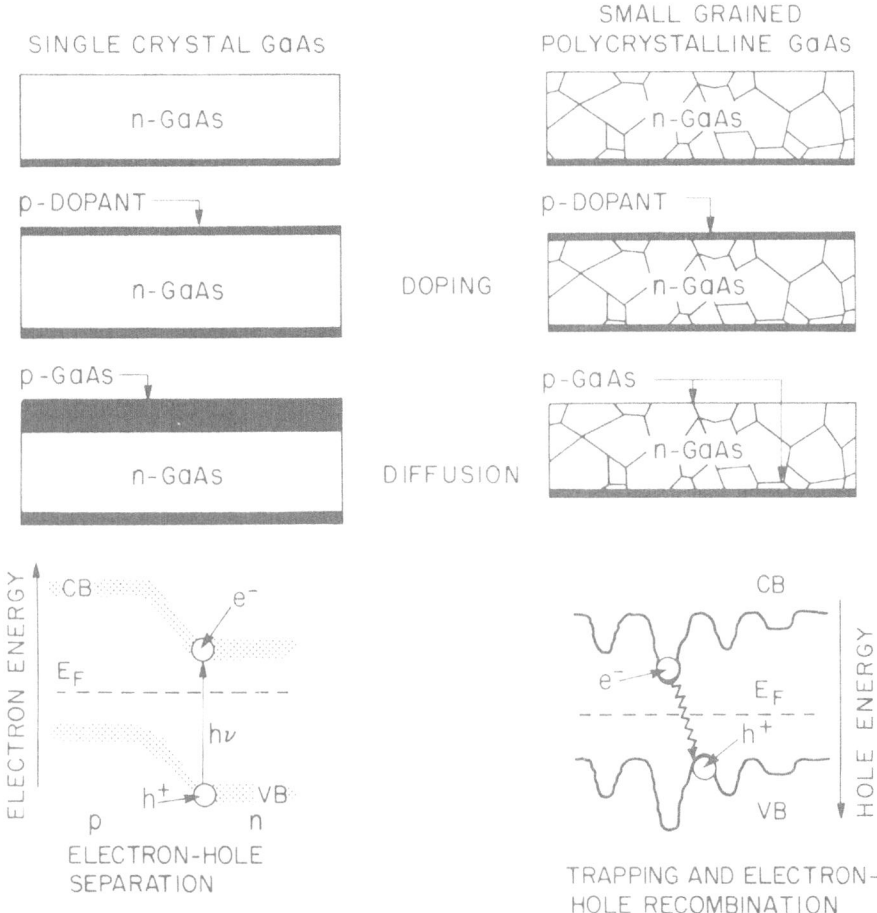

SINGLE CRYSTAL GaAs

SMALL GRAINED
POLYCRYSTALLINE GaAs

Figure 1.   p – n junction based solar cells, such as the GaAs cell
shown, are made by a process involving incorporation
of an appropriate impurity in the surface ("doping"),
followed by high temperature diffusion of the dopant.
In a single crystal the process produces a shallow
p – region, having a sharp boundary. An efficient
junction results: the photogenerated electrons and
holes are separated by the electric field and their
recombination is prevented. In small grained polycrystalline materials, produced by simple chemical
reactions, the dopant diffuses more rapidly through
grain boundaries than through the crystallites. This
leads to randomly oriented p – n junctions, which penetrate the films and shunt the cells. With random
junctions it is not possible to efficiently collect
electrons and holes; the carriers are trapped and
recombine non-radiatively.

semiconductors if a solution to the other major problem, that of electron hole recombination at grain
boundaries of covalent semiconductors, could be found.

One should note, that the shunting problem can
also be avoided in metal-semiconductor, metal-
insulator-semiconductor and heterojunction cells. In.
these, the junctions are formed by evaporation or
sputtering of a metal or of a semiconductor, by
electroplating of a metal, or by chemical reaction of
a semiconductor. Nevertheless, few of these cells
reach high efficiencies when made with chemically formed,
thin films of polycrystalline semiconductors. Metal-
semiconductor and metal insulator-semiconductor cells
require uniform 50-80Å thick metal films to simultaneously avoid excessive absorption of light by the
metal, and yet to provide sufficient surface conductivity. Such films are difficult to form on polycrystalline
semiconductors having complex surface topographies.
Heterojunction cells require properly matched electron
affinities of the two semiconductors forming the
junction, similar thermal expansion coefficients and
continuity of the lattice between the various crystal
faces of the two semiconductors. These conditions are
met only in a very small number of systems.

In 1974, the liquid junction cells appeared to
offer the advantages of metal-semiconductor junction
cells, without their disadvantages. It could be expected that the ionically conductive redox couple solutions, though many are colored, would absorb less light
than the metal layers and that the ionically conductive solutions would always conform to the surface
topography of the polycrystalline semiconductor films
to form high quality junctions.

To start the program on semiconductor liquid
junction solar cells, I joined Bell Laboratories in
August 1975. When we started our work, only large
band gap ($\doteq$ 3eV) ionic semiconductor electrodes such
as n-TiO$_2$ anodes[8] and photoanodes[12] were known to have
reasonable stability in photoelectrochemical cells.
These were of little use for solar conversion, since
they absorbed only in the ultraviolet, where there is
little solar radiation reaching the surface of our
planet. Efficient solar conversion requires materials
with bandgaps of (1.3 ± 0.3) eV, i.e. absorption of
light to $(10,000\pm2500)$Å. As the bandgap drops below
1.0eV, the photocurrent increases, but the voltage

declines and with it the power output and thus the
conversion efficiency. If the bandgap exceeds 1.6eV,
the photocurrent, the power output and the efficiency
decline, though the photovoltage may be high. In
general, bandgap and stability are correlated: The
lesser the bandgap the lesser the chemical stability and
the then existing, stable cells were necessarily
inefficient.[14,15]

Thus, two goals shaped our program on semiconduc-
tor-liquid junction solar cells. We wanted to confirm
the premise that these cells, in contrast with others,
remain efficient when made with small-grained polycrystalline semiconductors of appropriate band gap,
and we had to learn the nature of the chemical changes
at the semiconductor-electrolyte interface in order
to control them.

Principles and Definitions

When a semiconducting photoelectrode and a
counterelectrode are immersed in a redox couple solution and shorted, the conduction band (CB) and the
valence band (VB) are bent by up to E_f-E_{redox}, E_f
being the Fermi level of the semiconductor and E_{redox}
the potential of the redox couple in solution. This
difference, the barrier height, represents the upper
limit of the open circuit voltage, V_{oc} that can be
achieved under high irradiance. V_{oc} cannot exceed
$|E_f-VB|$ for photoanodes and $|E_f-CB|$ for photocathodes.[13]
Solar cells, under a fixed irradiance, operate most
efficiently with an external resistance for which the
output power reaches a maximum. The maximum power
output is a function of the open circuit voltage,
V_{oc}; of the short circuit photocurrent, I_{sc}, which is
proportional to the quantum yield of oxidation or
reduction at the photoelectrode; and of the fill factor,
ff, defined as $I_{max}V_{max}/I_{sc}V_{oc}$, where I_{max} and V_{max}
are the current and the voltage at the maximum power
point. The solar conversion efficiency is, in an
electrical power producing solar cell, the ratio of
this maximum power and the solar irradiance, integrated over the entire solar spectrum. In hydrogen
producing cells the efficiency is the ratio of the
electrical power saved in electrolyzing water and the
solar irradiance.

The process by which semiconductor liquid junction
solar cells convert sunlight into electrical power

resembles that of the p-n junction silicon cell,[3,4]
except that the circuit involves redox reactions at the
photoelectrode and at the counterelectrode. These are
equal but opposite in direction, and thus do not
introduce a net chemical change.

Cells with either photoanodes or photocathodes
can also be operated, in the absence of a redox couple,
to oxidize or to reduce a substrate instead of producing power. Thus protons can be reduced at photocathodes to produce hydrogen, and hydroxide anions can
be oxidized at photoanodes to produce oxygen. When
an external potential is required to supplement the
photopotential to cause electrolysis, the process is
termed photoassisted electrolysis.

Early Results: Overcoming Photocorrosion and Forming Thin Film, Polycrystalline Cells by Anodization

In his first paper on regenerative systems,
Gerischer studied the n-CdS$|$K$_4$Fe(CN)$_6$-K$_3$Fe(CN)$_6$-KOH$|$Pt
solar cell. He observed photocorrosion by holes
oxidizing the semiconductor to Cd^{2+} ions and elemental
sulfur, an insulator which covered the surface of the
photoanode and stopped it from functioning.[13] It was
obvious to me, as it was to others, that this layer
can be dissolved in a sulfide solution and that a
sulfide solution would suppress the dissolution of
Cd^{2+}. More importantly, however, B. Miller argued
that the hole transport kinetics to surface adsorbed
sulfide ions would be favored over that to crystal
bound species. The $\sim$1V difference between the
Fe(CN)$_6^{4-}$/Fe(CN)$_6^{3-}$ and the S^{2-}/S$_2^{2-}$ redox potentials
would buy kinetic stability, i.e. all the holes would
be captured by the oxidizable ion well before these
could oxidize the semiconductor.

To prove that one can achieve stability to
photocorrosion and also to establish our premise that
the new cells can be made with very small grained,
polycrystalline semiconductors, we chose to form our
n-CdS$|$Na$_2$S-Na$_2$S$_2$-NaOH$|$C cell simply by anodizing a
sheet of cadmium metal in the redox couple solution.
We observed, in our first experiment, the expected
improvement in stability as well as substantial
photovoltages and photocurrents.[16] Because CdS has a
bandgap of 2.4eV, larger than the 1.0-1.6eV optimum
for solar conversion, we repeated the experiment with
a bismuth metal electrode, forming 1.3eV bandgap

n-Bi_2S_3.[16,17] Independently, two other groups also
stabilized cadmium sulfide and selenide photoanodes
against photocorrosion by using the S^{2-}/S_2^{2-} couple.[18,19]
None of us was, however, the first to conceive such
stabilization. It was G. C. Barker, an esteemed
contributor to modern electroanalytical chemistry, who
first suggested the idea of stabilizing the n-CdS
photoanode by a sulfide solution.[20]

Bands, Bonds, Recombination and Pinning at Photo-electrodes: 7.5% Efficient Single Crystal and 5% Ceramic Cells

The results on cells made with semiconductors
formed by anodizing metals convinced us that kinetic
solutions to photocorrosion problems can be found.
The question that we faced in the spring 1976 was
whether the stabilized cells could be efficient. We
knew that there is no point in testing photoanodes for
efficiency until we addressed the issue of surface and
near surface states, in which the photogenerated
electrons and holes are trapped and recombine. In
extreme cases, such recombination results in losses of
photocurrent or quantum efficiency. Otherwise losses
in photovoltage and fill factor, i.e., power at the
maximum power point, result.

To explain the role of surface states and recombination losses, it is useful to start with concepts
of bonding in a lattice. As shown schematically in
Figure 2 (a,b,c), a valence band forms in a semiconductor when molecular bonding orbitals merge upon the
creation of a periodic lattice of identical bonds. A
conduction band forms when antibonding orbitals merge.
Since the distance between the highest bonding and
lowest antibonding orbital measures the energy required
to undo the binding of the most weakly bound electron,
there is a qualitative correlation between the strength
of the weakest bond and the band gap (unless the
transitions involve non-bonding electrons). At a
semiconductor surface, at a lattice dislocation, at
a single crystal defect, as well as at grain
boundaries of a polycrystalline semiconductor, bonding
is weaker and states between the valence and conduction
bands are introduced. These states trap and recombine
photogenerated electrons and holes (Fig. 2f,g,h). Thus,
for example, cleavage of diamond in vacuum results in
the formation of higher energy carbon atoms at the
surface. These are less strongly bound than those in

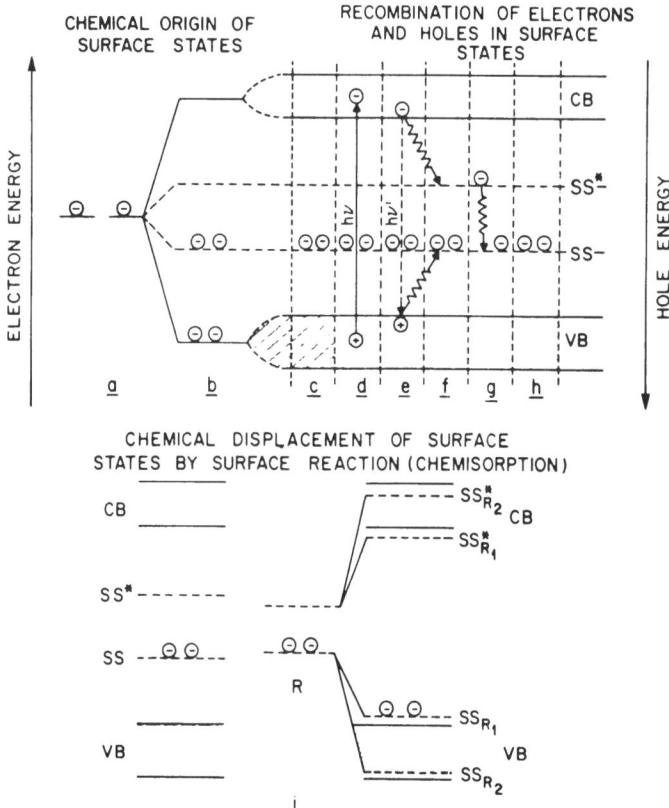

Figure 2. Bands in covalently bound semiconductors are formed when
           atoms (a) combine to form bonding and antibonding orbi-
           tals, (b), which coalesce upon evolution of a lattice to
form, respectively, valence (VB) and conduction (CB) bands, (c).
Weaker bonding at a surface or at a grain boundary leads to less
splitting between the bonding and antibonding orbitals (b,c) and
introduces surface states (SS,SS*) between VB and CB. Electrons
and holes produced by absorption of light, (d) may radiatively recombine (e), or recombine in a non-radiative process involving SS*
and SS (f,g,h). Non-radiative recombination leads to losses in current efficiency near the maximum power point (low "fill factor")
and in open circuit voltage ("pinning"). We discovered that these
losses can be reduced by chemical reaction of the surface or grain
boundary with a strongly bound reagent R (g). The reaction eliminates the weaker bonds and thus moves the surface and grain boundary states toward the bands, away from the center of the band gap.
"Deep" states, near the center of the band gap are most efficient
and "shallow" states near VB or CB (SS_{R1}, SS^*_{R2}) are least efficient
electron-hole recombination centers. When moved into CB or VB, the
surface states (SS_{R2}, SS^*_{R2}) no longer trap electrons or holes and
do not cause recombination.

the bulk and may trap electrons to form carbanions
trap holes to form carbonium ions. The two are
annihilated by surface recombination.

In a semiconductor crystal, a few percent of the
photogenerated electrons or holes decay by radiative
recombination, leading to luminescence at wavelengths
corresponding to the band gap, $h\nu'$, (Fig. 2e) but the
vast majority recombine non-radiatively at surfaces,
grain boundaries, defects or other sites associated
with weaker chemical bonding (Fig. 2 (f,g,h).
Recombination is evidenced by a decrease in the intensity and lifetime of the bandgap luminescence, $h\nu'$,
(Fig. 2e). In a semiconducting photoelectrode recombination is evidenced by a loss in quantum efficiency
for oxidation or reduction. The loss is associated
with an increase in the recombination current flowing
in a direction opposite to the photocurrent. Since
the open circuit voltage is the potential at which the
rate of photogeneration of electrons or holes equals the
rate of their recombination, i.e. when the opposing currents balance, recombination results in lower open circuit voltage. The electron-hole recombination rate, S_R,
is

$$S_R = \sum_i \sum_j \sigma_{ij} N_e^i N_h^j$$

where N_e^i is the electron population in the ith state,
N_h^j is the hole population in the jth state and σ_{ij} is
the cross section for their recombination. The
recombination current is an exponential function of
the electrode potential, i.e., Fermi level of the
semiconductor, since the populations N_e^i and N_h^j depend
(by Fermi statistics) on the energy difference between
the fermi level and the trapping state. The potential
at which the recombination current reaches the solar
photogeneration current defines a limit to the achievable open circuit voltage. When this limit is reached,
the open circuit voltage no longer depends on the redox
potential of the solution and is said to be "pinned"
by the (surface, lattice defect or grain boundary)
state causing the recombination. When the recombination
current is substantial near the maximum power point,
a loss in fill factor (vide supra) is observed. In
some cases recombination persists even at the solution
potential, i.e. when the electrodes are shorted, and
a loss in short circuit current is seen.

The first probe introduced to observe and eliminate
surface and near surface defects states was two-beam
photocurrent (and photovoltage) spectroscopy.[21,22]

With an intense laser beam populating the traps and a
second variable wavelength beam measuring the incre-
mental photocurrent (or photovoltage), we could observe
trapping and recombination and also identify some of
their causes. This allowed us to develop chemical
etching methods which eliminated near surface defects
and opened the way to the first efficient semiconductor
liquid junction solar cells.[21,22]

The initial phase of our work centered on the
n-CdSe|Na$_2$S-Na$_2$S$_2$-NaOH|C cell, the subject of earlier
studies at MIT and at the Weizmann Institute.[18,19]
By removing trapping states, we achieved 7.5% solar-
to-electrical conversion efficiency.[22] To prove that
there is little loss in efficiency when polycrystalline
materials are used, we prepared ceramic n-CdSe by hot
pressing CdSe powder, made by precipitation from
solution. In these first experiments, recombination
at grain boundaries was reduced by cadmium vapor
annealing of the material. Excess cadmium makes CdSe
n$^+$ type. Thus, when the metal diffuses along grain
boundaries, it dopes the edges of the grains heavily,
creating n$^+$ zones. The n$^+$/n junctions, formed along
the grain boundaries, repel the photogenerated holes,
forcing these to the solution interface and thus
reduce trapping and recombination (Fig. 3).

By late 1976, we reached 5.3% solar-to-electrical
conversion in a cell made with the resulting ceramic
material.[23] This represented an unprecedented reten-
tion of 75% of the single crystal efficiency of a
solar cell made with small grained (10-20μm)
crystallites. The experiment established that it is
possible to approach single crystal efficiencies with
polycrystalline materials and confirmed our premise
of the feasibility of using these simple materials
in semiconductor liquid junction solar cells.

Since 1976, polycrystalline cadmium chalcogenide
based cells have been substantially improved,
particularly by researchers at the Weizmann Institute
in Israel. By replacing the n-CdSe photoanodes with
n-Cd(Te,Se), Hodes et al. recently reached a 7.5-8%
solar conversion efficiency.[24]

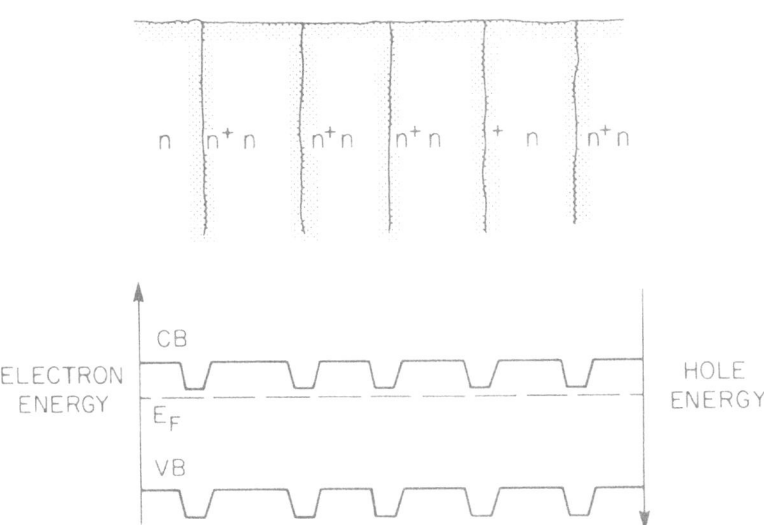

Figure 3. Preferential diffusion of dopants can be used to
           reduce carrier recombination at grain boundaries
           if n^+/n or p^+/p junctions are created at the edges
           of the grains (top). The first 5.3% efficient
           cells with polycrystalline n-CdSe photoanodes were
           made by exposing hot pressed n-CdSe powder to cadmium
           metal vapor. Excess cadmium creates heavily n-doped
           (n^+) zones along the boundaries. The electric
           field associated with the n^+/n junction repels
           holes from the boundaries (bottom). After removal
           of the n^+ region at the surface exposed to the
           redox couple (not shown), the photogenerated holes
           are channeled to the solution interface.

Problems of Ion Exchange: Stabilizing the n-CdSe|Na$_2$S-Na$_2$S$_2$-NaOH|C Cell

Having confirmed that reasonably efficient polycrystalline CdSe based cells could be made, we concentrated our effort on our second objective, that of identifying causes for instability and degradation.

Rapid ion exchange in the n-CdSe|Na$_2$S-Na$_2$S$_2$-NaOH|C cell creates a n-Cd$_x$Se$_{1-x}$ surface layer, [25,26,27,28] which introduces a barrier for the transport of holes to the interface.[25] The layer is created in a process similar to that of photocorrosion, except that the Cd^{2+} ions reprecipitate on the surface as CdS:

$$CdSe + 2h^+ \rightarrow Cd^{2+} + Se^\circ$$

$$Se^\circ + S^{2-} \rightarrow SeS^{2-} \qquad \text{photocorrosion}$$

$$Cd^{2+} + S^{2-} \rightarrow CdS \qquad \text{reprecipitation}$$

Photocorrosion takes place only when the kinetics of hole transport to the S^{2-} ions in solution are inadequate to cope with all the photogenerated holes. Thus, ion exchange damages performance only at or above rates of photogeneration of holes corresponding to normal solar irradiance. We solved the ion exchange problem by dissolving small amounts of elemental selenium in the sulfide solution. In the presence of the SeS$_2^{2-}$ ion the CdSe$_x$S$_{1-x}$ surface retains a sufficient selenide concentration to avoid the barrier problem. The stabilization experiments (Fig. 4) led to the first efficient, i.e. 7% liquid junction cell, that operated for months without change in its electrical characteristics.[25]

The n-GaAs|K$_2$Se-K$_2$Se$_2$-KOH|C

GaAs is one of the few available, well characterized semiconductors. Its absorption rises to 10^{5}cm^{-1} near the 1.3 ev band gap, which is optimal for solar conversion. We undertook the study of n-GaAs based semiconductor liquid junction solar cells in the spring of 1976 searching, as we did on our initial work on CdS, for a redox couple that would be fast and adequately reducing to assure that the rate of oxidation of the couple will greatly exceed the rate of

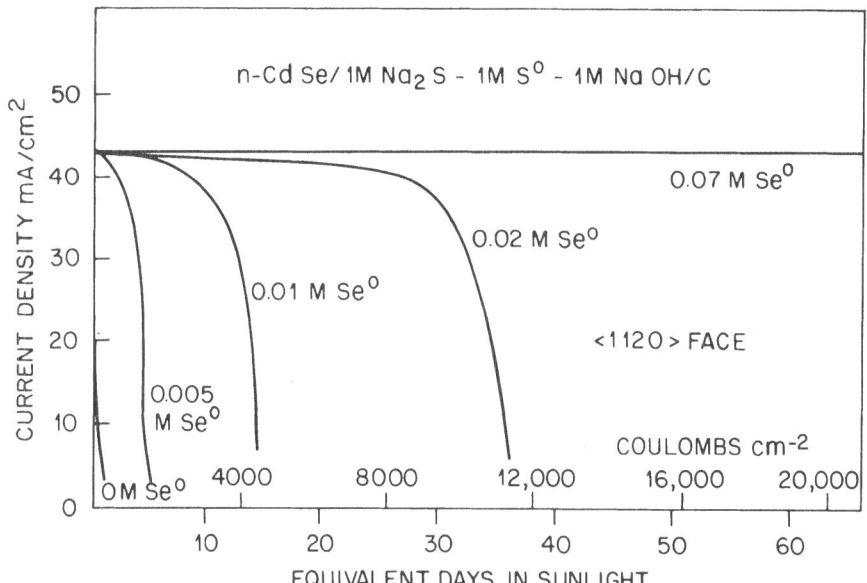

Figure 4. The first 7% efficient, stable semiconductor liquid
           junction solar cell was made by dissolving a small
           amount of selenium in the electrolyte of the 1M-
           Na2S-1MS-1M NaOH cell. In the absence of dissolved
           selenium, light assisted ion exchange between the
           solution and the surface of the semiconductor produces
           a CdS film, which blocks the transport of holes to
           the liquid interface.

photocorrosion. We found,[29] that while n-GaAs anodes photocorrode in basic $S^{2-}|S_2^2$ solutions, they do not photocorrode in $Se^{2-}|Se_2^{2-}$ at high concentrations of selenide. The selenide captures holes by the desired reaction

$$2Se^{2-} + 2h^+ \rightarrow Se_2^{2-}.$$

Wrighton's group carried out concurrent studies on n-GaAs photoanodes. Although they also observed oxidation of Se^{2-}, they did not realize[30] that the problem of photocorrosion has a kinetic solution and can be prevented by using a high enough selenide ion concentration to discharge the holes and thus prevent photocorrosion. As seen in Fig. 5, the ratio of the corrosion current to the total current declines with the inverse of the third power of $[Se^{2-}]$ Fig. 5.[29] At 2M K_2Se the rate of photocorrosion under 100 mw/cm^2 sunlight is estimated to be a few microns per year.

Chemical Modification of Surface and Grain Boundary States in Photoanodes: 12% Efficient Single Crystal and 7.8% Efficient n-GaAs Thin Film Cells

     Early in 1977, while experimenting with n-GaAs$|K_2Se-K_2Se_2-KOH|C$ cells, we observed variations in fill factors, and corresponding variations in efficiency between 7 and 10%. From the work of Boddy and Brattain in the 1960's it is evident that impurities, in submonolayer quantities, introduce surface states and change the electrical properties of semiconductor-electrolyte junctions.[31,32] Thus, the observed variations in efficiency were not unexpected and in 1977 we undertook a study of the effect of adsorbed ions on the performance on the n-GaAs$|K_2Se-K_2Se_2-KOH|C$ cell. We found that in a manner consistent with the model presented in the section on Bonds, Bands and Recombination (vide infra) cations could be divided into three groups: some, like Ca^{2+}, Sr^{2+} and Ba^{2+} were not adsorbed and had little effect on cell characteristics. Weakly chemisorbed ions, such as Bi^{3+}, caused deterioration in cell performance and very strongly chemisorbed ions, like Ru^{3+} and Pb^{2+}, improved the performance.[33,34] By chemisorbing Ru^{3+} and etching a textured structure into the surface of the n-GaAs crystals (to reduce light reflection losses), we increased the solar-to-electrical conversion efficiency of the n-GaAs cell to 12% (Fig. 6).[33-35]

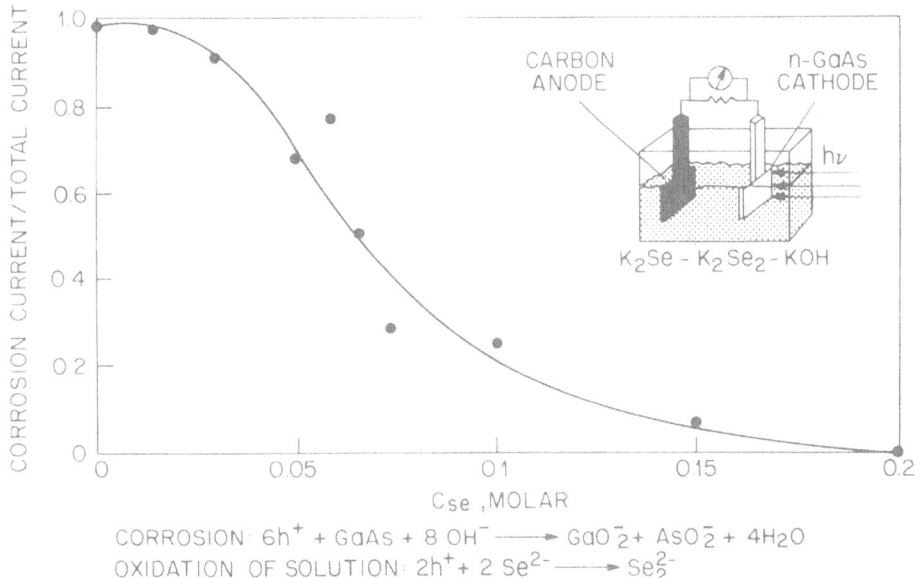

Figure 5.   Corrosion of photoanodes, caused by reaction of photo-
            generated holes with the surface of a semiconductor,
            competes with the transport of holes to the redox
            couple.   If the rate of transport of holes to the
            redox couple is much faster than the rate of photo-
            corrosion, i.e., the holes are captured by the redox
            couple before they can react with the semiconductor,
            the photoanode becomes kinetically stable.   In the
            n-GaAs|K$_2$Se–K$_2$Se$_2$-KOH|C cell the hole transport
            current to the solution increases with the concentration
            of selenide ions and the competing corrosion current
            declines.   The figure shows the ratio of the corrosion
            current, i_c, to the total photocurrent, i_t, as a function
            of the selenide ion concentration in solution.

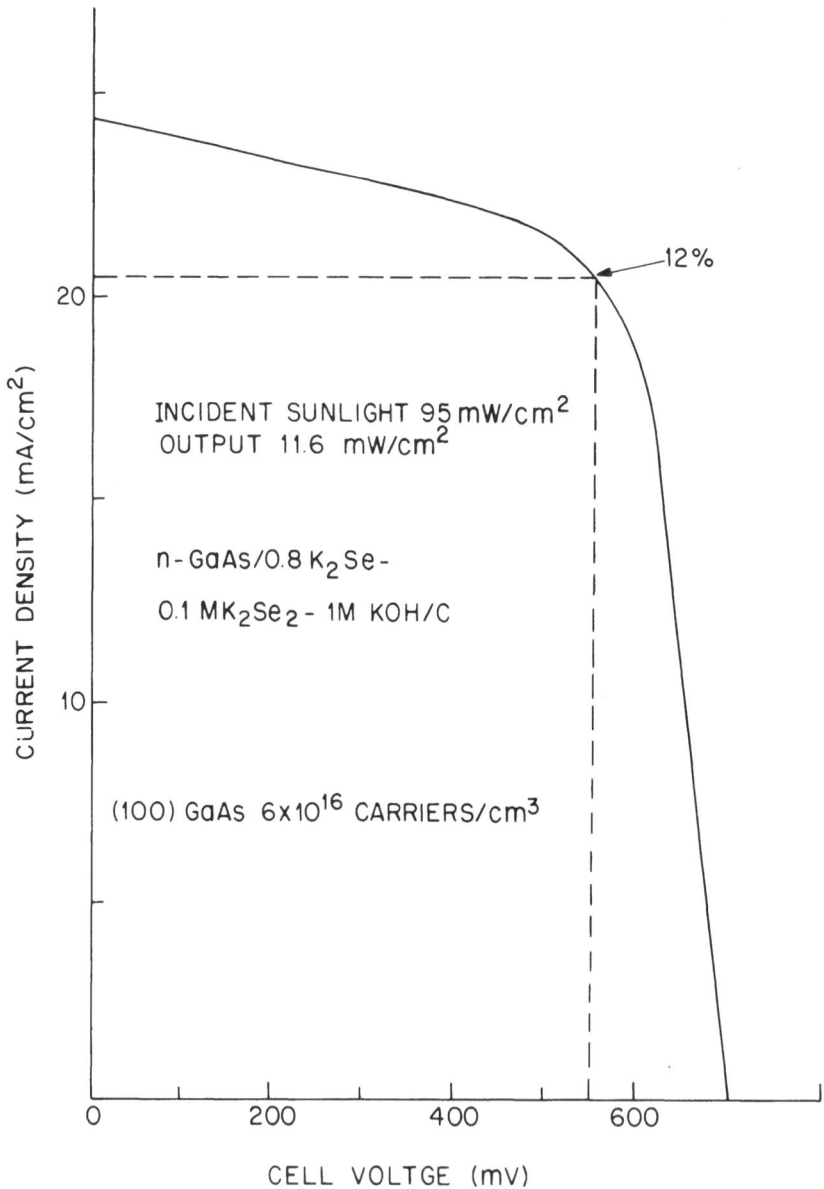

Figure 6.   12% solar-to-electrical conversion efficiency is
reached in the single crystal n-GaAs|0.8M K_2Se-
0.1M K_2Se_2-1M KOH|C cell upon reducing the electron-
hole recombination velocity at the solution interface. The figure shows the current-voltage characteristics of the cell.

The effect of strongly chemisorbed ions such as Ru^{3+}, was explained by our model[36-38] according to which chemisorption shifts the position of the surface states closer to, or even into, the nearer band of the semiconductor, as shown in Fig. 2i. This reduces the rate of recombination because σ_{ij}, the cross section for electron hole recombination, is decreased when the largest energy gap to be non-radiatively crossed increases. Such a decrease is expected from perturbation theory and a correlation between σ_{ij} and the largest gap has been borne out in studies by Lax on Si and Ge.[39-41] The relevant physical picture, common to radiationless transitions in these, as well as in other types of systems, is classical: The more energy that one must dissipate in a single step of a radiationless process, the more collisions, vibrations or phonons are needed for the transition to take place and the less likely the transition becomes.

The idea of shifting the surface states by chemical reaction of the surface in order to reduce recombination was revolutionary. It had been assumed that impurities on a surface only add damaging surface states and aggravate recombination problems. The object was to keep surfaces clean. We proposed, however, that the cause of surface and grain boundary recombination is in the weaker chemical bonding, (Fig. 2a-h) and that the solution to the problem is in strengthening the weak chemical bonds.[38]

We proved that ruthenium acts primarily by reducing the surface recombination velocity, rather than by catalyzing electrode reactions, in the following series of experiments.[42] We first measured, by Rutherford backscattering, the depth and surface density of the adsorbed ruthenium atoms. The ruthenium was found to be exclusively on the surface, forming on the (100) face 1/3 of a monolayer. Next, we measured the luminescence intensity and luminescence decay time due to radiative recombination of electrons in the conduction band with holes in the valence band. We found that upon chemisorption of Ru^{3+} both the intensity and the decay time increase (Fig. 7) and that the surface recombination velocity decreases from 10^6 cm/sec to 3×10^4 cm/sec. Since the sample was in air, not in contact with an electrolyte, this proved that the improvement upon Ru^{3+} chemisorption was not associated with any electrochemical reaction, and

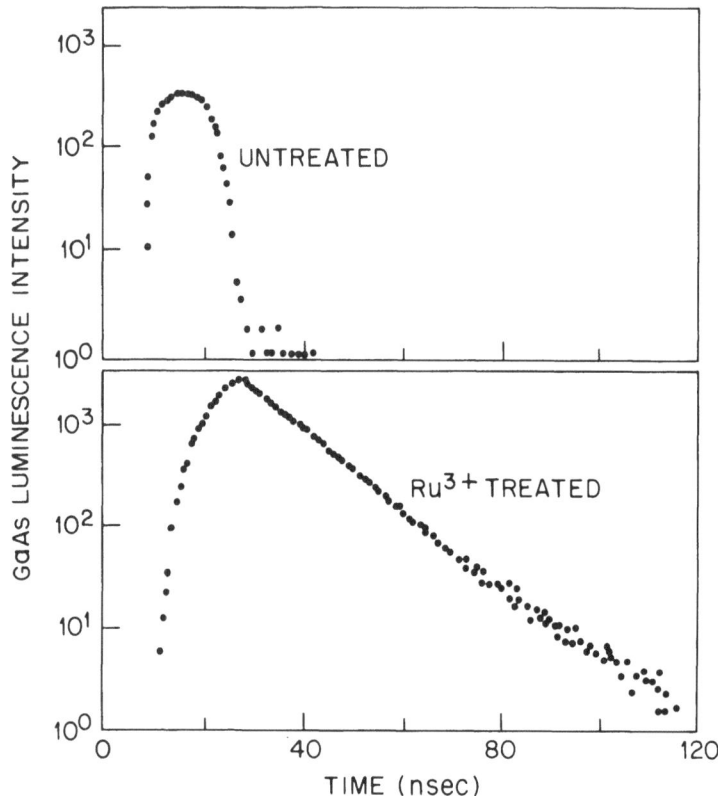

Figure 7.   The rate of non-radiative recombination of electrons
            and holes at the air n-GaAs interface is reduced
            thirtyfold upon chemisorption of 1/3 of a monolayer
            of Ru^{3+}.   The figure shows the rise and decay times
            of the band gap luminescence for the same crystal
            before and after dipping into a Ru^{3+} solution.

could not be due to electrocatalysis. It could only
be explained by the chemical redistribution of surface
states.

Reduction of Recombination at n-GaAs Grain Boundaries by Chemisorbed Ions: Thin Film Chemically Vapor Deposited Photoanodes

The minimization of recombination of electrons and
holes at grain boundaries necessitates, to make the
most efficient solar cells, the use of high quality
single crystals. Were it possible to solve this
recombination problem, the cost of materials for
photoelectrochemical cells would be lowered by at least
an order of magnitude. By building on our model for
reducing trapping and recombination in surface states
by strongly chemisorbed ions, we were able to increase
the efficiency of the n-GaAs|K_2Se-K_2Se$_2$-KOH|C solar cell
made with thin, polycrystalline chemically-vapor-
deposited films of n-GaAs on graphite. The concept
behind the efficient polycrystalline cells follows from
the reduction of the surface recombination velocity
by strongly chemisorbed ions. The boundary between
two crystallites is a region in which some of the
chemical bonds are weaker than in the bulk. Thus,
the principles that we established in our analysis
of the problem of surface states could apply also to
grain boundaries. We were able to predict that
diffusion of strongly chemisorbed species into grain
boundaries should improve the efficiency of polycrystalline semiconductor based solar cells.

Since diffusion of ions in grain boundaries is
much more rapid than in the bulk, soaking of the
polycrystalline films in aqueous solutions of cations
proved adequate for reacting the grain boundaries in
the top 10^{-5}cm light absorbing region. By diffusing
Ru^{3+} into boundaries of a polycrystalline. chemically-
vapor-deposited film on n-GaAs on graphite we
quadrupled the solar to electrical conversion efficiency
(Fig. 8).[43] Subsequent experiments[44] on a superior GaAs
film led to 7.3% efficient cells, and by co-adsorbing
Ru^{3+} and Pb^{2+} we reached, for an average grain size
of 9µm, a 7.8% solar-to-electrical conversion, retaining
2/3 of the single crystal cell efficiency.[45] These
experiments showed for the first time that it is possible to improve the performance of a polycrystalline
semiconductor based solar cell by a simple chemical
treatment of the grain boundaries.

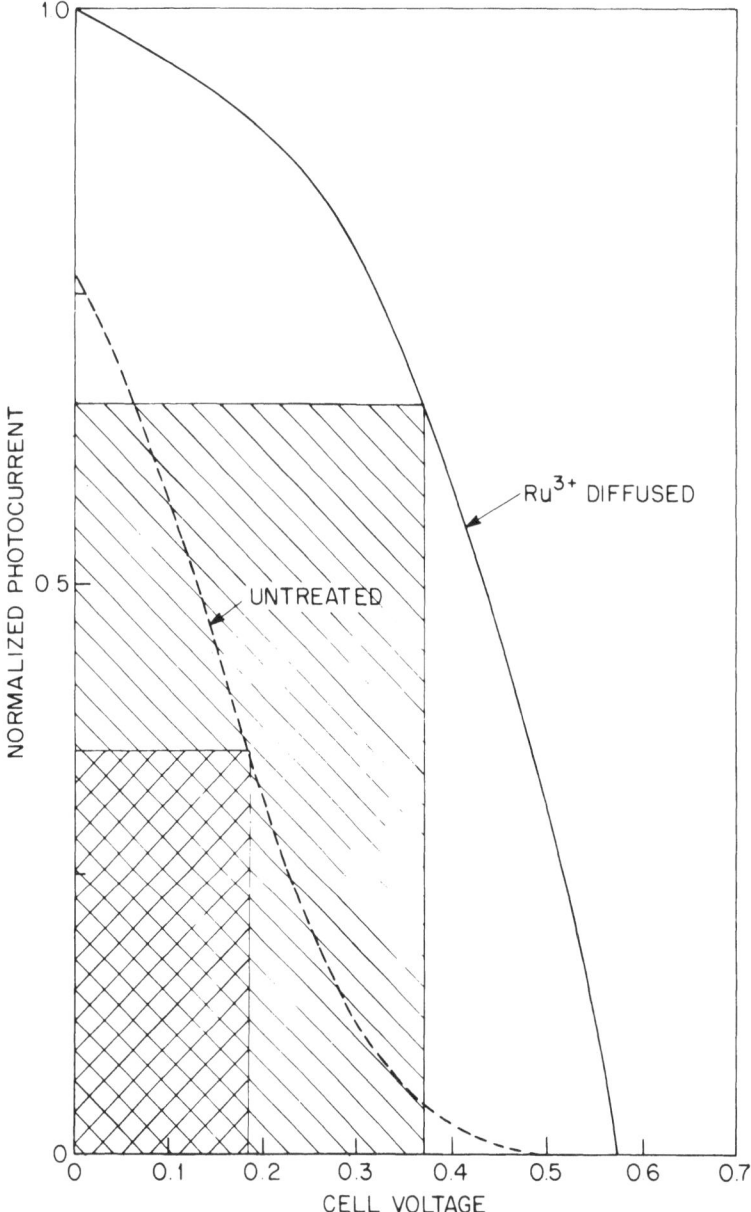

Figure 8.   First observation of an increase in the efficiency
of a polycrystalline solar cell by chemical treatment
of grain boundaries. The efficiency of the cell
n–GaAs|K$_2$Se–K$_2$Se$_2$–KOH|C (shaded area) quadruples when
Ru^{3+} is chemisorbed at the boundaries. A film of
n–GaAs on graphite, formed by reacting Ga(CH$_3$)$_3$ with
AsH$_3$, was used.

Experimental Confirmation of the Inverse Correlation: the
Strength of the Weakest Chemical Bonds and the Electron
Hole Recombination Velocity

    Our model correlating the surface recombination
velocity of electrons and holes with the strength of
the chemical bonds at a surface or a grain boundary was
confirmed in a charge collection electron microscopy
experiment on a layered compound semiconductor.[46,47]

    Layered compounds, like graphite, have Van der
Waals planes in which all the atoms at a surface are
strongly bound. Perpendicular to these planes the
bonding at the surface is weaker. According to the
model represented by Fig. 2, there should be a substantial difference in recombination velocity at the
two surfaces: when the surface bonding is weak,
recombination should be rapid; when it is strong,
recombination should be slow.

    To observe the difference we used charge collec-
tion electron microscopy (also known as electron beam
induced current microscopy, or EBIC).[46,47] The layered
semiconductor crystal was metallized on two opposing
Van der Waals faces. On the face exposed to the electron beam a $\sim 100 \text{Å}$ thick layer of gold was deposited to
form a Schottky junction. The opposite side was
metallized to form an ohmic contact. The 15keV electron
beam of the microscope could penetrate the thin gold
layer, to form a large number of electron-hole pairs.
When an appropriate potential was applied between the
two metal layers, the electrons and the holes were
collected and their flux measured as a current through
a circuit between the two metal layers (Fig. 9). By
scanning the sample with the electron beam and simultaneously measuring the current, one can determine the
efficiency of collection of electrons and holes at any
point. A decrease in current at a point proves that
the primary electron beam is residing on a site where
electron-hole recombination is rapid.

    Fig. 10 shows a scanning micrograph of a Van der
Waals plane of WSe_2, a layered semiconductor crystal
and a charge collection electron micrograph of the same.
Black areas in the charge collection micrograph represent high collection efficiency i.e., no electron-hole
recombination. White areas indicate sites of rapid
recombination. The relationship between the strength
of chemical bonds at the surface and the recombination

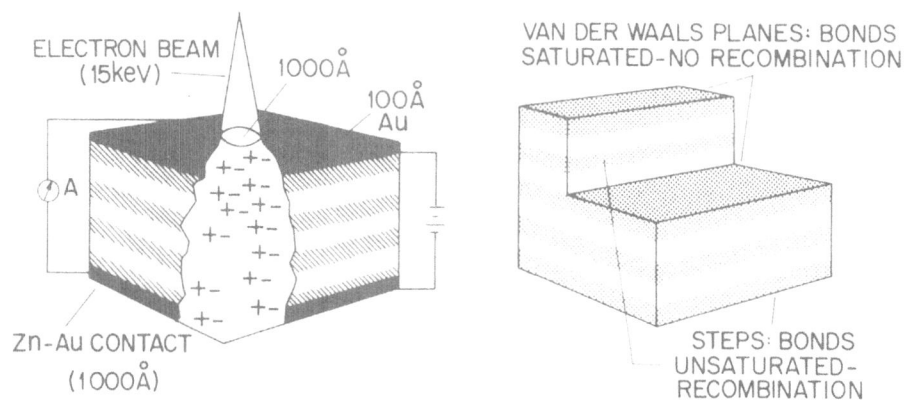

Figure 9: Schematic diagram of the experiment in which the
         dependence of the electron-hole recombination velocity
         on the strength of the chemical bonds at a semicon-
         ductor surface was observed. Left: A 15 keV, 1000 Å
         diameter electron beam of a scanning electron micro-
         scope penetrates through a 100 Å layer of gold into a
         layered-compound semiconductor (p-WSe$_2$) generating
         a track of secondary electron-hole pairs. A potential
         is applied between the gold layer on the top (which
         forms a Schottky junction) and the zinc-gold layer
         at the bottom (which forms on ohmic contact). If
         the secondary electrons and holes do not recombine,
         the electrons travel to the top and the holes to
         the bottom, causing a current to flow in the circuit
         on the left. If they do recombine, the current is
         reduced. Right: In layered compound semiconductors,
         the bonds at surfaces of steps are weaker than at the
         Van der Waals planes. If electrons and holes recombine
         at sites of weak chemical bonds, but do not recombine
         where the bonds are strong, recombination is expected
         to be rapid at steps and slow on the planes.

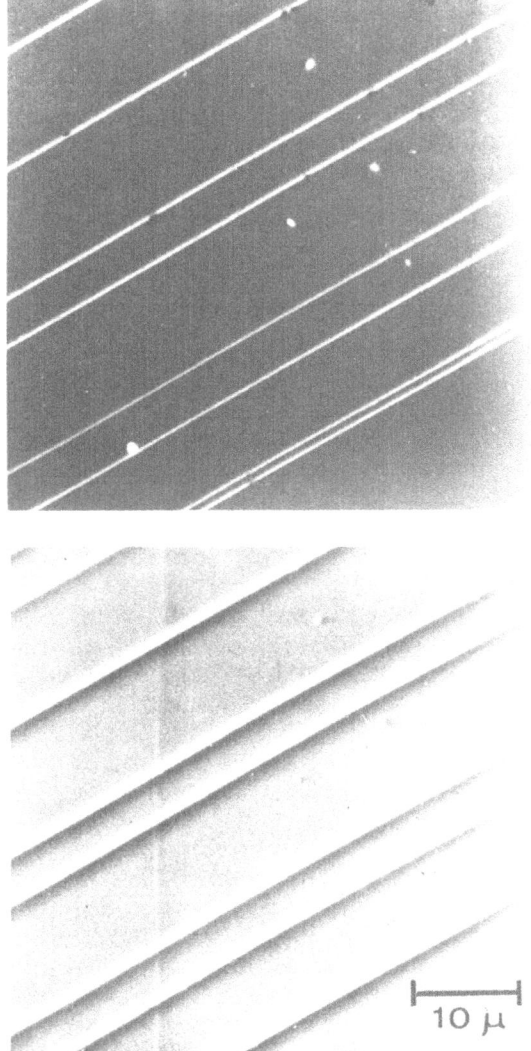

(Photoeffects at Semiconductor-Electrolyte Interfaces (47), p.28)

Figure 10.   Scanning micrograph of a Van der Waals plane of WSe_2, a
             layered semiconductor crystal, and a charge collection
             electron micrograph of the same. Black areas in the
             charge collection micrograph represent high collection
             efficiency, i.e., no electron-hole recombination. White
             areas indicate sites of rapid recombination.

velocity is obvious: there is little recombination at the Van der Waals plane, where the bonding is strong; at steps, with planes perpendicular to the surface, recombination of electrons and holes in rapid.[47]

As expected, there is also simple correspondence between the solar conversion efficiency of cells made with layered compound semiconductors and the height and density of steps on the surface. The smoother the surface, the higher the efficiency.[46-49]

Photocathodes: Removing the Obstacle of Pinning and Achievement of 11.5% Efficient Cells

Our success in solving the problems of surface and grain boundary states in n-GaAs and the confirmation of our model correlating the strength of surface bonds with the electron hole recombination velocity prompted us to address the issue of efficient photocathodes. Photocathodes have an important advantage over photoanodes. While photoanodes corrode under intense illumination (when hole transport kinetics to the redox couple cannot cope with all the photogenerated carriers), photocathodes are protected against oxidative corrosion by the photogenerated electrons arriving at the interface. The effect of light is thus equivalent to cathodic protection of the semiconductor. While reductive corrosion is possible, it is far less likely, because reduction products, such as hydrides, are less easily formed than oxidation products.

While the study of p-GaP photocathodes dates back to 1969,[50] the efficiency of the best of these had been only a few percent. It has been suggested[51-53] that the problem is in pinning by surface states. If pinned, the photovoltage achievable in a junction between a p-type semiconductor and a redox couple is no longer defined by the difference between the Fermi level and the redox potential, but by the difference between the bulk and the surface Fermi level of the semiconductor.

With the strong chemisorption model in mind, we looked for indications of extensive redistribution of surface states and reduction in surface recombination velocities upon chemisorption of oxygen.[38] We found evidence for these in the case of InP in the work of Casey and Buehler[54] and of Spicer and his colleagues.[55]

Indeed, our experiments showed that the open circuit voltage of the cells p-InP$|V^{n+}$-$V^{(n+1)+}$-$H^{+}|C$[57] and p-Si$|V^{n+}$-$V^{(n+1)+}$-$V^{(n+1)+}$-$H^{+}|C$[58] do vary with solution redox potential, over a >0.5V range, and that over this range surface states do not pin the photovoltage. We reached 9.3% solar-to-electrical conversion efficiency using either air exposed or anodized p-InP photocathodes in the p-InP$|VCl_3VCl_2$-HCl$|C$ cell.[57] Upon improving cleaning and oxidation of the surface, the efficiency of the cell increased to 11.5% (Fig. 11).[59] As predicted, the cells were quite stable. The simple electrode reactions, $V^{3+} + e^{-} \rightarrow V^{2+}$ at the photocathode and $V^{2+} \rightarrow V^{3+} + e^{+}$ at the carbon anodes, are fast at the concentrations employed. The first efficient semiconductor liquid junction cell that was no longer subject to oxidative corrosion upon illumination, however intense, was achieved.

Reduction of Grain Boundary Recombination in p-InP: Polycrystalline Photocathodes

If the p-InP$|VCl_3$-VCl_2-HCl$|C$ cell is made with a polycrystalline, chemically formed film of p-InP on graphite (prepared by reacting phosphine and indium chloride), its efficiency is of 0.1-1%. InP is a relatively covalent semiconductor and its grain boundaries abound in states that trap and cause recombination of electrons and holes. In contrast to the surface, the grain boundaries cannot be oxidized without generating insulated regions in the crystal. We have found, however, that not only oxygen but also silver ions react strongly with the surface of p-InP and improve the performance of both semiconductor liquid junction and semiconductor-metal junction cells.[60] To react the p-InP surface with silver the semiconductor is immersed, at room temperature, for 20 sec. in a dilute solution of a silver salt such as $KAg(CN)_2$-KCN. Such treatment causes 5-20% of the surface of be covered with silver atoms. Coverage depends on the crystal plane. Of the (100), (111) indium and (111) phosphorus planes, it is highest for the (111) phosphorus plane, suggesting that the surface states that are displaced upon chemisorption of silver involve phosphorus. The chemisorbed silver drastically reduces the electron-hole recombination at p-InP surfaces and grain boundaries[60] yet avoids the formation at Ag/p-InP Schottky junctions. The effect of silver on p-InP grain boundaries is more dramatic than the effect of ruthenium ions on grain boundaries of n-GaAs. With

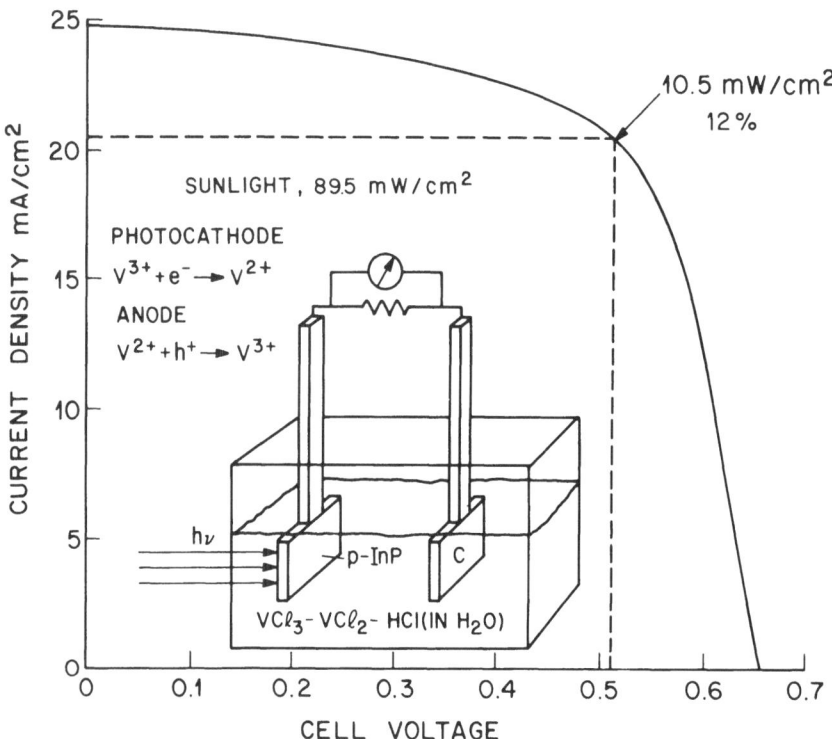

Figure 11. Current-voltage characteristics of the first efficient
photocathode based cell, p-InP|VCl₃-VCl₂-HCl|C. It
converts 11.5% of the incident sunlight into electrical
power. The photocathode is stabilized against
oxidative corrosion by the light generated electrons
arriving at the semiconductor liquid interface. An
oxide monolayer or submonolayer on the surface of the
photocathode prevents electron-hole recombination.

samples of 1-3μm grains, we have observed over a hundredfold improvement in both the p-InP|VCl$_2$-VCl$_2$-HC|C efficiency cell and in the collection of electrons and holes in charge collection electron microscopy experiments on p-InP Ti Schottky junctions. The charge collection experiments rule out the possibility that silver acts primarily as an electrocatalyst for reduction of vanadium cations at the photocathode. They prove that the rate of electron-hole recombination is reduced.[60]

Figure 12 shows the current-voltage characteristics of the thin film, polycrystalline p-InP|VCl$_2$-VCl$_2$-HC|C cell before and after a 20-second dip in a 0.1M KAg(CN)$_2$-0.1M KCN solution. A sixfold improvement brings the efficiency of the silver treated polycrystalline photocathode to about 7%.[60]

The experiments on the chemical modification of grain boundaries of n-GaAs by Ru^{3+} and of p-InP by Ag give us confidence in our ability to chemically reduce the recombination of carriers in other direct band gap polycrystalline semiconductors, made of abundant elements, on which future solar cells will be based.

Hydrogen: 12% Efficient Sunlight-Assisted Electrolysis of Water

The efficient p-InP photocathode opened an avenue to the long sought objective to efficient photoassisted electrolysis of water. Much of the worldwide effort had centered on photoanodes. These were inefficient because stability at the potentials required to generate oxygen could be achieved only in either large band-gap materials, absorbing only a small fraction of the sunlight, or in materials with an inadequate hole diffusion length/light absorption length ratio, where the quantum efficiencies were low. While photocathodes such as p-Si[61] and p-GaP[62-66] and p-GaAs[67] have been used for generation of hydrogen, their conversion efficiencies were low.

With pinning no longer a critical problem, we chose p-InP photocathodes for generating hydrogen, because its 1.35eV band-gap is ideal for solar conversion. InP is also a direct band-gap semiconductor, with an absorbance of ∿10^{5}cm^1 in the visible and in the near infrared. We have found in our initial experiments that surface oxidized p-InP is a poor electro-

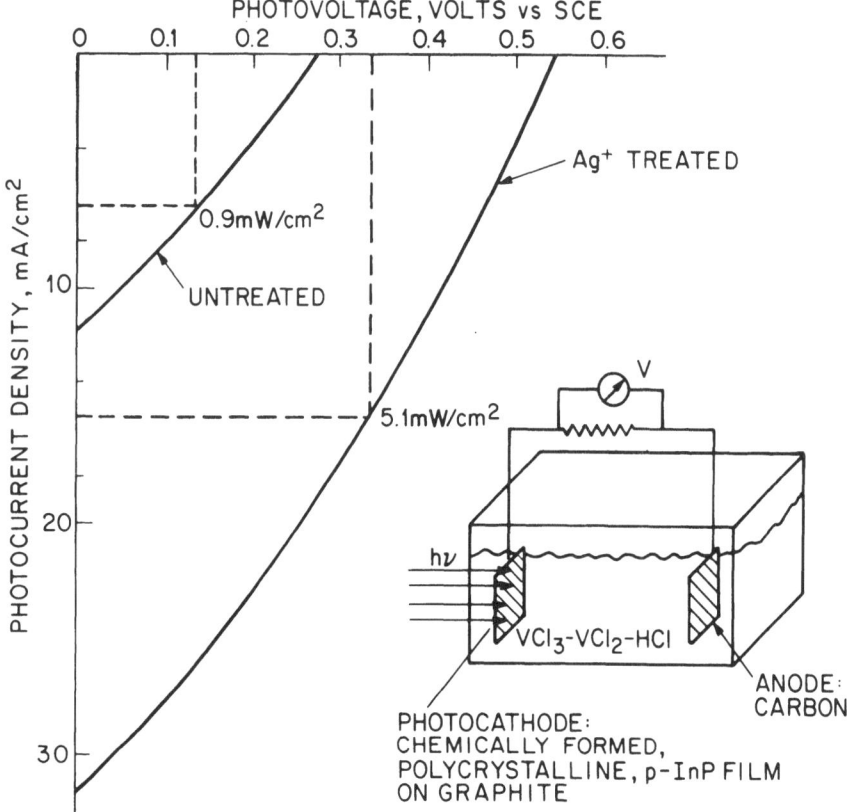

Figure 12. Reaction of grain boundaries of polycrystalline p-InP
with silver ions reduces the electron-hole recombination velocity and increases the solar conversion
efficiency of the $p-InP|VCl_3-VCl_2-HCl|C$ cell. The
figure shows the current-voltage characteristics of a
cell made with a thin chemically formed film of p-InP
on graphite, before and after immersion in a
0.1M $KAg(CN)_2$-0.1M KCN solution for 20 sec. Silver
ions diffuse into and react with the boundaries,
drastically increasing the photocurrent and photovoltage.

catalyst for hydrogen evolution. The hydrogen generation efficiency can, however, be increased by four
orders of magnitude by electrodepositing less than
100 A of Ph, Ru or Pt,[68,69] etching away most of the
metal, so that microscopic islands of the catalyst
are left, then oxidizing the p-InP surface by in situ
anodization to reduce recombination.

The mechanism of hydrogen production on the
p-InP(Pt), p-InP(Rh) or p-InP(Ru) photocathodes is
similar to the classical mechanism for platinum electrodes: protons adsorbed on the Pt, Rh, or Ru islands
react with photogenerated electrons to produce adsorbed
hydrogen atoms. These recombine to form hydrogen
molecules, which desorb. The anode reactions are controlled by the anode material and the anions in solution. If the electrolyte is aqueous hydrochloric acid
and an anode which catalyzes electron transport from
chloride anions (such as RuO_2-TiO_2 on Ti, RuO_2 on Pt
or IrO_2-$Ir(TaO_3)_4$ on Ti)[70] is chosen, hydrogen and chlorine
are produced. If perchloric acid is used, with an
anode such as IrO_2-$Ir(TaO_3)_4$-Ti,[70,71] hydrogen and
oxygen are produced.[68,69] Hydrogen evolution starts
on either p-InP(Rh) or on p-InP(Ru) under $84.7 mW/cm^2$
of sunlight at +0.36V versus the saturated calomel
electrode. On platinum, hydrogen is evolved only at
-0.23V. Thus, 0.59V are saved whether chlorine and
hydrogen are produced from hydrochloric acid, or
oxygen and hydrogen from water. At this level of
illumination the hydrogen and chlorine producing
p-InP(Ru)|2MHCl-2M$_2$KCl|Pt (RuO$_2$) cell yields a current
density of $24 mA/cm^2$ at a voltage saving of 0.43V
with respect to the platinum electrode (Fig. 13).[68,69]
The saving in voltage and current density for the
oxygen and hydrogen producing p-InP(Rh)|4M HClO$_4$|IrO$_2$-
Ir(TaO$_3$)$_4$-Ti cell are similar. The electric power
conserved upon substitution of the platinum cathode
by p-InP(Ru) or p-InP(Rh) amounts to 12% of the solar
irradiance.

The 12% engineering efficiency of these cells is
the highest for any direct scheme that uses sunlight
to generate fuels or chemicals. Plants are able to
convert only 1-3% of the sunlight to combustible fuels.

ACKNOWLEDGEMENTS

The contributions of my colleagues with whom I
had the privilege to work are evident from the
references.

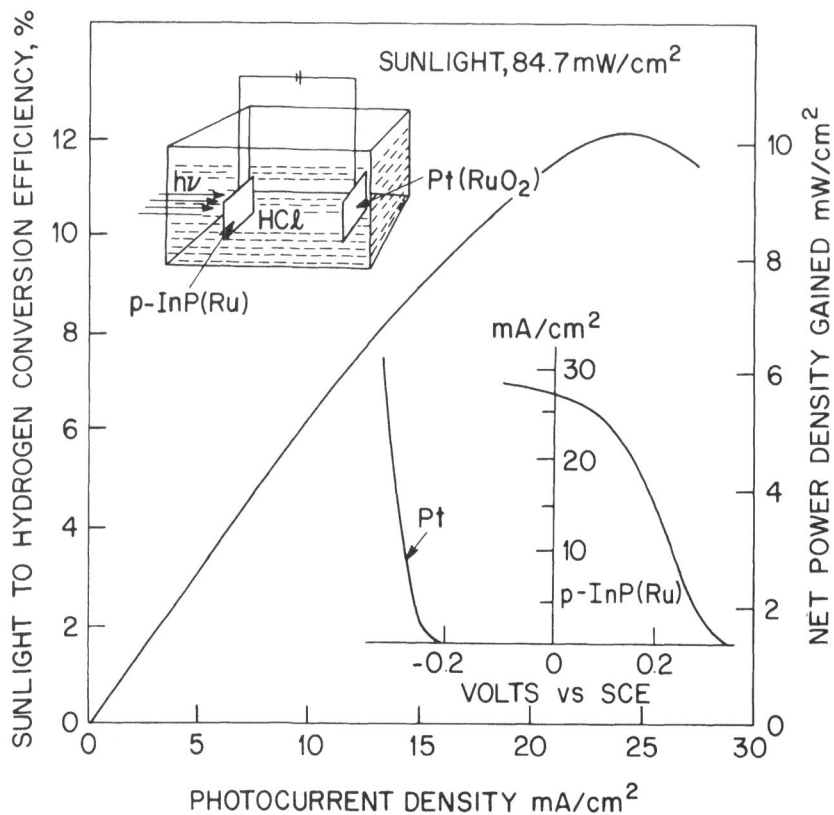

Figure 13.   12% of the incident solar energy is converted to hydrogen and chlorine by photoassisted electrolysis of aqueous
hydrochloric acid in the cell p-InP(Ru)|HCl-KCl|Pt or to
hydrogen and oxygen by photoassisted electrolysis of
water in the cell p-InP(Rh)|HClO$_4$|IrO$_2$-Ir(TaO$_3$)$_4$-Ti.
These cells represent the most efficient systems for the
conversion of sunlight into chemicals and fuels. The
current-voltage characteristics of a cell with a platinum
cathode and with a p-InP(Ru) cathode (under 84.7 mW/cm^2
sunlight) are compared in the insert. With p-InP(Rh) or
p-InP(Ru) photocathodes hydrogen evolution starts at
+0.36V vs. the saturated calomel electrode (SCE), or
0.64V vs. either the oxygen electrode or the chlorine
electrode. Platinum requires -0.23V vs. SCE or 1.23V
vs. oxygen or chlorine. (Chlorine is evolved below the
+1.36V standard potential for the reaction HCl→½ H$_2$ +
½ Cl$_2$ because its partial pressure above the solution
is less than 1 atm.)

Table 1.   REGENERATIVE CELLS

CELL	MATERIAL	SOLAR TO ELECTRICAL CONVERSION EFFICIENCY
n–CdSe $\mid$ Na$_2$ S–S–Se°–NaOH$\mid$C	Single Crystal	7.2%
n–CdSe $\mid$ Na$_2$ S–S–Se°–NaOH$\mid$C	Hot Pressed Powder	5.0%
n–GaAs $\mid$ K$_2$ Se–K$_2$Se$_2$ –KOH $\mid$C	Single Crystal	12.0%
n–GaAs $\mid$ K$_2$ Se–K$_2$Se$_2$ –KOH $\mid$C	Chemically Deposited Polycrystalline Film on Graphite	7.8%

Table 2.   REGENERATIVE CELLS

CELL	MATERIAL	SOLAR TO ELECTRICAL CONVERSION EFFICIENCY
P–InP $\mid$ VCl$_3$ – VCl$_2$ – HCl$\mid$C	Single Crystal	11.5%
P–InP $\mid$ VCl$_3$ – VCl$_2$ – HCl$\mid$C	Chemically Deposited Polycrystalline Film on Graphite	7.0%

Table 3.   ELECTROLYTIC CELLS

CATHODE	ANODE	ELECTROLYTE	PRODUCTS	SOLAR TO CHEMICAL CONVERSION EFFICIENCY
p–InP(Rh)	Ir (TaO$_3$)$_4$ –IrO$_2$ –Ti	4M HClO$_4$	H$_2$, O$_2$	12%
p–InP(Rh)	Pt(RuO$_2$)	4M HCl	H$_2$, Cl$_2$	12%

REFERENCES

1. Becquerel, E., Compt. Rend. 9, 561 (1839).
2. Adams, W. G., Day, R. E., Proc. Royal Soc., London,
    A25, 113 (1877).
3. Ohl, R. S., US Patent 2,402,662, filed May 27, 1941,
    issued June 25, 1946.
4. Chapin, D. M., Fuller, C. S., Pearson, G. L.,
    Appl. Phys. 25, 676 (1954).
5. Reynolds, D. C., Leies, G., Antes, L. L.,
    Marburger, R. E., Phys. Rev., 96, 533 (1954).
6. Brattain, W. H., Garrett, G. C. B., Bell Syst.
    Tech. J., 34, 129 (1955).
7. Dewald, J. F., Bell Syst. Tech. J., 39, 615, (1960).
8. Boddy, P. J., J. Electrochem. Soc., 115, 199
    (1968).
9. Boddy, P. J., Brattain, W., J. Electrochem. Soc.,
    109, 1053 (1962).
10. Turner, D. R., J. Electrochem. Soc., 108, 561,
    (1961).
11. Williams, R., J. Chem. Phys., 32, 1505 (1960).
12. Honda, K., Fujishima, A., Nature, 238, 37 (1972).
13. Gerischer, H., J. Electroanal. Chem. 58, 263, (1975).
14. Mavroides, J. G., Kafalas, J. A., Kolesar, D. F.,
    Appl. Phys. Lett., 28, 241 (1976).
15. Mavroides, J. G., Tchernev, D. I., Kafalas, J. A.,
    Kolesar, D. F., Mater. Res. Bull., 10, 1023
    (1975).
16. Miller, B., Heller, A., Nature, 262, 680 (1976).
17. Miller, B., Menezes, S., Heller, A., J. Electro-
    anal. Chem., 94, 85,(1978).
18. Ellis, A. B., Kaiser, S. W., Bolts, J. M.,
    Wrighton, M. S., J. Am. Chem. Soc., 98, 1635,
    (1976).
19. Hodes, G., Manassen, J., Cahen, D., Nature, 260,
    312 (1976); 261, 403.
20. Barker, G. C., discussing paper by Gerischer, H.,
    J. Electrochem. Soc., 113, 1182 (1966).
21. Heller, A., Chang, K. C., Miller, B., J. Electro-
    chem. Soc., 124, 697 (1977).
22. Heller, A., Chang, K. C., Miller, B., J. Amer.
    Chem. Soc., 110, 684 (1978).
23. Miller, B., Heller, A., Robbins, M., Menezes, S.,
    Chang, K. C., Thomson, J., J. Electrochem. Soc.,
    124, 1019 (1977).
24. Hodes, G., Manassen, J., Cahen, D., J. Amer. Chem.
    Soc., 102, 5962 (1980).
25. Heller, A., Schwartz, G. P., Vadimsky, R. G.,
    Menezes, S., Miller, B., J. Electrochem. Soc.,
    125, 1156 (1978).

26. Cahen, D., Hodes, G., Manassen, J., J. Electro-
    chem. Soc., 125, 1623 (1978).
27. Noufi, R. N., Kohl, P. A., Rogers, J. W., Jr.,
    While, J. M., Bard, A. J., J. Electrochem.
    Soc., 126, 949 (1979).
28. Gerischer, H., Gobrecht, J. Ber. Bunsenges. Phys.
    Chem., 82, 520 (1978).
29. Chang, K. C., Heller, A., Schwartz, B., Menezes,
    S., Miller, B., Science, 196, 1097 (1977).
30. Ellis, A. B., Bolts, J. M., Kaiser, S. W.,
    Wrighton, M. S., J. Am. Chem. Soc., 99, 2848
    (1977).
31. Boddy, P. J., Brattain, W. H., J. Electrochem.
    Soc., 109, 819 (1962).
32. Brattain, W. H. Boddy, P. J. Proc. Intl. Conf. on
    Phys. of Semiconductors, Exeter, 1962, London,
    Inst. Phys. and Phys. Soc., 1962, p. 797-806.
33. Heller, A., Parkinson, B. A., Miller, B., Proc.
    13th IEEE Photovoltaic Specialists Conf.,
    June 5-8, 1978, Washington, D. C., p. 2353-1254.
34. Parkinson, B. A., Heller, A., Miller, B., J.
    Electrochem. Soc., 126, 954 (1979).
35. Parkinson, B. A., Heller, A., Miller, B., Appl.
    Phys. Lett. 33, 521 (1978).
36. Heller, A., Miller, B., Adv. Chem. Ser., 184, 215
    (1980).
37. Heller, A., Miller, B., Electrochimica Acta, 25,
    29 (1980).
38. Heller, A., "Chemical Control of Surface and Grain
    Boundary Recombination in Semiconductors" in
    "Photoeffects at Semiconductors Electrolyte
    Interfaces", A. J. Nozik, Ed., ACS Symp. Series,
    Vol. 146, Am. Chem. Soc., Washington, D. C.,
    1981, p. 57-77.
39. Gummel, H., Lax, M., Annals of Phys. 2, 28, (1957).
40. Lax, M., J. Phys. Chem. Solids, 8, 66 (1959).
41. Lax, M., Phys. Rev., 119, 1502 (1960).
42. Nelson, R. J., Williams, J. S., Leamy, H. J.,
    Miller, B., Casey, H. J., Jrs., Parkinson,
    B. A., Heller, A., App. Phys. Lett., 36,
    76, (1980).
43. Johnston, W. D., Jr., Leamy, H. J., Parkinson, B.
    A., Heller, A., Miller, B., J. Electrochem.
    Soc., 127, 90 (1980).
44. Heller, A., Miller, B., Chu, S. S., Lee, T. Y.,
    J. Amer. Chem. Soc., 101, 7633, (1979).
45. Heller, A., Lewerenz, H. J., Miller, B., Ber.
    Bunsenges. Phys. Chem., 84, 592 (1980).
46. Lewerenz, H. J. Heller, A., DiSalvo, F. J., J.
    Am. Chem. Soc., 102, 1877 (1980).

47. Lewerenz, H. J., Heller, A., Leamy, H. J.,
    Ferris, S. D., in "Photoeffects at Semicon-
    ductor Electrolyte Interfaces", Amer. Chem.
    Soc. Symp. Ser., Vol. 146, A. J. Nozik,
    ed., Amer. Chem. Soc., Washington, D. C.,
    1981, p. 17-35.
48. Kautek, W., Gerischer, H., Tributsch, H., Ber.
    Bunsenges, Phys. Chem., 83, 1000 (1979).
49. Kline, G., Kam, K., Camfield, D., Parkinson,
    B. A., Solar Energy Materials, 4, 301 (1981).
50. Beckman, K. H., Memming, R., J. Electrochem. Soc.,
    116, 368 (1969).
51. Bard, A. J., Bocarsly, A. B., Fan, F. R. F.,
    Walton, E. G., Wrighton, M. S., J. Amer.
    Chem. Soc., 102, 3671 (1980).
52. Fan, F. R. F., Bard, A. J., J. Amer. Chem. Soc.,
    102, 3677 (1980).
53. Bocarsly, A. B., Bookbinder, D. C., Dominey,
    R. N., Lewis, N. S., Wrighton, M. S., J.
    Amer. Chem. Soc., 102, 3671 (1980).
54. Casey, H. C., Buehler, E., Appl. Phys. Lett., 30,
    247, (1977).
55. Spicer, W. E., Chye, P. N., Sheath, P. R., Su,
    C. Y., Lindau, I., J. Vac. Sci. Technol. 16,
    1422, (1979).
56. Spicer, W. E., Lindau, I., Skeath, P. R., Su,
    C. Y., Chye, P., Phys. Rev. Lett., 44,
    420, (1980).
57. Heller, A., Miller, B., Lewerenz, H. J. Bachmann,
    K. J., J. Amer. Chem. Soc., 102, 6555 (1980).
58. Heller, A., Lewerenz, H. J., Miller, B., J. Amer.
    Chem. Soc., 103, 200 (1980).
59. Heller, A., Miller, B., Thiel, F. A., Appl. Phys.
    Lett., 38, 282 (1981).
60. Heller, A., Johnston, W. D., Jr., Leamy, H. J.,
    Miller, B., and Strege, K. E., Proc. Fifteenth
    IEEE Photovoltaic Specialists Conference,
    Orlando, Florida, May 12-15, 1981.
61. Bookbinder, D. C., Bruce, J. A., Dominey, R. N.,
    Lewis, N. S., Wrighton, M. S., Proc. Nat.
    Acad. Sci., 77, 6280 (1980).
62. Yoneyama, H., Sakamoto, H., Tumura, H., Electro-
    chim. Acta. 20 341 (1975).
63. Nozik, A. J. Appl. Phys. Lett., 29, 150 (1976).
64. Tomkiewicz, M., Woodall, J. M., Science, 196,
    991, (1977).
65. Bockris, J. O'M, Uosaki, K., J. Electrochem. Soc.,
    124, 98 (1977).

66. Tamura, H., Yoneyama, H. Iwakura, C., Sakamoto, H., Murakami, S., J. Electroanal. Chem. 80, 357 (1977).
67. Fan, F. R. F., Reichmann, B., Bard, A. J., J. Amer. Chem. Soc., 102, 1488 (1980).
68. Heller, A. and Vadimsky, R. G., Phys. Rev. Lett., 46, 1153 (1981).
69. Heller, A., Accts. Chem. Res., 14, 154, (1981).
70. Beer, H. B., Netherland Pat. Appl. 6,606,302, Nov. 14, 1966, Chem. Abst., 67, 17379, (1967).
71. Smith, G. C., and Okinaka, Y., Abst. No. 374, Extended Abstracts, 158th Meeting of the Electrochemical Society, Fall 1980, Hollywood, Florida.

DISCUSSION

Dr. A. Gordon, Diamond Shamrock Corp., Painesville. OH:

Your p-sp PEC devices achieved only about 3% efficiency, as compared to the 10-12% for your other cells. Why?

Dr. Heller: Silicon photocathodes in aqueous acid solutions have a surface silicon dioxide layer, which becomes sufficiently thick to prevent the free tunneling of electrons to the oxidizing member of our redox couple, V^{3+}. This introduces a voltage dependent impedance into the system and thus a loss.

Dr. Elton Cairns, University of California, Berkeley:

Looking to the future, can you comment on which materials may be practical from a cost and performance point of view, assuming the use of thin film technology?

Dr. Heller: The materials will be direct band gap semiconductors made of abundant elements. Direct band gap semiconductors absorb most of the sunlight in a layer which is only a few thousand angstroms thick. This is important not only because thinner films can be used, but also because the carriers on their way to the interface traverse fewer grain boundaries and, most importantly, because it is much easier to diffuse a strongly chemisorbed species a few thousand angstroms than tens of microns, as would be necessary in order to passivate boundaries within the absorption length in indirect band gap materials like silicon.

RECENT PROGRESS IN PHOTOVOLTAIC/ELECTROCHEMICAL ENERGY

SYSTEM APPLICATION

E. L. Johnson

Texas Instruments Incorporated
P.O. Box 225303, M/S 158
Dallas, TX 75265

INTRODUCTION

Texas Instruments has developed a unique approach to solar energy utilization, combining the energy conversion and storage functions in a novel way.

If some form of energy storage is provided, the sun can supply power continuously or on demand. It can be generated at the point of use and be considered a replacement source rather than a supplement.

The amount of storage required is a function of the end use of the system and is strongly dependent on the nature of the application. With sufficient storage, energy can command retail price; without storage, it is worth a fraction of the wholesale price. Off-peak rates need be only slightly greater than the cost of the fuel to the utility, ranging from a few tenths of a cent per kilowatt hour for a nuclear plant to about one cent per kilowatt-hour for a coal facility at today's prices.

It is possible, of course, to use solar energy as a supplemental source, reducing the fuel requirements of the utility during the hours of sunshine. If this is done, the value of the energy to the utility is again equal to the value of the fuel saved: about a cent per kilowatt-hour.

We have chosen to apply the TI Solar Energy System (TISES) to residential applications. Although this choice may seem quite ambitious, it has a number of factors in its favor:

- The end application is one for which design data are readily available.
- It takes advantage of the distributed nature of solar energy.
- There is a good match between the roof area and the demand.
- No separate support structures are required, and the array can serve as a portion of the roof.
- The home represents one of the smaller significant units in the energy field, with far lower development and prototype costs than those required for larger installations.
- Since the TI technique provides storage, the energy produced for a home will command retail prices.
- Most important, the home market represents one of the few that is large enough to justify the costs of full-scale development.

The cost estimates made to date indicate that products of the TI system type have a good chance of being able to serve the residential market. The TI system concept is decribed in the following section.

SYSTEM DESCRIPTION

The need for supplanting fossil fuels as an energy source is well known. Many systems and techniques have been proposed for the conversion and utilization of energy from the sun. Although significant amounts of solar energy reach the earth, relatively large areas must be used to collect and generate significant amounts of power. About 10 kilowatt-hours (kWh) per square foot per month reach the earth at latitudes of 35°.

Since the radiation received is not continuous, means must be provided for storage of the energy received. Arrays of large area solar cells connected to banks of storage batteries have been proposed for this purpose, but four major problems are encountered in practical systems.

(1) Conventional techniques for the manufacture of efficient cells require large amounts of semiconductor material to support and otherwise make accessible a relatively small volume of active material.

(2) Semiconductor cells are basically low voltage devices requiring interconnections capable of handling thousands of amperes for kilowatt capacities.

(3) Because it is necessary to connect many cells receiving solar radiation in parallel to obtain usable outputs from large areas, flaws or leakages in a single cell can seriously degrade the performance of a large array. This causes added costs in manufacture and problems in the reliability of the finished array.

(4) Since the energy from the sun must be stored to make it available when required, conventional solar cell systems are used to
charge storage batteries. Such batteries are expensive, bulky,
and difficult to maintain.

Systems have been proposed in which conventional solar cells
are used to operate a cell for the electrolytic dissociation of
water. The hydrogen from the cell is then stored for later use.
Although this procedure aids in the storage problem mentioned above,
it does not contribute to a solution of the first three problems.
Other attempts have been made to simplify the interconnection problem by using large sheets of single crystal silicon. In a radical
departure, the present concept is embodied in a system in which all
the above problems are avoided.

The concept involves solar energy conversion by operation of
small silicon solar cells immersed in an electrolyte, and the current generated by the cells is used to separate the electrolyte into
its constituent parts. These parts can be stored separately until
needed and then recombined in a fuel cell to produce electrical
energy. The concept encompasses not only the efficient conversion
of solar energy (the solar chemical convertor), but also the efficient storage and use of the converted solar energy and the efficient
use of materials for that purpose.

The system block diagram, Figure 1, is a closed-loop system
capable of continuous operation. The system is composed of four
main components: the solar chemical convertor (SCC), hydrogen
storage, a fuel cell, and a heat exchanger. The only input is solar
radiation, and the output is electrical and thermal energy.

The SCC array is represented in a schematic cross section in
Figure 2. In this structure, separate photovoltaic elements, somewhat spherical, are fixed in a glass matrix, electrically inter-
connected, and immersed in an electrolyte to form the SCC. The
spherical solar cells, both n-on-p and p-on-n types, are provided
with metal electrodes to form cathodes and anodes, respectively.
An advantage of this structure is that defective solar cells can be
tolerated. Short or open circuit cells are not fatal to the SCC
operation.

When the system is used with an electrolyte such as an aqueous
solution of hydrobromic acid, the reaction within the SCC may be
written as:

$$2 \text{ HBr} + H_2O + \text{Electrical Energy} \rightarrow H_2 + Br_2 + H_2O$$

The Br_2 thus prduced may be considered to exist as tribromide (Br_3^-)
ions that form by reaction with the bromide (Br^-) ions from HBr.
The products of this reaction, hydrogen and tribromide ions, are

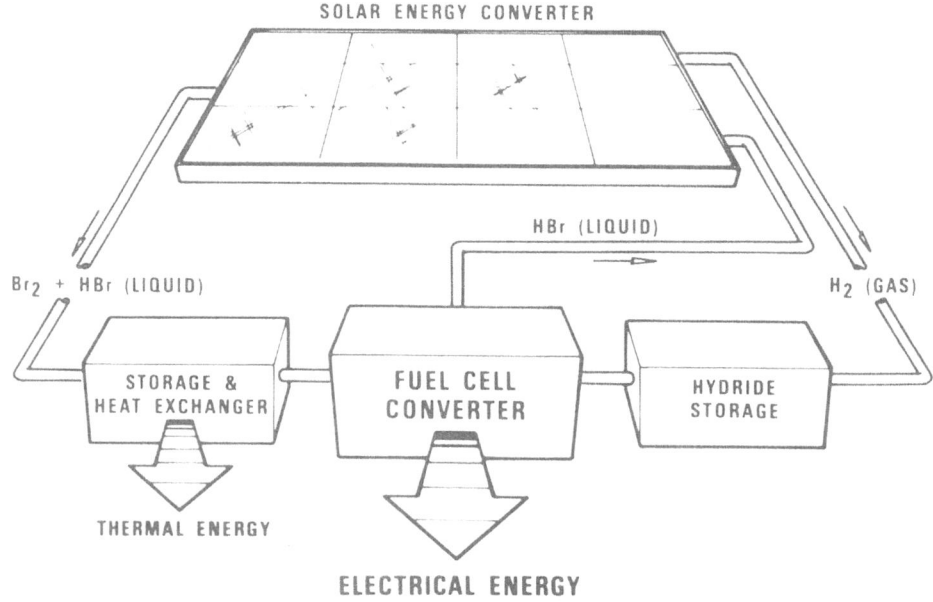

Figure 1.   Closed loop solar energy conversion system.

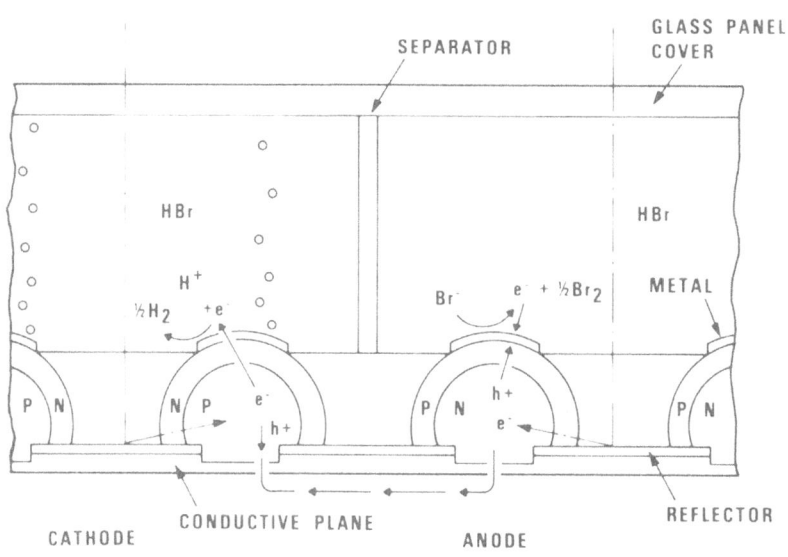

Figure 2.   Silicon solar conversion array.

removed from the SCC panel. The reaction products are readily phase-separated as hydrogen gas and the tribromide in aqueous solution with unreacted HBr. The hydrogen is transported to a storage unit where it may be stored in the form of a metal hydride. A suitable hydride storage process has been described by a number of authors such as Wiswall and Reilly of Brookhaven National Laboratory (1).

Electrical energy can be obtained from the system by recombining the hydrogen and bromine in a fuel cell. In the fuel cell the reaction will then be:

$$H_2 + Br_2 + H_2O \rightarrow 2\ HBr + H_2O + \text{Electrical Energy}$$

Because the solution in the array will also be heated by the rays of the sun, energy in the form of heat may be taken from the system by a conventional heat exchanger. The recombined products from the fuel cell can then be returned to the SCC panel for reelectrolysis.

The closed-loop system, operated with an electrolyte such as hydrobromic acid, is efficient because the charge reaction can be carried out without detectable electrode overpotential, and the discharge reaction can be carried out without a significant electrode polarization. As a result, the reaction

$$H_2 + Br_2 + H_2O \rightarrow 2\ HBr + H_2O + \text{Electrical Energy}$$

can be made to approach thermodynamic reversibility very closely, thus providing an efficient energy storage and supply system.

Applications for the system, such as a residential installation, will require hundreds or even thousands of square feet of solar cell strips. Efficient fabrication of this element of the system is of paramount importance. Such strips must be producible on an economically feasible basis with efficient utilization of the materials.

SUMMARY

In summary, this system has the following advantages:

• Small single crystal silicon cells can be made inexpensively, at a small fraction of the cost of an equivalent area prepared by conventional or advanced crystal growing techniques. TI has developed inexpensive processes similar to those used in making lead shot, simply and inexpensively. No silicon is wasted.

• It is anticipated that these small cells can be made and cast into very large sheets by continuous processes. Sheets as large as four feet by eight feet may be practical. The largest sheet cast to date is more than an order of magnitude larger than the largest single crystal grown to date, and no barriers to even larger sizes are apparent.

• Silicon cells cast in a glass sheet serve as an efficient light collector, so that complete coverage of the exposed area by silicon is not required. Sheets with 13% electrical efficiency have been built, and these efficiencies appear feasible for future routine production.

• The electrolyte is an efficient interconnection mechanism, drawing the full current that each cell can deliver. Shorted or open cells are readily tolerated. No external electrical connections to the sheet are required.

• Many electrolytes could be used with the system. One has been identified that is completely reversible and that has a gas, hydrogen, as one of its constituent parts. This simplifies the problem of separating the reaction products.

• A fuel cell suitable for use with the chosen electrolyte has been designed, built, and operated with acceptable electrical performance.

• The hydrogen will be stored as a metal hydride, using techniques similar to those developed at Brookhaven Laboratories of DOE. This is a safe and inexpensive form of storage for this application.

• Extensive cost studies and computer modeling have shown that the TI system should be able to produce more than 90% of the energy required in a typical home at a cost less than four cents per kilowatt-hour, the present cost of electrical energy. Although these costs are a function of geographical location, locations as diverse as Washington, D.C., Dallas, and Phoenix appear favorable.

Although a substantial amount of effort has been expended on the proposed system, major efforts will be required to make it a commercial reality. Feasibility of each of the major elements and a system test bed have been demonstrated. Additional effort will be required to further develop the components, complete the system design, and commercialize the product. Advances in the state-of-the-art will be required in three technologies: semiconductors, fuel cells, and hydrogen storage.

DISCUSSION

Dr. R. P. Frankenthal, Bell Laboratories: What is the metallization between the Si and the electrolyte? How thick is it? Assuming it is a platinum-group metal, does not the voltage pulse caused by the solar radiation going on and off corrode the metallization? It is known that voltage pulses greatly accelerate the corrosion of these metals.

Dr. E. L. Johnson: It is a Pt-group metal, 50 - 500 Å thick. At the voltages we use, this effect should not come into play. We would need to push the anodes to the oxygen potential. (The maximum voltage between anode and cathode is 1.2 V.)

CORROSION - NEW DIRECTIONS

R. P. Frankenthal

Bell Laboratories
Murray Hill, NJ

Although generally not recognized, metallic corrosion is one
of society's costly blights. The damages can be measured in
dollars, in human life, and in the quality of life. Corrosion
wastes material resources and pollutes our environment. It
affects every person, business, and industry.

Let us examine a few of these points. The economic costs
are large. A study by the National Bureau of Standards
indicated that in 1975 the cost of corrosion in the United
States was 70 billion dollars, or almost 4.2% of the gross
national product (1). What is perhaps worse is that about 15%
of this cost or 10 billion dollars could have been saved
simply by using the best available technology. Assuming that
most people wish to avoid the cost of corrosion, this implies
that many are not properly educated about the dangers and the
cost of corrosion and the technology available for minimizing
or preventing it.

Our safety is endangered by corrosion. For example, in
1967 the Silver Bridge across the Ohio River collapsed,
allegedly due to stress corrosion cracking, with the loss of
46 lives.

Our health is affected by corrosion. Prosthesis, such as
body implants, and cardiac pacemakers are subject to attack by
highly corrosive body fluids. As an example, Figure 1
illustrates pitting of a platinum-iridium electrode tip from a
pacemaker. Parsonnet et al. (2) have found corrosion on most
explanted electrodes, even in normally functioning pacers.

307

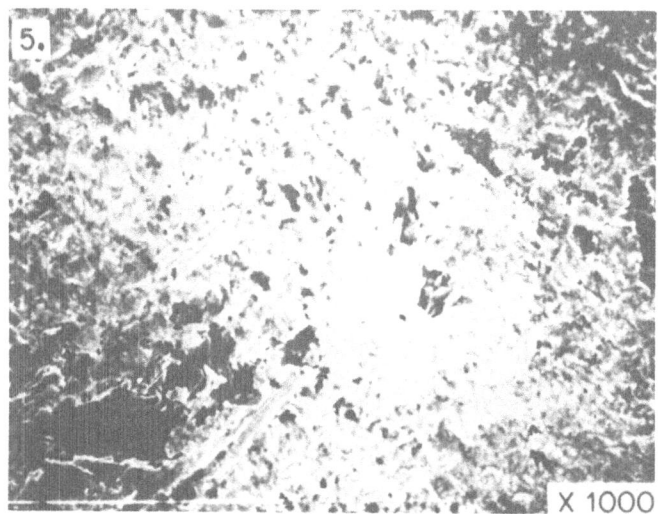

Figure 1.   Pitting corrosion on a platinum-iridium electrode
            tip from a cardiac pacemaker (courtesy of
            V. Parsonnet).

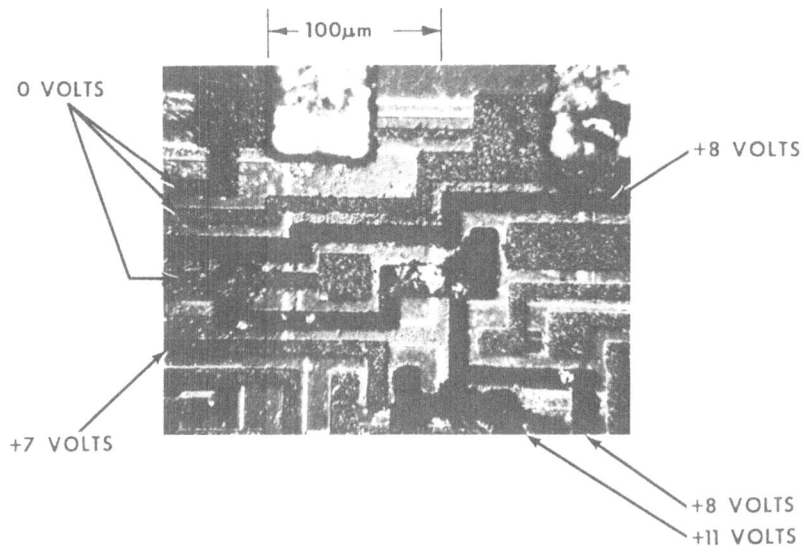

Figure 2.   Corrosion failure of a silicon integrated circuit due to
            formation of $Au(OH)_3$ on the positively biased conductor
            paths (3). Applied voltages shown next to conductor paths.
            (Reprinted by permission of the publisher, The Electro-
            chemical Society, Inc.).

The development of many technologies is impeded by corrosion. For example, many advanced energy systems, such as batteries, solar cells, and solar thermal heating systems, are encountering corrosion problems. The development of magneto-hydrodynamic energy generators is limited by the lack of materials to withstand the corrosivity of the hot gases that drive the system. At the other end of the size scale, we have the world of microelectronics, which is continuously undergoing further miniaturization. An example of corrosion here is shown in Figure 2, in which the positively-biased gold metallization on a silicon integrated circuit has been badly corroded. The presence of a high relative humidity and a sufficiently large applied voltage was all that was necessary for corrosion to occur. The problem becomes more severe as the spacing between conductor stripes is decreased (3).

Art and metal artifacts are also subject to corrosion. Many statues and other metallic objects exposed to polluted environments have been severely damaged by corrosive attack. Museum conservators and corrosion scientists are working together to prevent further corrosion and to restore corroded objects (4).

Not all corrosion is bad, however. Metallic corrosion may be used constructively. For example, electrochemical machining, which is used to make difficult to form metal objects, is a corrosion process; and in the electronics industry, circuits are generated by etching metal to obtain the desired design on integrated or thin film circuits or on printed wiring boards. In the world of art and architecture, the weathering steels are frequently used for sculpture and for building structures ranging from buildings and bridges to highway guard rails. These steels corrode in the initial stages to form a protective brown-purple patina. Also, many fine buildings have roofs made of copper, which exhibit with time a beautiful green patina.

What are the new directions in corrosion research and corrosion prevention? They are many. In what follows, we will discuss some specific cases of corrosion and then will indicate means by which it might be mitigated. In this manner, some of the more promising and intriguing methods of corrosion protection will be illustrated.

Underground utility cables and pipes for electricity, telephone, water, and gas are generally buried in a common trench for economic reasons. For safety reasons, the power cables are enveloped by a neutral ground wire. This wire, commonly referred to as the concentric neutral, is subject to corrosion (Fig. 3) not only by the soil but more severely from currents induced by the alternating current in the cable (5,6).

Figure 3. Corrosion of concentric neutral surrounding a
           power cable induced by alternating current in
           cable (courtesy of G. Schick).

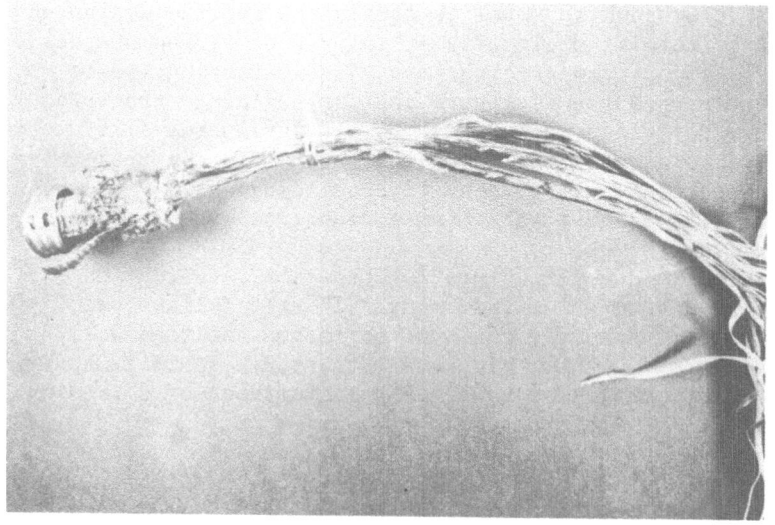

Figure 4. Burned-out utility cable resulting from ac-induced
           corrosion failure of concentric neutral on a
           nearby power cable (courtesy of G. Schick).

If corrosion should cause an open circuit in the concentric
neutral, the induced current must find another path to ground,
such as a nearby cable or pipe. This may result and indeed has
resulted (Fig. 4) in corrosive attack, sometimes explosively
fast, on that cable or pipe. Clearly this is a safety hazard.
The utility companies have investigated this problem. One
solution being tried at the present is the extrusion of a
carbon-filled conductive polyethylene coating over the concentric
neutral (7). This coating provides continuous grounding and
will not corrode, while preventing catastrophic corrosion of
the concentric neutral. However, the large area of carbon
black serves as an effective cathode, and any metal in contact
with it is likely to corrode. This problem must still be solved.

When a metal is composed of more than one phase, galvanic
corrosion between the phases may be a problem. Since the choice
of an alloy is generally governed by numerous requirements, such
as its mechanical and wear properties, as well as its corrosion
resistance, it is not always possible to avoid a corrosion
problem. However, it is possible to treat the surface of the
alloy to make it a single phase region, which avoids the problem
of galvanic corrosion. This may be done, for example, by a new
technique called laser surface melting (8), with which the
surface of the alloy is melted by a short pulse of a powerful
laser beam and then rapidly quenched by conduction of heat to
the bulk. The melted and quenched surface layer is homogenized,
while the bulk of the material retains its original structure
and properties (Fig. 5). For the aluminum bronze shown in
Figure 5, laser surface melting significantly improved the
corrosion resistance of the surface layer (9). It appears that
metastable phases can be formed by this technique.

Oxidation and tarnishing reactions may be greatly
inhibited by another surface modification technique called ion
implantation (10). In this technique, the species to be
implanted is ionized in a vacuum chamber and accelerated at
voltages ranging from 10 to 100 kilovolts to the target. The
depth to which the ionized species is implanted depends on the
energy of the ions. By varying the energy, different depth
profiles of the implanted species are obtained. Not only does
ion implantation change the composition of the near surface
region, but it also changes its defect structure and may make
it amorphous. Metastable alloys, which cannot be produced by
conventional metallurgical techniques, can be made by ion
implantation. The effects of implantation on the reactivity of
the surface are many. For example, it has been shown (11)
that yttrium implanted into stainless steel greatly reduces its
rate of oxidation. Copper implanted with boron tarnishes at a
significantly lower rate than copper in air or in the presence
of moist hydrogen sulfide (12). Not all implants, however,

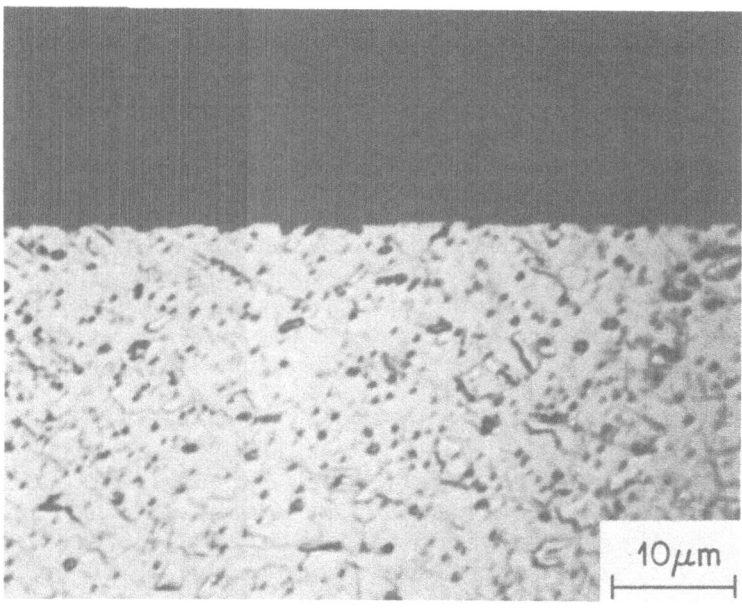

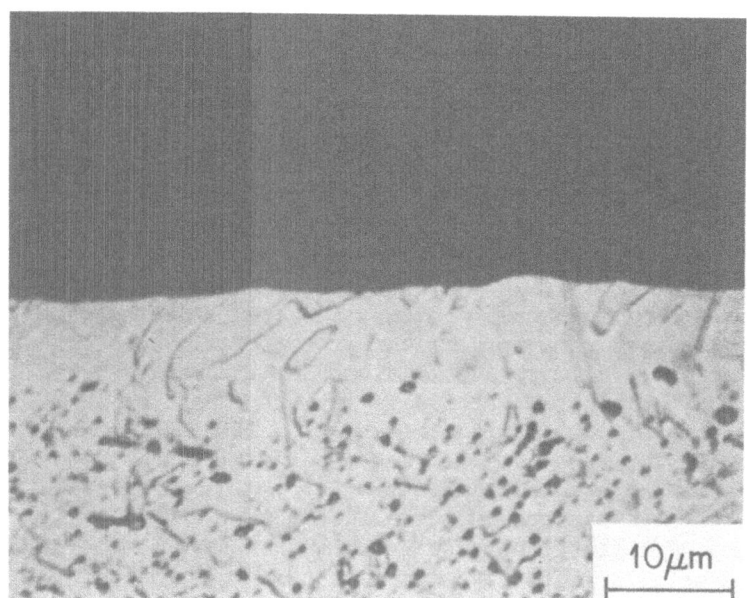

Figure 5.   Micrograph of an aluminum bronze. Top - as
           received showing two phases. Bottom - laser
           surface melting has homogenized the near surface
           layer (9).

reduce the rate of corrosion; some increase this rate (11).

     Low voltage, low force electrical contacts are usually
very sensitive to the presence of small quantities of tarnish
film. Films as thin as 100 to 200 Angstroms may give what is
effectively an open circuit, while even thinner films may
introduce unacceptable noise into the circuit. To avoid the
presence of tarnish films, it frequently is necessary to plate
the contact with a noble metal, such as gold. Since the price
of gold has increased 20 fold during the decade of the 70's,
it has become necessary to seek substitutes for this noble
metal. While many metals and alloys are under investigations
for this purpose, compounds that are good electronic conductors
are also being sought. One example of such a compound is
ruthenium dioxide. Recent investigations (13) of ruthenium
dioxide have shown that its electrical conductivity and
contact resistance are that of a good contact material and
that it is chemically stable in several corrosive atmospheres,
including moist hydrogen sulfide, at temperatures up to 100C.

     Dust and other contaminants have been shown to be
associated with the corrosion of fine instruments and electronic
devices (Fig. 6). These contaminants react in one of three
ways. First, they may be intrinsically corrosive, absorbing
moisture to form an electrolyte on the surface. Second, they
may be indirectly corrosive, absorbing both moisture and a
corrodant from the atmosphere or being conductive and serving
as a cathode in a corrosion cell. Third, they may be harmless
but may produce a geometry that is conducive to crevice
corrosion.

     These mechanisms are dependent on the critical relative
humidity, at which moisture is rapidly absorbed by the
contaminant. Thus, to predict the contaminant's corrosiveness,
it is first necessary to have techniques for measuring the
critical relative humidity (15). It is also helpful to
catalog the critical relative humidities for many inorganic
and organic compounds, for dusts, and for other contaminants
(15,16). Munier et al. (16) have applied these and other
techniques to determine qualitatively and quantitatively the
composition or the constituents of contaminants on electronic
equipment after an extended urban exposure; these results were
correlated with equipment degradation.

     Since corrosion is an electrochemical process, it should
be noted that surface modification can affect both the rate
and the mechanism of electrochemical reactions and thereby
of the corrosion reaction. It has recently been shown (17)
that platinum implanted into the surface of iron reduces the
rate of hydrogen permeation through the iron. The reduction

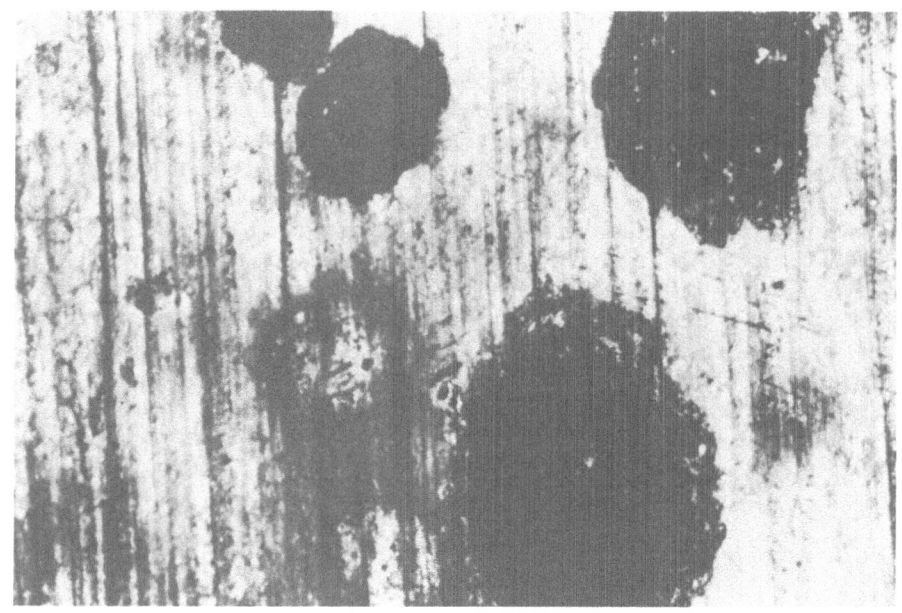

Figure 6. Localized corrosion that occurred under dust
            particles on phosphor bronze in an indoor
            environment at 70 F and 75% relative humidity (14).

in the permeation rate results from a change in the relative
rates of the different steps of the hydrogen evolution reaction.
This finding could have significant consequences for those
metallurgical properties of steel that are affected by the
presence of hydrogen in the metal. Among these are hydrogen
embrittlement and stress corrosion cracking.

It is well known that small additions of platinum or
palladium to titanium greatly reduce its corrosion rate in
chloride solutions, because the noble metal depolarizes the
hydrogen evolution reaction and shifts the corrosion potential
from the active to the passive region. However, as these noble
metals become more expensive, alloying titanium with them
becomes economically less feasible. The recently developed
technique of laser surface alloying (18, 19), as well as ion
implantation, may provide a method for protecting titanium and
other metals against corrosion while using much smaller
quantities of noble metals. This has recently been demonstrated
by Draper and co-workers (20), who vapor deposited 150 Å
of palladium onto a titanium surface and then laser melted and
rapidly quenched it. The resultant surface was an alloy of
palladium and titanium, which corroded about 30 times more
slowly than pure titanium in 2M HCl.

Conclusion

There are many areas in which corrosion affects all
persons, businesses and industry. It is important to
sensitize people to the dangers of corrosion and to the
necessity for controlling it. In the foregoing, we have
illustrated several types of corrosion and methods currently
being developed to minimize corrosion in the future. Continued
research and development in this area will provide real
benefits to society.

References

1. "Economic Effects of Metallic Corrosion in the United
        States", NBS Special Publication 511-1, SD Stock No.
        SN-003-003-01926-7, 1978.
2. V. Parsonnet, A. Villanueva, Y. Kresh, and J. Driller,
        Abstract No. 527, Sixth World Symposium on Cardiac
        Pacing, October 2-5, 1979, PACE, $\underline{2}$(5), A-110 (1979).
3. R. P. Frankenthal and W. H. Becker, J. Electrochem. Soc.,
        $\underline{126}$, 1718 (1979).
4. "Corrosion and Metal Artifacts - A Dialogue between
        Conservators and Archaeologists and Corrosion Scientists",
        NBS Special Publication 479, SD Stock No. C13-10-479, 1977.

5. D. W. McLellan and G. Schick, IEEE Conference Record -
      Supplement, 1974 Underground Transmission and
      Distribution Conference, IEEE, New York, 1974, p. 95.
6. K. Compton, Materials Performance, 14(8), 14 (1975).
7. G. Schick, Private communication.
8. E. M. Breinan, B. H. Kear, and C. M. Banas, Phys. Today,
      29(11), 44 (1976).
9. C. W. Draper, R. E. Woods, and L. S. Meyer, Corrosion, 36,
      405 (1980).
10. "Applications of Ion Beams to Metals", S. T. Picraux,
      E. P. EerNisse, and F. L. Vook, Eds., Plenum Press,
      New York, 1974.
11. J. E. Antill, M. J. Bennett, R. F. A. Carney, G. Dearnaley,
      F. H. Fern, P. D. Goode, B. L. Myatt, J. F. Turner, and
      J. B. Warburton, Corrosion Sci., 16, 729 (1976).
12. G. W. Kammlott, C. M. Preece, T. E. Graedel, J. P. Franey,
      E. N. Kaufmann, and A. Staudinger, Corrosion Sci.,
      Submitted for publication.
13. R. G. Vadimsky, R. P. Frankenthal, and D. E. Thompson,
      J. Electrochem. Soc., 126, 2017 (1979).
14. G. B. Munier, unpublished results.
15. J. D. Sinclair, J. Electrochem. Soc., 125, 734 (1978).
16. G. B. Munier, L. A. Psota, B. T. Reagor, B. Russiello,
      and J. D. Sinclair, J. Electrochem. Soc., 127, 265 (1980).
17. M. Zamanzadeh, A. Allam, H. W. Pickering, and G. K. Hubler,
      J. Electrochem. Soc., 127, 1688 (1980).
18. F. D. Seaman and D. S. Gnanamuthu, Metal Progress, 108(3),
      67(1975).
19. C. W. Draper, C. M. Preece, D. C. Jacobson, L. Buene, and
      J. M. Poate, in "Laser and Electron Beam Processing of
      Materials", C. W. White and P. S. Peercy, Eds.,
      Academic Press, New York, 1980, p. 721.
20. C. W. Draper, L. S. Meyer, D. C. Jacobson, L. Buene, and
      J. M. Poate, Thin Solid Films, in press.

DISCUSSION

Gholamabbas Nazri, Case Western Reserve University: Corrosion
studies in general, particularly wet corrosion, can be studied
by in situ technique such as Raman spectroscopy and infrared
spectroscopy, thus avoiding the need to transfer the passive film
from the electrochemical environment to the vacuum of such instruments as SIMS, ISS, ESCA, etc. Furthermore, the nature and type
of adsorption of inhibitors on the surface can be studied by Ramana
spectroscopy.

Dr. Robert Frankenthal: I agree with the comment. However, the
purpose of my presentation was to discuss techniques for controlling or preventing corrosion, not to discuss techniques for
studying corrosion mechanisms.

PASSIVITY AND BREAKDOWN OF PASSIVITY

Jerome Kruger

Center for Materials Science
National Bureau of Standards
Washington, D.C. 20234

Some metals and alloys form a film with special properties
when interacting with a corrosive environment. The formation of
this film results in a metal surface becoming resistant (passive)
to corrosive attack. This phenomenon of passivity underlies the
successful use of non-noble metals in all the industries that support every modern society; the breakdown of passivity is responsible for many of the localized corrosion failures of these non-
noble metals.

Passivity is an old subject. The first metal that was observed to exhibit this phenomenon was iron. Uhlig[1] in a review
of the history of passivity found three instances in the 18th
century of the observation of the formation of an unreactive surface on iron after immersion in concentrated nitric acid, by the
Russian Lomonsov in 1738, the German Wenzel in 1782, and the Eng-
lishman Keir in 1790. A major contributor to the study of passivity in the next century was the great Faraday who referred to the
passivity of iron as "this very beautiful and important case of
voltaic condition presented to us by the metal iron."[2] For a
comprehensive treatment of the subject of passivity see a recent
book edited by Frankenthal and Kruger.[3]

Using iron and its alloys as examples of metals exhibiting
passivity, a very brief and non-comprehensive description will be
given of the latest ideas on the nature of the passive film and
the way in which its protective properties are lost when breakdown of passivity occurs. This discussion while brief will provide a number of references to the literature for those who wish
to pursue the subject more deeply. I will then suggest avenues
that may be pursued in the future that will produce surfaces with

317

films that are more effective in resisting corrosion.

DEFINITION OF PASSIVITY

Uhlig[4] has set down definitions for two types of passivity
that are still in force today:

Type 1 - A metal active in the EMF series is passive
         when its electrochemical behavior becomes
         that of a metal noble in the EMF series (low
         corrosion rate, noble potential).

Type 2 - A metal is passive while still at an active
         potential when it exhibits a low corrosion
         rate (low corrosion rate, active potential).

This discussion will concern itself only with Type 1
passivity. Examples of metals or alloys exhibiting such
passivity are nickel, chromium, titanium, iron in oxidizing
environments, stainless steels, and many others. Examples of
Type 2 passivity are lead in sulfuric acid and iron in an inhibited pickling acid. A major characteristic of a Type 1
passive system is the existence of a polarization curve of the
sort shown in Fig. 1. The curve illustrates well a restate-
ment of the definition of Type 1 passivity as first proposed
by Wagner[5]. He suggested that a metal becomes passive when
upon increasing its potential in the positive or anodic direction, a potential is reached where the current (rate of anodic
dissolution) sharply decreases to a value less than that observed at a less anodic potential. This decrease in anodic
dissolution rate, in spite of the fact that the driving force
for dissolution is brought to a higher value, is the result of
the formation of a passive film.

NATURE OF THE PASSIVE FILM

A major controversy in past years has centered around
the question as to whether the passive film is two-dimensional
or three-dimensional. The advocates of the two-dimensional
or adsorbed layer picture of passive, for example, Uhlig[1],
proposed that the passive film on iron was a monolayer of
chemisorbed oxygen that slowed down the rate of anodic dissolution because of the formation of strong bonds between
the iron atoms on the surface and the oxygen. The proponents
of the three-dimensional or oxide film theory, most notably
U. R. Evans[6], proposed that passivity was the result of the

formation of an oxygen lattice film that served as a barrier
for diffusion of metal ions into aqueous solutions.

In recent years these two concepts have merged and become considerably more detailed and complicated. The passive

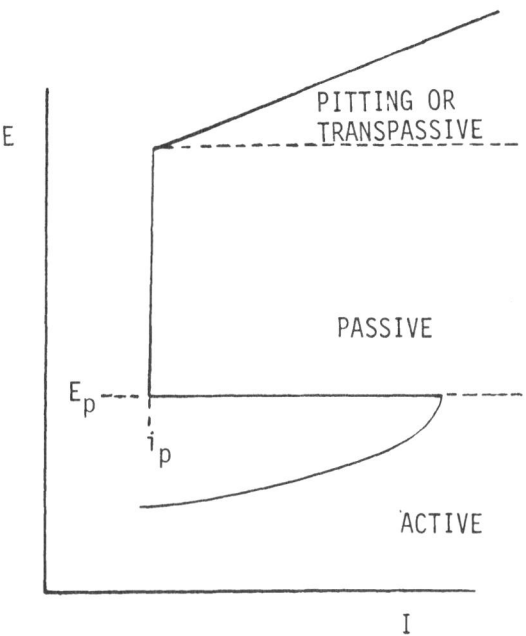

Fig. 1 An idealized anodic polarization curve for
        a system exhibiting passivity. Three dif-
        ferent potential regions are identified.
        E_p is the potential above which the system
        becomes passive and exhibits the passive
        current density, i_p.

film on Cr may indeed be only a monolayer thick upon attain-
ment of passivity[7]. However, in-situ optical studies of iron
surfaces prepared in an ultra-high vacuum system prior to
passivation gave evidence for a three-dimensional film[8].
Electron diffraction studies of passivated iron foils[9] have
also given strong indications that the passive film is a three-
dimensional crystal lattice. The studies have also identified
the oxides that form on an iron surface when it is passivated.

The prepassive iron surfaces were found to have films that were either $\alpha FeOOH$, or Fe_3O_4, the passive films were found to be γFe_2O_3 and the films formed in the transpassive region were found to be Fe_3O_4. Subsequent studies[10] showed by using a radiotracer technique that the passive film contains hydrogen in its outer layers. This does not rule out γFe_2O_3 since Bloom and Goldenberg[11] proposed that this oxide is stabilized by the entry of hydrogen into its lattice and ultimately becomes HFe_5O_8 with no change in its spinel structure as it goes through a number of stoichometric changes. A recent Mössbauer spectroscopic study, however, by O'Grady[12] found that the passive film on iron is amorphous or made up of polymeric chains which lose water upon introduction into the vacuum of an electron diffraction apparatus. The film thus becomes γFe_2O_3 (or, my interpretation, HFe_5O_8). Sato and others[13,14] have also found evidence for the existence of water ("bound water") in the passive film on iron and have proposed various structural and compositional models for the passive layer on iron which involve the incorporation of anions, the existence of nonstoichiometric compositions, and the possibility of a bipolar film[15] consisting of an anion-selective layer on the metal side and a cation-selective layer on the solution side which can rectify ionic migration.

How do these structures and compositions change when iron is alloyed with chromium? McBee and Kruger[16] determined the structure of the passive film on a series of iron-chromium alloys using electron diffraction and obtained the results given in Table 1. As Table 1 shows, the degree of crystallinity of the passive film decreases with chromium content in an alloy. Moreover, Table 1 shows that as the passive film becomes noncrystalline with increasing chromium content the ability of the alloy to resist corrosion increases. These results and that from the work of others[17,18] have led to the suggestion, perhaps first proposed by Hoar[19], that noncrystalline passive films provide superior corrosion resistance. Recently, Revesz and Kruger[20] have developed a description of the characteristics of effective noncrystalline films and ennumerated the factors that promote such noncrystallity. They have suggested that passive films that are noncrystalline or "glassy" are the most effective films to provide the properties just described that are needed to promote successful kinetic strategies. This is so because these noncrystalline films in contrast to crystalline ones do not have grain boundaries which provide paths of easy diffusion toward the metal interface for damaging species from the environment or easy diffusion out into the environment for metal ions from the metal interface. Revesz and

Table 1 - Variation in Limiting Film Thickness as a Function of Chromium Content in Iron Chromium Alloys and Crystallinity of the Passive Films Formed on these Alloys (from 20).

%Cr	Degree of Crystallinity of Passive Film	Limiting Film Thickness $\overset{\circ}{A}$
0	Well oriented spinel	36
5	Well oriented spinel	27
12	Poorly oriented spinel	21
19	Mainly non-crystalline	19
24	Completely non-crystalline	18

Kruger have suggested that bond and/or structural flexibility is an important factor in determining the ability to form a non-crystalline structure. The flexibility of the noncrystalline structure ensures a good accommodation at the oxide/substrate interface without requiring an epitaxial relationship. Such an interface is less prone to chemical attack than an imperfect one. A high degree of short range order in glassy passive films also contributes to their chemical resistance. The existence of short range order suggests that these glassy films can be looked upon as large inorganic polymeric molecules. The structural bond flexibility and, hence, the tendency toward noncrystallinity can be increased by additives. The effect of Cr in promoting passivation behavior of stainless steel is an example. One can look upon these additives as being glass formers. The formation and stability of noncrystalline oxide films is a kinetic rather than a thermodynamic problem as demonstrated, for instance, by the influence of substrate.

Besides having the compositions and structures just described, passive films are also thought to have certain electrical and mechanical properties. As is the situation with structure and compos-

ition, there is, as yet, no complete agreement about these properties of the passive film on iron or iron alloys. This is especially
so with regard to electronic conduction. Sato[15] in a review points
out that "Based on the present state of knowledge, it is highly
probable that for most metals the passive film is a poor electronic
conductor ranging from semiconductor to insulator. Redox reactions,
however, can occur by electronic tunneling if the passivating film
is sufficiently thin."

The mechanical properties of passive films of iron and a number of other metals and alloys that exhibit passivity have been
studied by Bubar and Vermilyea[21]. They found a wide range of
ductility for the passive films on the metals they studied. However, the noncrystalline films that form on 304 stainless steel
was more ductile than the passive film on iron. The greatest
ductility for all the metals they studied was exhibited by the
noncrystalline films that form on tantalum. Thus, it appears that
noncrystalline passive films may be more ductile than crystalline films.

MECHANISM OF FORMATION

A number of mechanisms of passive film formation have been
proposed (see Fromhold[22] for a review). The two most widely suggested mechanisms have been either some form of field assisted ion
conduction based on the oxidation theory developed by Cabrera and
Mott[23] or the place exchange mechanism of Sato and Cohen[24]. A
major problem in deciding which mechanism is operative for
iron in neutral solutions is that the kinetics of passive film
growth follow equally well either inverse logarithm or direct
logarithmic rate laws. Fig. 2 shows the extent of this problem for
iron in a neutral borate-buffer solution. The inverse logarithm

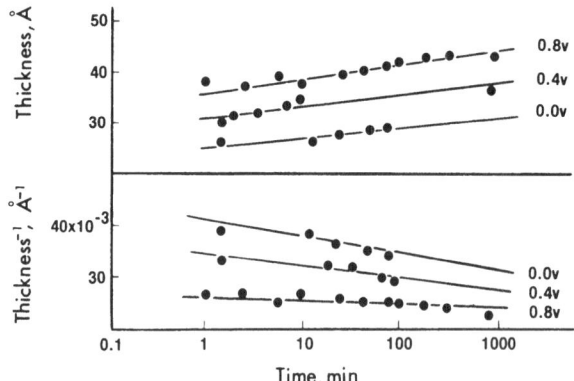

Figure 2. Direct and indirect logarithmic plots of the growth of a
           passive film on iron by potentiostatic anodic polarization
           at different potentials in a pH 8.4 borate buffer solution
           (from 25). (Reprinted by permission of the publisher,
           The Electrochemical Society, Inc.)

law indicates a field assisted ion conduction mechanism; the direct logarithmic law can be expected from a place exchange mechanism. Fromhold[22] has indicated that neither mechanism can explain all aspects of the kinetics. Thus, the situation is both more complicated that the two most popular mechanisms would suggest and other, perhaps, non-kinetic measurements, e.g. spectroscopic, electrochemical, and electrical[26,27,28] ones may be needed to obtain a sounder picture of the mechanism of passive film formation.

BREAKDOWN OF PASSIVITY

As stated at the beginning, breakdown of passivity leads to the most damaging kinds of corrosion, such localized forms of corrosion as pitting, crevice corrosion, intergranular attack. This brief discussion will be concerned only with chemical breakdown--those processes that are brought about by the chemical alteration of the passive film or the environment so that the film can no longer effectively prevent destructive anodic currents from flowing. Mechanical breakdown of passivity which results from the application of externally applied stress will not be discussed. More detailed discussions of breakdown have been given by Kruger[29], Galvele[30], and, most recently, by Janik-Czachor, Wood and Thompson[31].

Phenomenology and Models of Breakdown

Four major phenomena have been observed that are characteristic of breakdown:

a. A certain critical potential E_c must be exceeded, called by some the critical potential for pitting.

b. Damaging species (for example, Cl^- or the higher atomic weight halides) are needed to initiate and propagate breakdown.

c. There exists an induction time separating the initiation of the breakdown processes by the introduction of conditions conducive to breakdown and the completion of the process when pitting or crevice attack commences.

d. Breakdown occurs at highly localized sites.

A number of models have been proposed that seek to incorporate the above phenomena in the breakdown process. They will be described but a discussion of how well they encompass the phenomenology of breakdown is beyond the scope of this paper. Such discussion is given elsewhere.[29,30,31]

1. Adsorbed ion diplacement models. Two mechanisms have been suggested under this general model. First, there is the mechanism

suggested by Kolotyrkin[32] and Uhlig.[33] This model considers the
passive film to be an adsorbed film (probably a monolayer) of
oxygen. Breakdown occurs when a more strongly adsorbing damaging
anion (e.g. chloride ion which is highly polarizable and hence
capable of adsorption) displaces the oxygen forming the passive
film. Once the Cl⁻ ion is adsorbed at the surface, breakdown
commences. When other anions are present, they can compete with
the Cl⁻ for sites and inhibit breakdown.

The second model that can be considered as falling under the
general heading of an adsorbed ion displacement model is that described by Hoar and Jacob[34]. It has, however, elements in common
with the ion migration models also. In this model, a small number
(3 or 4) of halide ions jointly adsorb on the surface of the passive film around a lattice cation. The probability of formation
of such a high energy complex is small and thus requires a high
activation energy for formation. Once formed, however, the halide
ions will readily remove the cation from the passive film lattice
(this model assumes a three dimensional passive film). The film
is thus made thinner at the site where the complex first formed and
the stronger anodic field at the thinned site will rapidly pull
another cation through, where it will meet more halide ions, complex with them, and thereby enter the solution. Thus, once started,
the passive film is locally thinned and breakdown proceeds, using
Hoar's word "explosively." Recent work[35,36,37] provides electrochemical evidence for passive film thinning at localized sites.

    2. Ion Migration of Penetration Models. The theoretical
models that can be grouped under this heading all require the
penetration of damaging anions through the passive film, the breakdown process being complete when the anion reaches the metal-film
interface. Most of these models consider the passive film to be
three dimensional. The models differ widely, however, in the mode
of penetration.

At one extreme is a model that is quite old and was suggested
by Evans[38] and co-workers. More recently, Richardson and Wood[39]
have revised the pore mechanism and conclude that breakdown of the
passive film does not occur at all because there always exist pores
or defects in the film that allow instantaneous penetration and
hence instant localized corrosion.

At the other extreme of the types of penetration models are
those models involving migration of the damaging anion through a
lattice, via defects, or via some sort of ion exchange process.
The processes which involve ion migration in a lattice can occur
in a variety of ways. Hoar, et al[40] suggested anion entry without
exchange to produce "contaminated" passive films. The site where
this occurs becomes a path of high conductivity where high cation
currents and hence the initiation of pitting are possible. Pryor

and co-workers[41] also proposed the production of an anion contaminated film but suggested that cation vacancies are created which lower ionic conductivity and promote breakdown. Another ion migration scheme can involve exchange of $O^=$ or OH^- with the damaging anion.[42] Equation (1) shows that such a process can also create anion vacancies that will further enhance the migration of damaging anions to the film-metal interface.

$$Cl^-_{solution} + \square 0^= + 20H^-_{lattice} = Cl^-_{lattice}$$

$$+ 2 \square _{OH^-} + 20H^-_{solution} \tag{1}$$

3. Chemico-Mechanical Models. This group of theoretical models which involves chemically induced mechanical disruption of the passive film is perhaps the newest. One version was briefly described by Hoar.[43] He postulated that the adsorption of a damaging anion on the surface of the passive film lowers the interfacial surface tension (surface free energy) at the solution interface until "peptization" occurs. This "peptization" results from the mutual repulsion of the adsorbed anions. When the repulsive forces are sufficient, the passive film cracks and damaging anions can attack the metal exposed.

Sato[44] has extended this concept of chemical mechanical breakdown by providing a more detailed examination of the model. He has suggested that high fields could lead to mechanical rupture of thin films by electrostriction pressures exceeding the compressive fracture strengths of the film. The so-called film pressure this generates is given by:

$$p - p_0 = \frac{\varepsilon(\varepsilon-1)E^2}{8\pi} - \frac{\gamma}{L} \tag{2}$$

where p = film pressure, p_0 = atmospheric pressure, ε = film dielectric constant, E = electric field, γ = surface tension, and L = film thickness.

Sato pointed out in his description of his mechanical breakdown theory that whether a pit nucleates or dies depends not only on breakdown but also on film reformation. This concept, the interplay between breakdown and repair, underlies the last model of the chemico-mechanical group to be discussed. Videm[45] has recently proposed that it is the interplay between the breakdown (although he does not specifically say so, this could be mechanical as Sato has suggested) and repassivation that determines whether a passive film will suffer permanent breakdown that leads to pitting. Recent measurements by Bertocci[45] of small current fluctuations ("corrosion noise") found that the level of these fluctuations increase markedly when breakdown conditions are introduced (chlor-

ide ions and a potential above E_c) suggest that a dynamic break-
down/repair process of the sort suggested by Videm may be occurring
when pitting is initiated.

   4. Localized Acidification. This model most strongly advoca-
ted by Galvele[30] proposed that breakdown initiates as a result of
the localized acidification that takes place at the passive film-
solution interface and by so doing causes local dissolution and
also hinders reformation of the passive film. The dissolution of
a metal in a solution containing damaging anions (those of a strong
acid) will result in the following reactions:

$$Me = Me^{z+} + Ze^- \qquad (3)$$

and then

$$Me^{z+} ZH_2O = Me(OH)_z + ZH^+ \qquad (4)$$

at potential below E_c only the reaction

$$Me + ZH_2O = Me(OH)_z + ZH^+ + Ze^- \qquad (5)$$

will occur and the metal will repassivate. But above E_c the acidity
created at a localized site produced according to reaction (5) can
create conditions for reaction(3). Reaction (3) followed by reaction (4) will maintain localized acidity at a metal surface.

   All of the models for breakdown just described contain common
elements. A unified model, if such is possible, which incorporates
all the phenomenlogy of breakdown in a convincing way remains to be
built.

Alloy Composition and Structure Factors.

   It is not possible within a limited space to discuss the many
complex and contradictory results concerned with the effect of alloy composition and structure factors that influence breakdown
initiation processes. A excellent review has been given by
Smialowska[47]. Instead, I will describe briefly how some of these
factors relate to the theoretical models for breakdown.

   1. Alloy Composition. Kolotyrkin[48] points out in a review
of pitting that the tendency to breakdown decreases with an increase in the content of N, Ni, Cr, and Mo, especially for the
latter two. The main effect of these beneficial alloying elements
is to shift E_c in a noble direction. This explains phenomenologi-
cally why they are beneficial but does not actually tell us how
they work. One way in which they can retard initiation processes
is by producing a passive film that is more difficult to penetrate

because it provides fewer diffusion paths. Revesz and Kruger[20] have suggested that some of these beneficial alloying elements can do this by producing (glassy) films. McBee and Kruger[16] have found experimental evidence for this when Cr is added to Fe and also have found reduced rates of penetration.[49]

2. Alloy Structure. Only the surface structure of alloys will be considered here. Three kinds of possible variations in surface structure have been studied. First, Kruger[50] has looked at the role of crystallographic orientation on breakdown tendency for iron. He found that the tendency to pit goes up as the surface approaches the [110] orientation. He also found that pitting density varied with crystallographic orientation. Another example is the single crystal work by Smialowska and colleagues[51] on Fe-16Cr. They found that nucleation occurred at crystal subboundaries as well as at other sites where there were not necessarily any metal inhomogeneities.

Also, surface heterogeneities play a big role in initiating breakdown. Smialowska and co-workers[51] have found extensive evidence for the nucleation of pits at inclusions such as sulfides or chromium oxide for a number of alloys. Bond and co-workers[52] showed that very small amounts of segregated metallic impurities in quite pure aluminum single crystals could produce breakdown nucleation sites.

The obvious, but sometimes contradictory, effect of surface structure or breakdown can be explained by all the theoretical models discussed because it is reasonable to expect adsorption, penetration, complex formation, or repassivation kinetics to be affected by surface heterogeneities.

NEW DIRECTIONS

The foregoing discussion of passivity and breakdown of passivity suggests that in order to produce passive films that will resist breakdown it will be necessary in the future to develop new alloys and new environments (inhibitors) that will form films that:

a) Have structures that resist diffusion of damaging species

b) Have high ductility to resist mechanical breakdown

c) Have low solubility

d) Have low electron conductivity

e) Have rapid rates of repassivation

The preceding discussion has hinted at ways to achieve some of
these ends, for example, the development of alloys whose structures
and compositions promote the formation of noncrystalline (glassy)
passive films.

ACKNOWLEDGEMENT

I am grateful to the Office of Naval Research, which supported
this work under contract NAONR 18-89 NRO 36-082.

DISCUSSION

Dr. Myron A. Coler, Coler Engineering Co., New York, New York: I'd
like to point out that I had occasion over forty years ago to verify
experimentally work that had been done on the effect of a direct
magnetic field parallel to the D.C. electrical field in E-I active-
passive-transpassive studies. Strong magnetic fields produced major
shifts. Removal of the field removed the effect reversibly. This
suggests a study tool - other reagents can, of course, be added.
This also might lead to protective treatments. "Pure" soft Armco
Fe was used.

I add to my remarks that there have been tremendous advances in
magnetic field equipment in the interim. Superconducting, highly
homogeneous field or high gradient field magnets are commercially
available, etc.

Dr. Jerome Kruger: I am not too surprised at the interesting results
you describe. Since passive film growth may involve a field assisted
diffusion of ions, it seems reasonable that a magnetic field could
affect such a process. Using a magnet to protect iron from corrosion is an intriguing idea.

REFERENCES

1. H.H. Uhlig, in "Passivity of Metals," Frankenthal and Kruger,
    eds., Electrochem. Soc., Princeton (1978) p. 1.
2. M. Faraday, "Experimental Researches in Electricity", Vol. 2
    p. 250 London (1844).
3. "Passivity of Metals," R.P. Frankenthal and J. Kruger, eds.,
    Electrochem. Soc., Princeton (1978).
4. H.H. Uhlig,"Corrosion and Corrosion Control", John Wiley, New
    York (1963) p. 58.
5. C. Wagner, described in reference 4 above.
6. U.R. Evans, J. Chem. Soc. (London), 1927, 1020.
7. M.A. Gershaw and R.S. Sirohi, J. Electrochem. Soc., 118, 1558,
    (1971).
8. J. Kruger, J. Electrochem. Soc., 110, 654 (1963).

9. C.L. Foley, J. Kruger and C.J. Bechtodt, J. Electrochem. Soc.,
    114. 994 (1967).
10. H.T. Yolken, J. Kruger, J.P. Calvert, Corros. Sci.,8,103 (1968).
11. M.C. Bloom and L. Goldenberg, Corrosion Sci, 5, 623 (1965).
12. W.E. O'Grady, J. Electrochem. Soc., 127, 555 (1980).
13. T. Noda, K. Kudo and N. Sato, Z. Phys. Chem. N.F., 98, 271 (1975).
14. K. Kudo, T. Shibata, G. Okamoto and N. Sato, Corros. Sci.,8,
    809 (1968).
15. N. Sato, in "Passivity of Metals," R.P. Frankenthal and J. Kruger,
    eds., Electrochemical Soc., Princeton (1978) p. 29.
16. C.L. McBee and J. Kruger. Electrochemica Acta, 17, 1337 (1971).
17. G. Okamoto, Corros. Sci., 13, 471 (1973).
18. J.S.L. Leach, Surf. Sci. 53, 257 (1975).
19. T.P. Hoar, J. Electrochem. Soc.,117.,17c (1970).
20. A.G. Revesz and J. Kruger, in "Passivity of Metals," Frankenthal
    and Kruger, eds., Electrochem. Soc., Princeton, (1978) p. 137.
21. S.F. Bubar and D.A. Vermilyea, J. Electrochem. Soc., 113, 892 (1966).
22. A.T. Fromhold, in "Passivity of Metals," Frankenthal and Kruger
    eds., Electrochem. Soc., Princeton (1978) p. 59.
23. N. Cabrera and N. Mott, Rep. Progr. Phys., 12, 163 (1949).
24. N. Sato and M. Cohen, J. Electrochem. Soc., 111,52 (1963).
25. J. Kruger and J.P. Calvert, J. Electrochem. Soc., 114,43 (1967).
26. D.J. Wheeler, B.D. Cahan, C.T. Chen and E. Yeager in "Passivity
    of Metals," Frankenthal and Kruger, eds.; Electrochem. Soc.,
    Princeton (1978) p. 546.
27. J.W. Schultze, ibid., p. 82.
28. F.N. Delnick and N. Hackerman, ibid., p. 116.
29. J. Kruger in "Passivity and its Breakdown on Iron and Iron
    Base Alloys," Staehle and Okada, eds., Nat. Assoc. Corros.
    Engineers, Houston (1976) p. 91.
30. J.R. Galvele in "Passivity of Metals," Frankenthal and Kruger,
    eds., Electrochemical Soc., Princeton (1978) p. 285.
31. M. Janik-Czachor, G.C. Wood and G.E. Thompson, Brit. Corros. J.,
    in press.
32. Ya. M. Koloyrkin, J. Electrochem. Soc., 108, 209 (1961).
33. H. Bohni and H.H. Uhlig,J. Electrochem. Soc., 116 906 (1969).
34. T.P. Hoar and W.R. Jacob, Nature,216, 1299 (1967).
35. K.E. Heusler and L. Fischer, Werkstoffe u. Korros., 27, 697 (1976).
36. M. Janik-Czachor, Werkstoffe u. Korros., 30, 255 (1979).
37. H.H. Strehblow, B. Titze and B.P. Loechel, Corros. Sci.,19,1047
    (1979).
38. U.R. Evans, L.C. Bannister, and S.C. Britton, Proc. Roy. Soc.,
    A131, 367 (1931).
39. J.A. Richardson and G.C. Wood, J. Electrochem. Soc.,120, 193 (1973).
40. T.P. Hoar, D.C. Mears, and G.P. Rothwell, Corrosion Sci.,5, 279
    (1965).
41. M.A. Heine, D.S. Keir, and M.S. Pryor, J. Electrochem. Soc.,
    112, 29 (1965).
42. C.L. McBee and J. Kruger "Localized Corrosion" Staehle, Brown,
    Kruger, and Agrawal, eds., National Association of Corrosion
    Engineers, Houston, (1974), p. 252.

43. T.P. Hoar, Corrosion Sci., 7, 335 (1967).
44. N. Sato Electrochica Acta, 19, 1683 (1971).
45. K. Videm, Kjeller Report KR-149, Institutt for Atomenergi, Kjeller, Norway (1974).
46. U. Bertocci, J. Electrochem Soc, 127, 1931 (1980).
47. Z. Szklarska-Smialowska, "Localized Corrosion," Staehle, Brown, Kruger, and Agrawal, eds., National Association of Corrosion Engineers, Houston, 312 (1974).
48. Ya. M. Kolotykin, Corrosion 19 261 (1963).
49. C.L. McBee and J. Kruger, "Passivity and its Breakdown on Iron and Iron Base Alloys," Staehle and Okada, eds., National Assoc. of Corros. Engineers, Houston (1976) p. 131.
50. J. Kruger, J. Electrochem. Soc., 106, 736 (1959).
51. A. Szummer, Z. Szklarska-Smialowska, and M. Janik-Czachor, Corrosion Sci, 8 ,827 (1968).
52. A.P. Bond, G.F. Bolling, and H.A. Domian, J. Electrochm. Soc., 113 ,773 (1966).

NEW DIRECTIONS IN THE ALUMINUM INDUSTRY

Theodore R. Beck

Electrochemical Technology Corp.
3935 Leary Way N.W.
Seattle, Washington 98107

Like Case Institute, now celebrating its centennial, the
electrolytic aluminum industry will soon celebrate its 100th anniver-
sary--that of the independent inventions of Hall and of Heroult in
1886 of an economic electrolytic process to produce aluminum metal.
Up to that time aluminum was available only in limited quantities
at a high price from sodium reduction of aluminum chloride. The
early history of aluminum production is described in several references (1-3). Hall and Heroult electrolyzed alumina dissolved in a
cryolite "bath" in carbon-lined crucibles using carbon anodes.
Charles Martin Hall produced his first globules of aluminum metal
on February 23, 1886, in his home laboratory at Oberlin, Ohio, not
far from Cleveland.

The Pittsburgh Reduction Company (later to become the Aluminum
Company of America) was formed in 1888 to exploit Hall's patent. The
first two cells, or "pots," had cast iron shells 24 inches long, 16
inches wide, and 20 inches deep and were lined with 3 inches of
baked carbon. The cells had six to ten 3-inch diameter carbon
anodes, operated at 1700 to 1800 amperes, and together produced
about 50 pounds of aluminum per day. The technology has made enormous quantitative strides since that time; cell size has increased
to over 250 kA, specific energy consumption has decreased from about
22 kWh/lb to about 7 kWh/lb, and worldwide production has increased
to over 15 million tons per year. Cell design and operation, however, is still qualitatively the same; alumina dissolved in a cryolyte "bath" is electrolyzed between a carbon anode and an aluminum
"metal pad" in a carbon-lined pot.

The growth of aluminum production and electrical energy production in the United States since 1900 is shown in Fig. 1. An inter-

331

esting parallel is seen between the rates of growth of the two
industries; over a 50-year span from 1920 to 1970 they each grew at
the average rate of about 7% per year. Aluminum production had
greater perturbations during the depression of the 1930s and the war
effort of the 1940s. The aluminum industry on the average has used
about 4% of the electrical energy in the United States, peaking at
about 7% during World War II.

Consistent with the theme of this symposium, one may ask what
are the future directions of aluminum and electrical energy production in the United States? The large growth rate of electrical
energy production until about 1970 can be attributed to decreasing
costs relative to other energy forms due to economies of increasing
scale and to increased thermodynamic efficiencies of higher-tempera-
ture thermal plants. Sharp increases in fossil energy costs in the
early 1970s, however, caused a dramatic decrease in the growth rate
of electrical energy production. The utility industry, which once
believed the 7% growth rate to be a natural law, has been revising
its forecasts steadily downward to a present 3½% growth rate for the
next decade or two (4). The aluminum industry predicts a 3% per
year increase in primary aluminum supply worldwide and 1% per year
increase from recycled aluminum in the next five years (5).

Major markets for aluminum for a recent ten-year period (6)
are shown in Fig. 2. It is difficult to make accurate predictions
for new directions in the marketplace, but increasing petroleum
costs provide an incentive to increase aluminum usage in surface

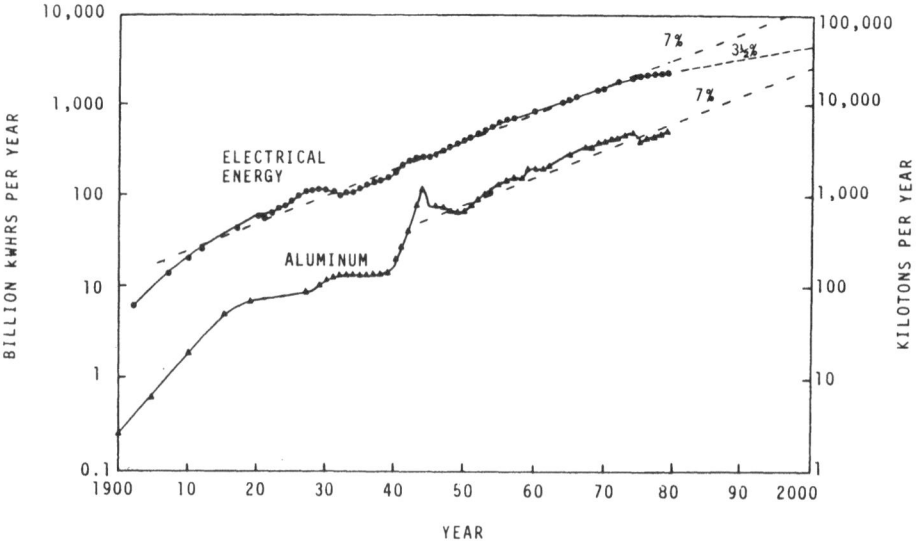

Fig. 1.   Growth of aluminum and electrical energy production
          in the United States.

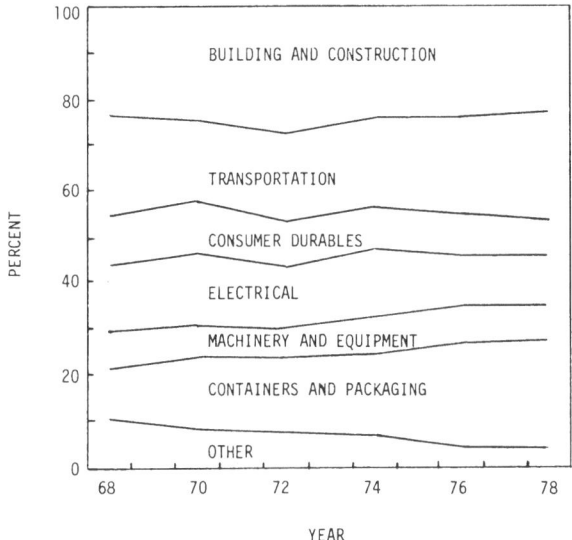

Fig. 2. Product shipment by major market (United States).

transportation equipment. Each pound of aluminum used in an automobile can displace about two pounds of steel resulting in a net
saving of one pound of weight for the vehicle (7). The energy
saving per pound of aluminum used may be approximated as follows.
Assume as a basis a 4,000-pound car which operates at 15 miles per
gallon of gasoline for a total of 100,000 miles and gasoline with a
density of 6 lb/gal and a heating value of 20,000 Btu/lb (8). These
numbers give a specific energy consumption of 200,000 Btu/lb of
vehicle. The average electrical energy cost to make aluminum is
presently about 7.5 kWh/lb in the United States (9). This number
translates to 79,000 Btu/lb aluminum based on 10,500 Btu of fossil
fuel energy consumed per kWh of electrical energy delivered to an
industrial customer (9). Each pound of aluminum used therefore
saves about 2½ times its fuel-propulsion energy for the life of an
automobile. The energy savings accrue to the owner, however, not
the manufacturer. The energy saving stimulus to increase use of
aluminum is less intense to the manufacturer and the impact on the

growth rate is not clear.   Furthermore, there is also competition
from plastics.

     Although prediction of markets and production is uncertain at
best, it is quite clear that increasing energy costs will exert
continued pressure on the aluminum industry to increase efficiency of
production.   Great strides have already been made in decreasing electrical energy consumption for production of aluminum as shown in Fig.
3 (9).   The improvements to date have been largely due to increase in
pot size and thus increased thermal efficiency as well as general
tightening of operating conditions.   The ultimate limit for a self-
sustaining pot operating at 975°C is 2.89 kWh/lb based on the enthalpy change for the overall reaction,

$$Al_2O_3(25°C) + 3/2\ C(25°C) \longrightarrow 2\ Al(975°C) + 3/2\ CO_2(975°C) \quad (1)$$

Methods for decreasing the actual specific electrical energy consump-

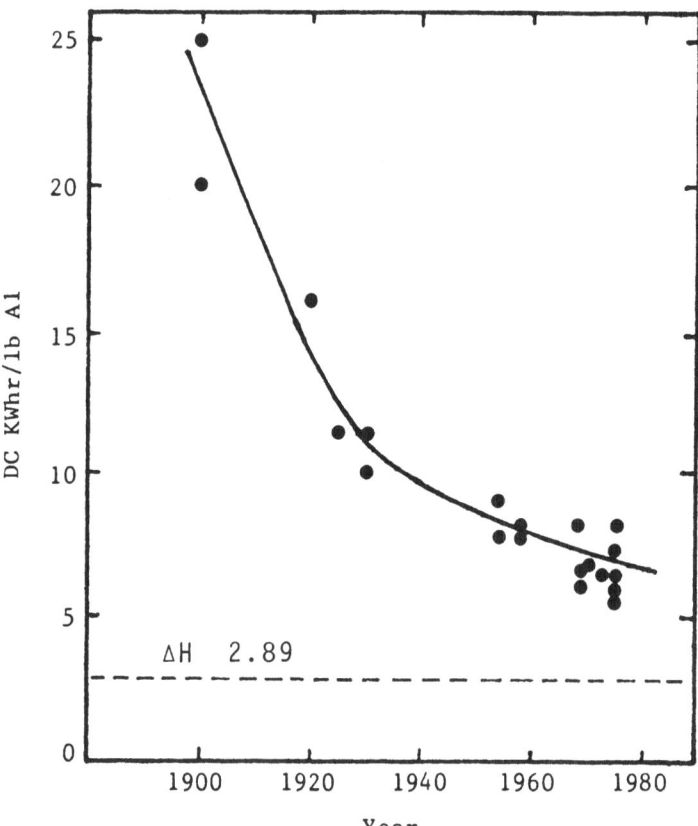

Fig. 3.   World historical trend in specific
           electrical energy consumption.

tion of aluminum production based on published information are described in this paper.

The Bayer-Hall-Heroult process shown in Fig. 4 is first described
in more detail. Bauxite ore containing 30% to 50% alumina is crushed,
ground, and digested with sodium hydroxide to make a sodium aluminate
solution. The iron oxides, silica, and other impurities are settled
as red mud and disposed. The solution is cooled and seeded with
aluminum hydroxide particles to precipitate aluminum hydroxide which
is separated, washed, and dried. The precipitate is calcined at
1,100°C to alumina which is shipped in bulk to the Hall-Heroult
reduction plants.

Petroleum coke and coal tar pitch are the other basic raw materials for the reduction plant. All materials have to be very pure in
respect to metals more noble than aluminum, because all such metals
would be codeposited with aluminum, decreasing its purity. The coke
and pitch are mixed hot and either pressed and then baked at 1,100°C
to make prebake anodes or fed as a paste to Soderberg anode cells.
Soderberg anodes are baked in place in the cells by excess heat generated by electrolysis.

A typical cross section of a Hall-Heroult cell with prebaked
anodes is shown in Fig. 5 (10). A steel shell containing a carbon
lining is supported on steel cradles. Alumina is used for thermal
insulation on the bottom. The carbon lining is either made of prebaked blocks or of a paste of anthracite coal and pitch that is
rammed and baked in place. Steel collector bars embedded in the
lining carry current out of the cell through flexible connectors to
the cathode ring bus. The cavity in the lining holds a molten aluminum cathode and the "bath" of molten cryolyte containing dissolved
alumina. Carbon anodes are suspended in the bath from the anode bus.
Current is carried from the anode bus through aluminum or copper
anode rods to steel stubs fastened by cast iron poured into holes in
the tops of the anodes. Current in the series potline is carried
from one pot to the next by a riser bus from the cathode ring bus of
one cell to the anode bus of the adjacent cell.

In operation, bath freezes on the carbon sidewall to form a
protecting "ledge" and over the top to form a "crust." Alumina is
periodically fed to the crust from an "ore" bin in the pot super-
structure. Crust is then broken along the centerline of the pot with
a steel rake which is also used to stir alumina into the bath. The
pots are deliberately underfed to avoid accumulation of alumina or
"muck" on the bottom. This procedure results in concentration polarization leading to high voltage (up to 100 V) "anode effect" about
once per day. Aluminum is siphoned periodically from the metal pad
into crucibles which convey it to the casting area. Enough metal is
always left in the pot to cover the lining and serve as the cathode.

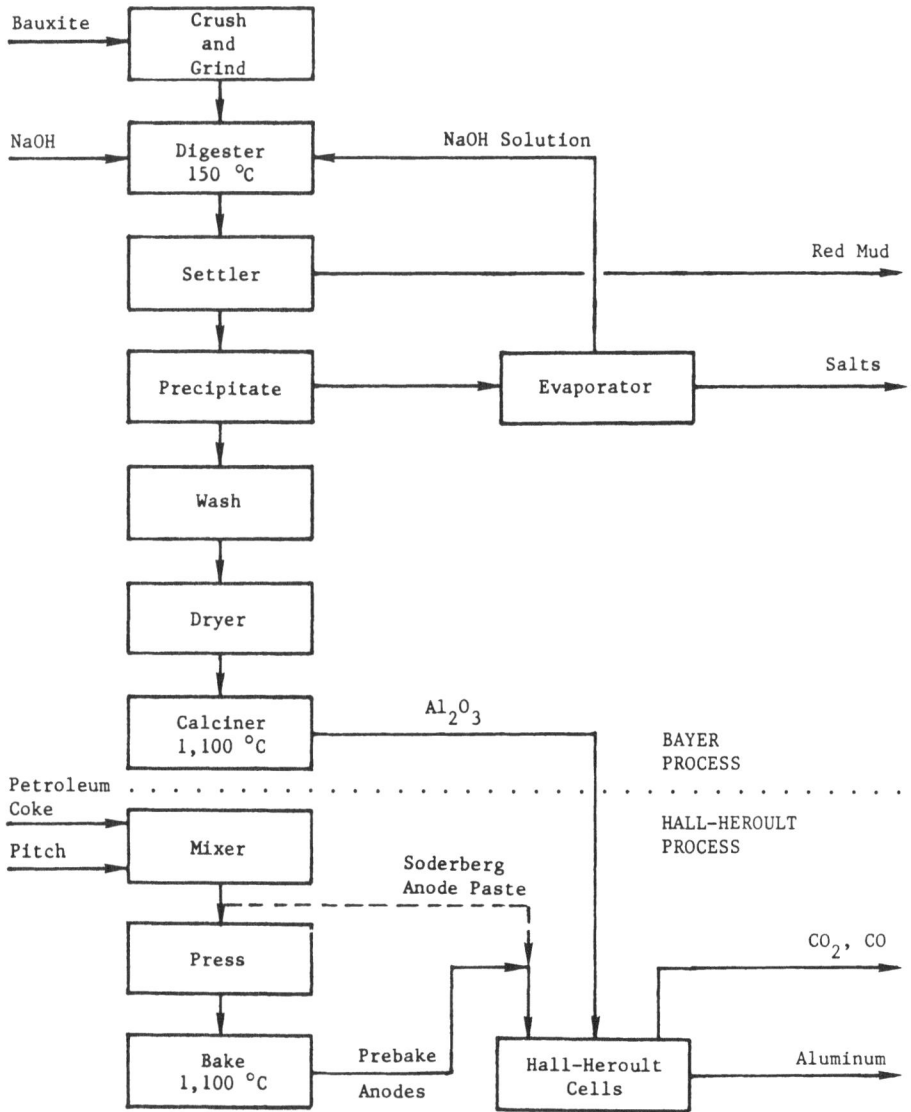

Fig. 4. Flowsheet for the Bayer–Hall–Heroult Process
to produce aluminum from bauxite.

The physical chemistry and electrochemistry of reactions in Hall-
Heroult cells have been extensively studied (11), but the experimental

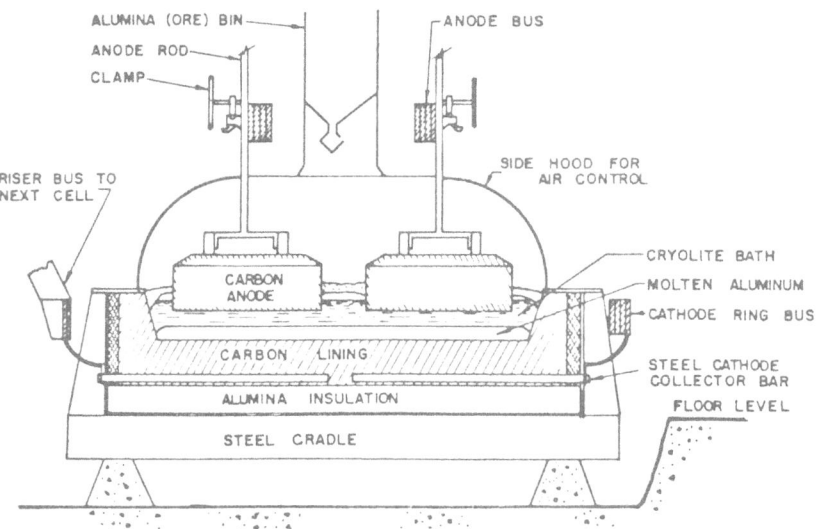

Fig. 5. Prebake aluminum reduction cell. (From the ENCYCLOPEDIA
OF ELECTROCHEMISTRY edited by C. Hampel. Copyright ©
1964 by Van Nostrand Reinhold Co. Reprinted by permission).

difficulties are great, and the detailed mechanisms are still not
completely understood. In general there is evidence that cryolyte,
Na_3AlF_6, dissociates to Na^+, $AlF_6{}^{3-}$, $AlF_4{}^-$ and F^- ions, and alumina
dissolves and reacts to form $Al_2OF_6{}^{2-}$ ions. The $AlF_6{}^{3-}$ ions are discharged at the cathode to form aluminum, and the $Al_2OF_6{}^{2-}$ ions discharge at the carbon anode to form CO_2 as the primary gaseous product.

Current efficiency is now usually in the range of 85% to 92%
(12). The inefficiency reaction is mostly due to reaction of dissolved aluminum and CO_2 in the bath.

$$2Al + 3CO_2 \longrightarrow Al_2O_3 + 3CO \qquad\qquad (2)$$

The reaction appears to be mass-transport limited, but it is unclear
whether aluminum or other reduced species is oxidized at the anode,
dissolved CO_2 is reduced at the cathode, or the reaction takes place
in the bath between dissolved components. Electromagnetic stirring
of the metal pad and bath caused by interaction of the current flow
through the cell with the magnetic field from the current in the busbar between cells promotes the inefficiency reaction 2. Electromagnetic forces also cause waves and distortions in the metal pad
giving nonuniform current distribution between anodes and shorting of
anodes to the pad if the anode-cathode distance is too small.

A typical distribution of voltage in a Hall-Heroult cell is
as follows (13). The thermodynamic reversible potential for reaction

1 based on ΔG of reaction for all components at 975°C is 1.2 V, which
is 1.0 V lower than the reversible potential to dissociate alumina
into aluminum and oxygen. The activation overpotential for the anodic formation of CO_2 is about 0.5 V, and the cathode overpotential
is less than 0.1 V. The largest voltage loss in a cell, about 2 V,
is due to ohmic drop in the bath between the anode and cathode. The
bath conductivity is high, but the current density is also high,
about 1 A/cm^2 (6 A/in^2), because an economic balance is made with
high capital costs. CO_2 gas bubbles near the anodes contribute
about 0.2 V to the bath resistance (12). Ohmic losses in the lining
(0.5 V), in the anodes (0.3 V), and buswork (0.2 V) add another
0.8 to 1.0 V. Time-averaged anode effects may add an average 0.1 V.

Specific energy for aluminum production is related to cell voltage and current efficiency by (9)

$$kWh/lb = \frac{135 \ V}{\% \ CE} \qquad\qquad (3)$$

Decreasing the anode-cathode distance (ACD) decreases voltage drop in
the bath and cell and also decreases current efficiency due to
reaction 2. The present economic balance between these two effects
occurs at an ACD of 1.5 to 1.75 inches. A pot voltage of 4.5 V and
a current efficiency of 90% give a specific energy of 6.75 kWh/lb in
equation 3.

Worldwide trends in cell current and current density since the
inception of the Hall-Heroult process are shown in Fig. 6 (9). The
approximately one-hundred-fold increase in cell current was made to
achieve economies in capital and labor costs. The upward trend is
likely to continue in the future; the major impediment to increase
in current has been electromagnetic effects. The early steep decrease in current density can be attributed to increasing cell size
and rapidly moving up the learning curve. The high surface-to-volume
ratio is critical for cells under about 10,000 A; a relatively high
heat generation rate is therefore required for small cells. The
recent rapid escalation in electrical energy costs has shifted the
economics of aluminum production to larger, lower-current-density
cells with lower cell voltage and specific energy consumption. In
the past five years the economic current density has changed from
about 6 A/in^2 to about 4.5 A/in^2 for new plants in the United States.

An energy flow diagram for the Bayer-Hall-Heroult process is
presented in Fig. 7 (9). The diagram is based on average production
data in the United States in 1975; the specific energy consumption is
7.6 kWh/lb in this diagram. A rectifier efficiency of 97% was
assumed. The fossil fuel energy to delivered ac electrical energy
at the plant was assumed to be 10,500 Btu/kWh. This number corresponds to an energy conversion and delivery efficiency of (3,413/
10,500)(100) = 32.5%. Although a large fraction of the aluminum

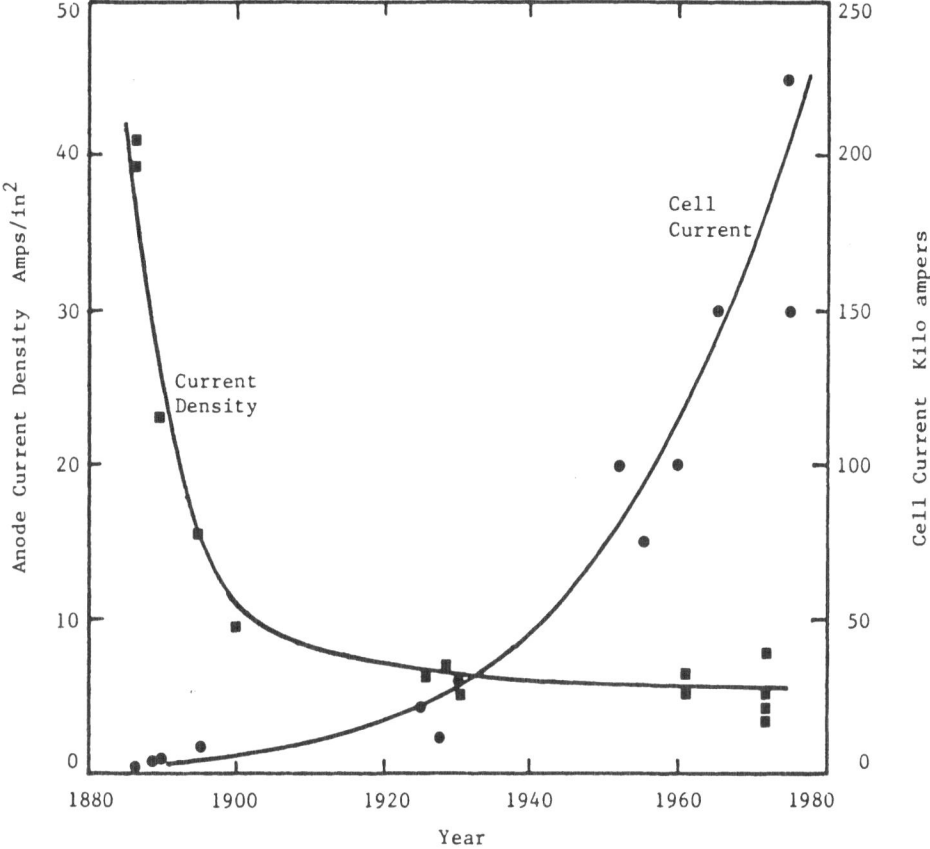

Fig. 6.   Trends in Hall-Heroult cell current capacity
          and anode current density 1886-1976.

industry in the United States now operates on hydropower, new production facilities will be largely based on thermal power sources,
and thus a fossil fuel basis is used in Fig. 7.

    Several new processes and potential process improvements are
being developed. The most successful alternative to date to the
Hall-Heroult process is the Alcoa aluminum chloride process (9, 14,
15). Operation of a 15,000 T/yr pilot plant began in Texas in June
1976 and is still under evaluation. A simplified flow sheet is
given in Fig. 8 (16). Technical details are limited to what can be
deduced from patents (17). Bayer alumina is reacted with chlorine
and petroleum coke at 700° to 900°C. Aluminum chloride is condensed
from the $AlCl_3$-CO/CO_2 product mixture at 70°C, and the CO/CO_2 gases
are separated out and rejected. Anhydrous aluminum chloride is fed
to the cells in which it is electrolytically decomposed to aluminum
metal and chlorine gas. This chlorine is recycled to the reactor.

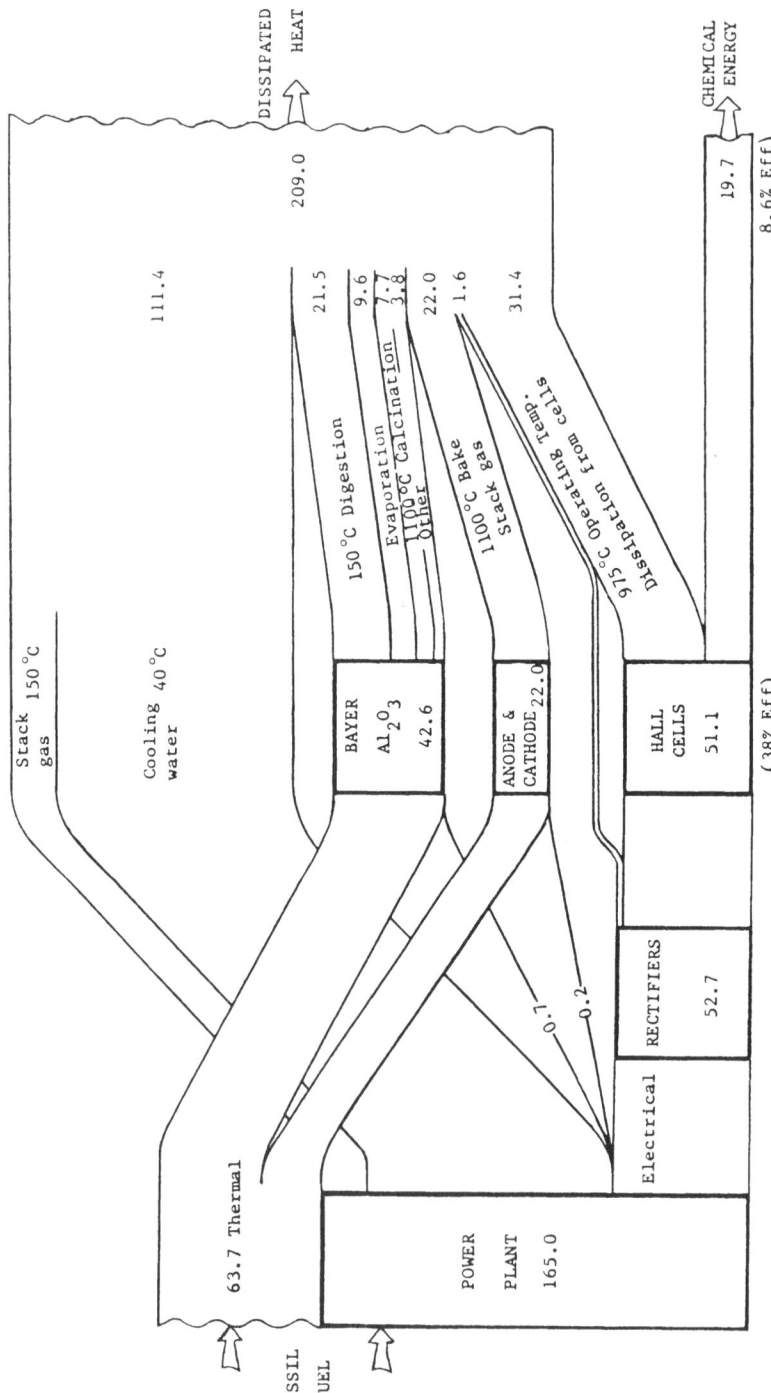

Fig. 7. Energy flow diagram for Bayer-Hall-Heroult process for production of aluminum.
Numbers = million Btu/Ton Al

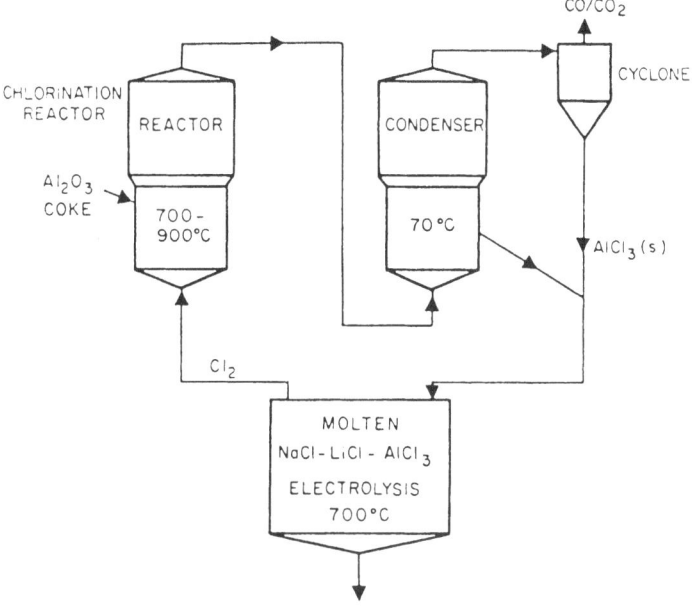

Fig. 8. Alcoa aluminum chloride process flow sheet.

Electrolysis is carried out at about 700°C with a 50% sodium chloride–45% lithium chloride–5% aluminum chloride electrolyte. A battery of bipolar graphite electrodes as shown in Fig. 9 is used. Aluminum is deposited on the upfacing cathodic surfaces, and chlorine is deposited on the downfacing anodic surfaces. The aluminum globules flow across the cathodes and drop to the bottom of the cell; the chlorine gas bubbles flow across the anodes and rise to the top of the cell. The electrodes are apparently inclined to promote convective flow. Aluminum is periodically siphoned out by inserting a metal tube through a port in the cell.

Anode-cathode distance is about 1/2 inch and electrolyte conductivity is greater than for cryolite, so that cell voltage is lower than for Hall-Heroult cells. Current densities are in the range of 5 to 14 A/in^2, and current efficiency was cited to be 86% for a small, monopolar cell. A specific energy as low as 4.5 kWh/lb is claimed for large cells. Cell capital cost is substantially decreased from that of conventional Hall-Heroult cells, but a part of the

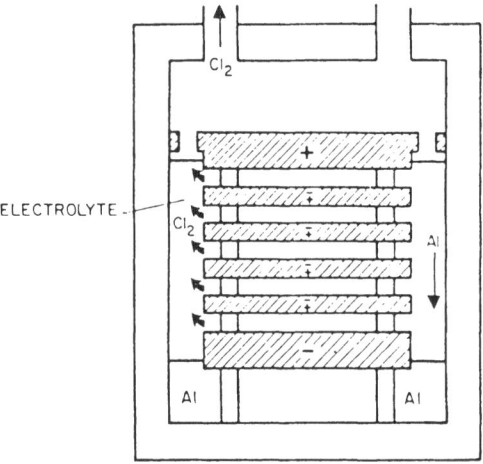

Fig. 9. Alcoa bipolar aluminum chloride cell.

trade is the requirement of the chemical plant to make AlCl₃. The
most serious technical problem appears to be avoiding or minimizing
air and moisture leaks which contaminate the AlCl₃ and result in
oxidation of the graphite anode surfaces.

An energy flow diagram estimated for a Bayer-aluminum chloride
process is given in Fig. 10. This diagram is based on information
disclosed in Alcoa patents and press releases and thermodynamic data.
Aluminum chloride is produced by chlorination of Bayer alumina with
petroleum coke reducing agent by the reaction

$$Al_2O_3 + 3C + 3Cl_2 \longrightarrow 2\ AlCl_3 + 3CO \tag{4}$$

The enthalpy of reaction at 800°C is +50 kcal/mol (18). No process
data are available, so a 50% thermal efficiency is assumed, which
gives an equivalent 6.8 x 10⁶ Btu/T of aluminum produced. The overall cell reaction is

$$AlCl_3 \longrightarrow Al + 1.5\ Cl_2 \tag{5}$$

At a working temperature of 700°C, the enthalpy change is +153.5
kcal/mol which is equivalent to a specific energy of 3.00 kWh/lb or
20.5 x 10⁶ Btu/T. The energy efficiency for the cells is 67% based
on an actual 4.5 kWh/lb. The overall process efficiency is 14.3% in

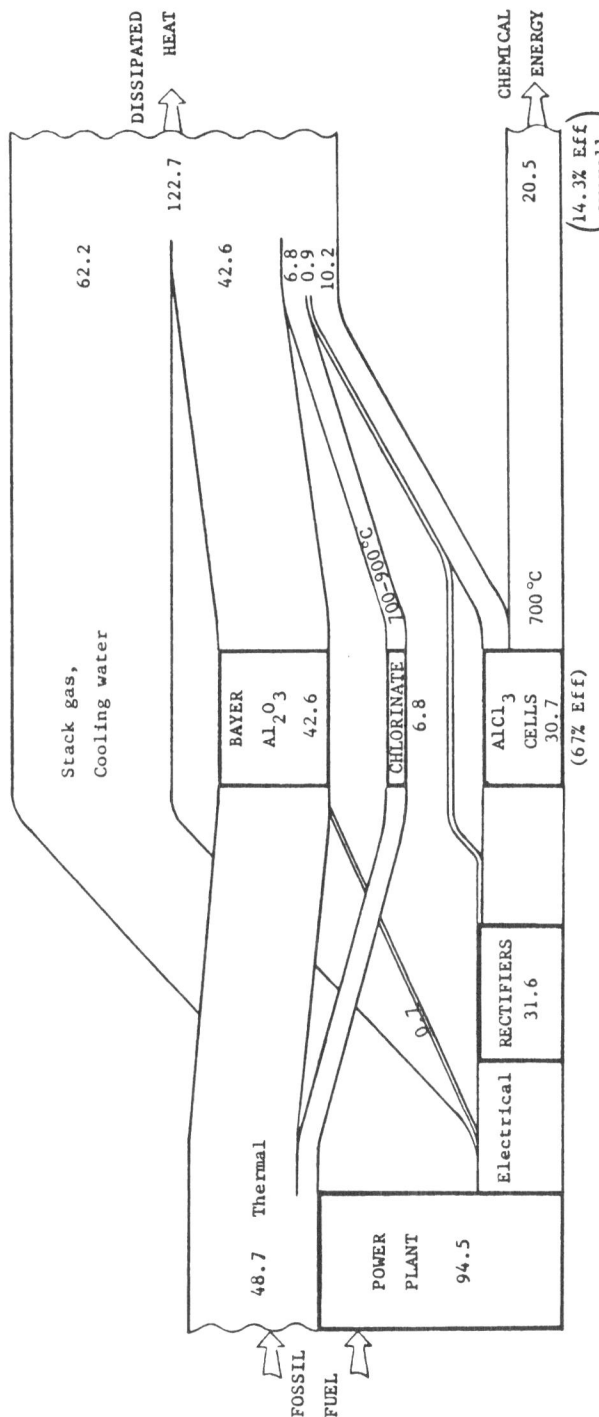

Fig. 10. Energy flow diagram for proposed Bayer—Alcoa AlCl₃ electrolysis process for production of aluminum.

Many attempts have been made to carbothermally reduce alumina to aluminum by reaction 1 using electric furnaces. To obtain pure aluminum requires over 3,000°C, and no such process has been able to compete with the Hall-Heroult process. It is possible to make aluminum alloys at a lower temperature, and aluminum bronze and other alloys have been made commercially by carbothermal reduction at various times. Recent increases in the cost of electrical energy have spurred interest in making aluminum-silicon casting alloys by carbothermal reduction.

Electrolytic reduction of aluminum sulfide and aluminum nitride require a lower free energy than reduction of aluminum oxide, but to date no practical processes have been developed using these cell-feed materials.

A potential improvement to the Hall-Heroult process that has received considerable attention in recent years is the use of refractory hard metal cathodes. Titanium diboride is the prime candidate because it has good electrical conductivity, it is resistant to attack by molten cryolite and aluminum, and it is wet by aluminum. The discovery that refractory hard metals might be used advantageously in aluminum reduction cells was made by Ransley (19) in the early 1950s. An excellent review of inert cathodes for aluminum electrolysis has recently been written (20).

The first attempted application of TiB_2 was as collector bars that directly contacted the metal pad through the sides of the cell. Several technical problems caused abandonment of this approach (21). The early-produced material was of insufficient purity (high oxygen), and the bars were subject to intergranular attack and stress corrosion cracking. Erratic current distribution resulted from a "pinch effect" in the liquid aluminum around the bars due to interaction of the current in the metal with strong magnetic fields from the busbars. It was also found that the current density in the bars had to be high enough that sufficient heat was generated within the bars to prevent bath from freezing on them during a pinch effect. To have the necessary zero-temperature gradient on the bars at their inboard ends required a current density in them which gave at least a 0.3 V potential drop. The TiB_2 collector bars thus gave no smaller potential drop than a conventional carbon lining when new. A final problem was that without generation of heat in the pot bottom, the muck tended to freeze on the bottom and build up.

Later it was realized that a more advantageous use of TiB_2 would be as a dimensionally-stable cathode as shown in Fig. 11 (22). In this case, TiB_2 tiles are laid in the carbon bottom of a cell under the two rows of anodes. The tiles are sloped so that the deposited

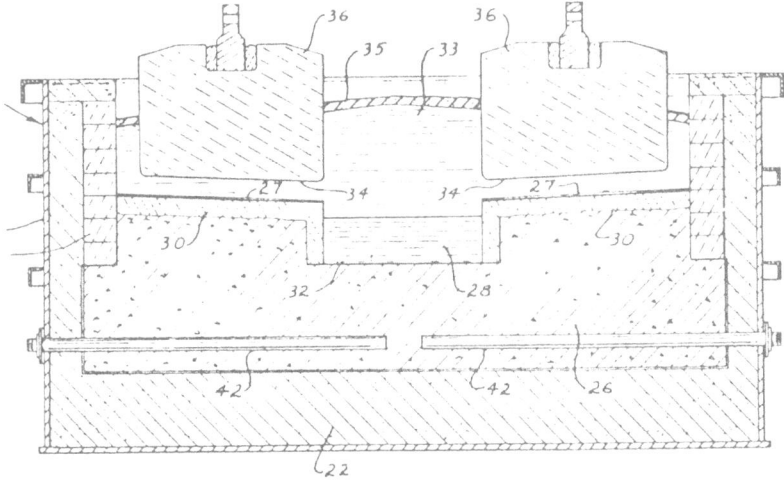

Fig. 11.   TiB$_2$ drained cathode plate cell adaptable
to conventional Hall-Heroult cell designs.

aluminum drains into a central channel in the cell where it collects
for siphoning. The cathodically-deposited aluminum wets the tiles
and forms only a thin layer. This cathode is dimensionally stable,
uninfluenced by electromagnetic effects, so that the anode-cathode
distance can be decreased without significant loss in current efficiency. The cell voltage and specific energy consumption are thereby
decreased. Breakage of tiles has been a severe problem.

An alternate method to use TiB$_2$ is tubes which fill with aluminum above the general metal pad level (23) as shown in Fig. 12. The
aluminum in these tubes is less influenced by electromagnetic effects
than the metal pad so the effective anode-cathode distance can be
decreased. A 10% to 20% decrease in specific energy consumption is
claimed by use of TiB$_2$ cathodes (9, 12).

Development of an inert, nonconsumable anode has long been a
goal of the aluminum industry (24). Such an anode would evolve oxygen rather than carbon dioxide so the overall cell reaction would be

$$Al_2O_3 \longrightarrow 2Al + 3/2\ O_2 \qquad\qquad (6)$$

The thermodynamic reversible potential for this reaction is 1.0 V
higher than for reaction 1, but there are other compensating effects.
The anodic overpotential for generation of oxygen in cryolyte is
negligible (12) compared to about 0.5 V for generation of CO_2, thus
the penalty for reaction 6 is only a net 0.5 V. The oxygen produced
may have other use in the smelter, whereas the CO_2 has no value. The

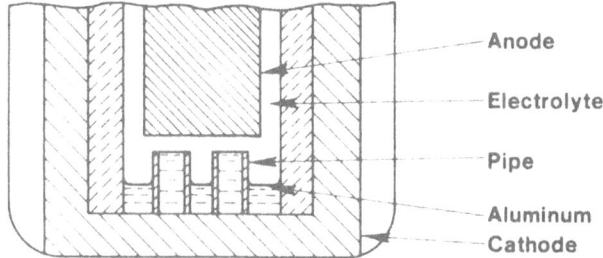

Fig. 12. Cell with tube cathodes, filled with molten aluminum.

process would be freed from dependence on petroleum coke, now in
increasingly tight supply, and the heat energy for anode baking
saved. The overall results could be near breakeven on total energy
used and some operating economies.

The long-range goal for nonconsumable anodes is to combine them
with a nonconsumable TiB_2 cathode to make bipolar electrodes. Close
anode-cathode spacing and economies in cell construction would accrue
similar to those for the $AlCl_3$ cells (Fig. 9). An example battery of
cells employing bipolar composite electrodes is shown in Fig. 13 (25).

Proposed materials for inert anodes have mainly been oxides or
mixtures of oxides. To date no completely satisfactory material has
been reported. Properties required are (24): insolubility in molten
cryolite, resistance to accidental contact by molten aluminum, resistance to oxidation, thermal stability, low specific resistance, low
oxygen overvoltage and no contamination of the aluminum product.

A need for inexpensive aluminum scrap reclaiming is increasing
as energy costs rise. Production of high-purity aluminum in a Hoopes
cell has long been practiced, but the energy cost is greater than for
making primary aluminum (26). In the Hoopes cell, illustrated in
Fig. 14, a dense aluminum-copper alloy is placed on the bottom of the
cell as the anode. The high-purity cathode floats on top of the
cell. The intermediate density bath is essentially cryolite with
barium fluoride added for density control. Metals lower than aluminum in the electromotive series remain in the cathode alloy. Metals
higher than aluminum in the electromotive series are not reduced in
the cathode. The specific energy consumption is high because a thick
bath layer is employed to avoid accidental contact of anode and
cathode.

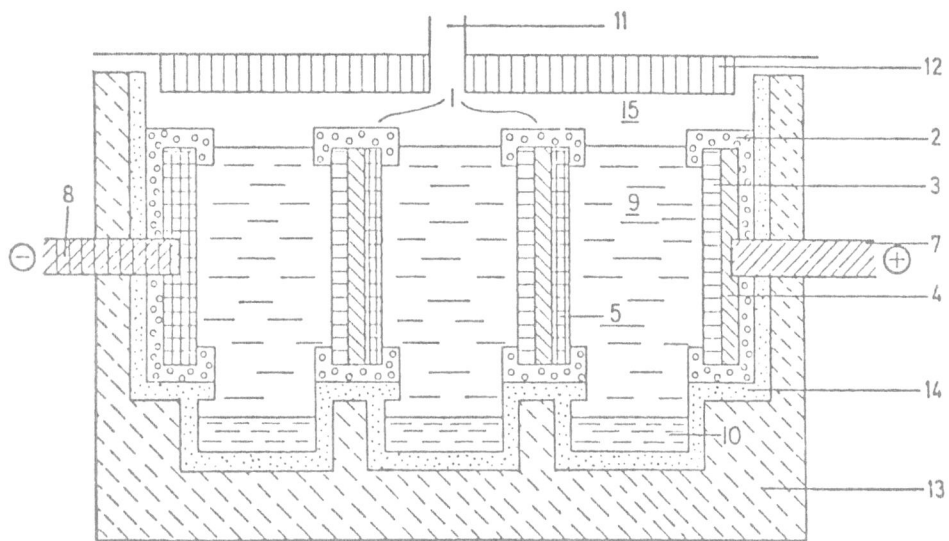

Fig. 13.   Battery with bipolar electrodes. 1. Vertical,
           bipolar electrodes, 2. Frame, non-conducting,
           3. Anode layer, ceramic oxide, 4. Intermediate
           layer, good conductor, 5. Cathode layer, carbon,
           TiB_2 or TiC, 7. and 8. Bus bar, 9. Electrolyte,
           10. Molten aluminum, 11. Opening for anode gas,
           12. Top of the cell, 13. Lining, 14. Electrical
           insulation, 15. Atmosphere.

     Recently there has been activity to develop refining cells with
thin diaphragms containing electrolyte in order to decrease the
specific energy consumption (27). One such cell under development
employs a graphite cloth diaphragm impregnated with molten salt bath
(28), illustrated in Fig. 14. The molten salt wets the graphite
cloth but the metal does not. Current densities of 30 A/in^2 and
specific energy consumptions of 3 kWh/lb have been claimed for laboratory cells.

CONCLUSIONS

1. The aluminum industry is under increasing pressure to improve
the energy efficiency of producing aluminum.

2. Three major areas are under active investigation for improving
energy efficiencies: an alternate aluminum chloride electrolysis
process, improvements to the Hall-Heroult process, and improved
scrap-refining cells.

3. The major improvements to the Hall-Heroult process presently
being developed include: still-larger lower-current-density cells,

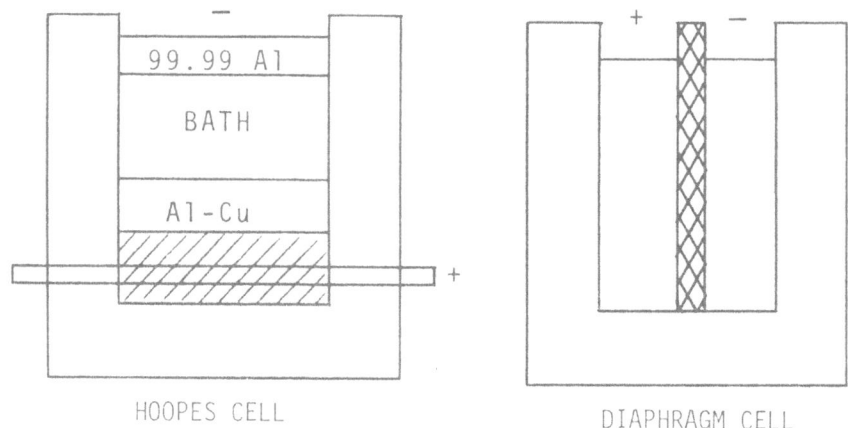

Fig. 14. Aluminum refining cells.

titanium diboride cathodes, inert anodes, and inert bipolar electrodes.

4. Significant further decrease in specific energy to produce
aluminum can be expected.

REFERENCES

1. J. W. Richards, "Aluminum, Its History, Occurance, Properties,
     Metallurgy and Applications Including Its Alloys," 3rd ed.,
     Henry Carey Baird & Co., Philadelphia (1896).
2. A. Minot, "The Production of Aluminum and Its Industrial Use,"
     (Trans.), first ed., John Wiley, New York (1905).
3. J. D. Edwards, F. C. Frary, and Z. Jeffries, The Aluminum
     Industry, McGraw-Hill, New York (1930).
4. E. Marshal, Science, 208:1353 (1980).
5. anon, Fortune, Oct. 20, 1980.
6. Aluminum Association, Aluminum Statistical Review 94:19 (1978).
7. A. Nag, Wall Street Journal, Nov. 18, 1980.
8. R. H. Perry and C. H. Chilton, "Chemical Engineers Handbook,"
     fifth ed., 3-145, McGraw-Hill, New York (1973).
9. T. R. Beck, Final Report, "Improvements in Energy Efficiency of
     Industrial Electrochemical Processes," to Argonne National
     Laboratory, ANL/OEPM-77-2 (1977).
10. C. A. Hampel, Ed., "The Encyclopedia of Electrochemistry,"
     Reinhold, New York (1964).
11. K. Grjotheim, C. Krohn, M. Malinovsky, K. Matiasovsky and
     J. Thonstad, "Aluminum Electrolysis," Aluminum-Verlag GmbH,
     Dusseldorf (1977).
12. N. Jarrett, "Aluminum Smelting, Past, Present and Future

Challenges," presented to AIChE, Portland, Oregon, August 18, 1980, to be in: AIChE Symposium Volume, Ed. R. C. Alkire and T. R. Beck.

13. J. M. Parry, A Survey of Potential Processes for the Manufacture of Aluminum, Contract No. 31-109-38-4210, Argonne National Laboratory to A. D. Little, ANL/OEPM-79-4 (1979).

14. Anon, "A Revolutionary Alcoa Process," Business Week, January 20, 1973.

15. J. Beizer, "Will Alcoa Process Deck the Halls?" Iron Age, January 18, 1973.

16. J. G. Peacey and W. G. Davenport, "Evaluation of Alternate Methods of Aluminum Production," Journal of Metals, p 25-28, July 1974.

17. U. S. Patents 3,755,099; 3,847,761; and 3,956,455; Aluminum Company of America.

18. K. Grjotheim, et al., "The Alcoa and Toth Processes for Aluminum Production - Outline and Comparison," Aluminium, 11:51, 697-699 (1975).

19. C. E. Ransley, British Patents 784,695; 784,696; 802,471; 802,905; U. S. Patent 3,028,324; British Aluminium Company.

20. K. Billehaug and H. A. Oye, "Inert Cathodes for Aluminum Electrolysis in Hall-Heroult Cells," (submitted to Aluminium).

21. Author's personal experience.

22. R. Hildebrandt and R. A. Lewis, U. S. Patent 3,400,061, Kaiser Aluminum & Chemical Corp.

23. P. A. Foster and S. C. Jacobs, U. S. Patent 4,071,420, Aluminum Company of America.

24. K. Billehaug and H. A. Oye, "Inert Electrolysis in Hall-Heroult Cells, Part II Anodes," (submitted to Aluminium).

25. H. Alder, U. S. Patent 3,930,967, Swiss Aluminium Ltd.

26. T. G. Pearson, "The Chemical Background of the Aluminum Industry,: The Royal Institute of Chemistry, Lectures, Monographs and Reports, No. 3 (1955).

27. D. J. Fray, "Potential for Fused Salt Electrolysis in Metal Winning and Refining, in Energy Consideration of Electrolytic Processes," D. S. Flett, Organizing Chmn., papers presented at Newcastle-Upon-Tyne, England, July 1-3, 1980, Society of Chemical Industry, London (1980).

28. K. A. Bowman, "Electrolytic Purification of Recycled Aluminum in Molten Salt Media," in: Extended Abs. No. 24, The Electrochemical Society, St. Lewis, Missouri, May 11-16, 1980.

DISCUSSION

Dr. R. W. Glazebrook, Shell Research, Ltd.: Will you elaborate on the overall efficiencies of the aluminum processing industry? In particular, does the data you presented take into account that the energy contained in the product (aluminum) is free energy whereas the energy input is partly free energy (as electricity) and partly enthalpic (as fossil fuel)?

Dr. Beck: The energy flow diagram, Fig. 7, is totally based on enthalpies. Further details may be found in reference 9.

THE ROLE OF ELECTROCHEMISTRY IN THE ELECTRONIC INDUSTRY[*]

N. B. Hannay

Bell Laboratories

Murray Hill, New Jersey

Electrochemistry plays a vital role in electronics. Classical processes (plating, anodization, electropolishing, corrosion, etc.) are relevant to most electronic components. Of particular interest, however, are the electrochemical processes used in the construction of integrated circuits and other micro-electronic devices. These processes are central to the fabrication of such structures, and can contribute to the continuation of the spectacular progress in the complexity and performance of solid-state circuits that has occurred over the last two decades. In addition, other new electronic components have an essentially electrochemical basis; these include electrochromic cells, lithium intercalation compounds for battery cathodes, solid electrolytes, and liquid-junction solar cells.

---

[*] Full manuscript not available

PANEL PRESENTATION AND DISCUSSION

FUTURE TRENDS IN MAJOR ELECTROCHEMICAL INDUSTRIES

Each of the panel participants has been asked to make a position statement as to what he sees as the future trends in specific major electrochemical industries. The industries covered in the panel discussion have been restricted to chlor-alkali, aluminum, batteries (both primary and secondary) and the electronics industry. Each of the speakers is a well recognized expert in his area. They are as follows:

      Jeff Cole: chlor-alkali
      Noel Jarrett: aluminum
      Robert Powers: primary batteries
      Alvin Salkind: secondary batteries
      Elton Cairns: automobile propulsion batteries
      N. B. Hannay: electronics industry

The panel discussion was moderated by Ernest Yeager of Case Western Reserve University.

The manuscripts for four of the panel participants were available and are reproduced herein.

FUTURE TRENDS IN THE CHLOR-ALKALI INDUSTRY

Jeff C. Cole - Consultant

The H. K. Ferguson Company and Department of Energy

777 Eastwood Drive
Painesville, Ohio 44077

INTRODUCTION

The chlor-alkali industry is one of the world's larger chemical processing industries in terms of product tonnage and value. The United States is the leading producing country with about 1/3 of the total; it is followed by West Germany with about 10 percent of the world's production, and then Japan and USSR in that order. Global production of chlorine is currently in excess of 31 million metric tons per year from a total annual capacity of about 40 million tons. The United States has an annual capacity of 13.2 million tons of chlorine.

Electrolytic production of chlorine and caustic soda is highly energy intensive with a total power expenditure ranking second to aluminum among the electrochemical industries. High energy costs along with environmental and other government regulations have spurred much technological development during the past decade. Equipment design and operations are presently changing from an art to a science.

GROWTH PROSPECTS

It is impossible to forecast with complete confidence either world-wide or domestic growth rates for chlorine and caustic. Many are convinced that with today's political turmoil, economic climate and regulatory restrictions, the industry will have a much lower average growth rate than in the early 70's. A few optimistic market researchers are forecasting average growth rates as high as 5 or 6 percent annually, but the more knowledgeable experts predict about 2 to 3 percent average annual growth in the United

355

States and perhaps a slightly larger percentage worldwide over the next five year period. This is a more realistic prediction.

The United States chlor-alkali industry operated at about 64 percent capacity in August, slightly better in September, and is currently at 73-75 percent capacity or about 10 million metric tons annually. There are no economic signs at this time pointing toward a complete recovery in the industry in the near future. The Japanese industry operating rate is still lower than the United States-- about 65 percent of capacity.

Vinyl chloride monomer remains one of the brighter prospects for increased chlorine consumption. Production of chlorinated solvents and other chlorinated organics will probably grow but their use is continuing to be challenged by environmentalists and OSHA type regulations. Reduced chlorine demand is contributing to a short supply picture in the caustic industry.

ENERGY SITUATION

While the industry has been very aggressive and successful in reducing the power requirements per unit of production with improved cell technology, energy is still the largest cost item in producing chlorine and caustic -- in some plants exceeding 50 percent of the total operating cost.

The chlor-alkali industry power cost per kilowatt hour in the United States is probably averaging a little better than 30 mills, in Germany about 35 to 40 mills, but in Japan it is up to 65 mills. High power costs and the necessity to import salt at the excessive cost approaching $35 per ton has created almost insurmountable economic problems for the industry in Japan. Several Japanese companies are considering building chlor-alkali plants in the United States to take advantage of lower power and raw material costs.

ENVIRONMENTAL AND SAFETY CONSIDERATIONS

The industry is plagued with many meaningless EPA and safety restrictions, and is continually threatened with still more burdensome regulations. Both the United States and foreign government regulatory agencies are looking more and more seriously at asbestos. The withdrawal of asbestos would be a severe blow to chlorine and caustic soda producers since probably one half of the global production is from diaphragm cells. While the chlor-alkali industry can safely design around the occupational hazards of using asbestos in chlorine cell diaphragms, stringent government manufacturing regulations directed at others could force asbestos producing companies out of business.

In some countries, mercury cell operators are being confronted with restrictions which limit their purchase of total mercury to a small quantity per ton of chlorine produced. If this practice should spread it could force the closing of many mercury cell chlorine plants in the United States and around the world. In Japan the remaining mercury cell plants are required by government edict to cease production in 1984.

TRANSPORTATION

Several accidents have occurred world-wide in transporting chlorine and usually the accidents have received unfavorable press coverage. The industry is greatly concerned that additional crippling regulations will be forthcoming on barge, rail car and truck transportation and storing of larger quantities of liquid chlorine. Scandinavian countries and other countries in Europe are imposing stringent regulations on transporting chlorine. In this country, Chicago and New York prohibit the transporting or use of liquid chlorine within the city limits.

CELL TYPES AND DEVELOPMENTS

Mercury Cells

The industry has demonstrated that mercury losses can be reduced to the point where they represent no hazard to health and are not an environmental problem. Unfortunately the mercury environmental problem long ago became a political one in almost every country. It will probably continue as a political consideration and unfortunately will eventually force the chlor-alkali industry to completely abandon the highly developed, efficient and economical mercury cell technology. This is really unfortunate for the chlor-alkali industry.

Diaphragm Cells

The real breakthrough in advancing diaphragm and mercury cell technology and energy conservation came with the development of dimensionally stable anodes. This was, of course, a tremendous breakthrough that revolutionized the industry. Unfortunately, we cannot foresee any more advancements of this magnitude and importance.

One development in diaphragm cells the industry is currently working on is a non-asbestos diaphragm. Several companies have announced technical success, but the economic viability is probably not there as long as asbestos is available and can be used.

Probably the technological improvement closest to commercialization in the diaphragm cell is the coated cathode to reduce the

hydrogen voltage overpotential. While the magnitude of energy
savings is nowhere comparable to the savings associated with coated
anodes, it is still significant. Diamond Shamrock has a commercial
cathode coating plant operating at its Battleground, Texas facility.
The first plants to have coated cathodes will be Diamond's Battleground and Deer Park diaphragm cell plants.

Logistics is a problem with coated cathodes because the economics depends on the coating plant being located in a concentrated
area of chlor-alkali producers -- otherwise shipping costs of the
cathode to and from the central coating plant would be prohibitive.
Favorable areas are the Gulf Coast, Japan and possibly Europe.

Another development that could apply to diaphragm cells is the
so called oxygen cathode which has been discussed earlier in this
symposium. This is in the R & D stage and in all likelihood will
show more favorable economics in the membrane cell than the diaphragm cell.

Membrane Cells

Without question the industry is tremendously excited about
membrane cells as a technology that produces high quality caustic
and avoids many of the known environmental problems. To date, only
small plants have been built, but several companies around the world
are considering large tonnage plants using membrane cells. At least
three installations world-wide ranging from 600 to 800 tons of
chlorine per day have been announced for engineering and construction. By 1985 there will probably be several very large chlorine-
caustic membrane cell plants operating. The technology will eventually displace diaphragm cells. One advantage of membrane cells
could be pressure operation eliminating the need for chlorine and
hydrogen compressors.

R & D work is in progress to discover better catalyzed cathodes
to reduce the hydrogen overpotential in membrane cells. Another
related development is the oxygen cathode which may show improved
economics over other uses for the hydrogen produced. It is generally
believed, however, that the oxygen cathode for membrane cells is at
least five years into the future. One company that offers membrane
cells for license reports its geometry of design is such that the
oxygen cathode when developed can be installed without a prohibitive
added cost. This is important and was one of the reasons the dimensionally stable anode was such a quick commercial success.

We at The H. K. Ferguson Company have made a number of com-
parative economic studies on chlor-alkali plants employing various
designs of membrane and diaphragm cells. In many cases we have
found the membrane cell installations are lower in both operating
and capital cost considerations.

PLANT TRENDS, SIZES AND INSTALLATIONS

Over the past several years a number of small, inefficient
and high energy cost plants built many years ago have been closed.
This trend is likely to continue.

Transportation difficulties in chlorine will tend to favor
the producing plant being located adjacent to the consuming point.
Energy considerations will make a limited number of large scale
petrochemical complexes economically viable. A major one recently
announced is the $3 billion complex in Saudi Arabia which is a
joint venture with Shell Oil. Others are under consideration.

The membrane cell makes possible small chlor-alkali plants
producing 15 to 30 T/D of chlorine which are economically viable
for operation in the developing countries. Also, unlike diaphragm
cells, small membrane cell plants may offer viable economics even
in industrial countries.

THE ALUMINUM INDUSTRY IN THE 1980's

Noel Jarrett, Asst. Director, Metal Production
Laboratories
Aluminum Company of America
Alcoa Laboratories
Alcoa Center, PA 15069

About four years ago, two years after the first OPEC oil
embargo, economic projections into the future showed four major
problems facing the aluminum industry in the 1980's and beyond:

1. Ever increasing cost and probable decreasing availability
    of all energy, more pointedly, electrical energy.

2. Dependence on foreign supplies of raw materials--bauxite
    and petroleum, from which anode coke is made.

3. High capital intensity--ever increasing capital investment
    per annual ton of aluminum.

4. Ever tightening environmental controls of evolutions
    and waste disposal.

It is difficult to separate these issues for discussion,
but in general today they boil down to a sort of bad news/good
news type of report.

Let's begin with the bleak parts of the outlook. Major
increases in domestic aluminum production will continue to be
constrained by a slower demand growth and the availability and
high cost of electrical power. By 1985 demand for petroleum coke
and coal tar pitch needed for our smelters is projected to exceed
supplies.

The good news is, there will continue to be plenty of
affordable bauxite on the international market. Cost/benefit/

361

risk analysis will become the dominant equation in environmental
regulatory agencies. The demand for aluminum is expected to
essentially balance capacity during the early part of this decade
but to exceed capacity perhaps by as much as 2×10^6 MT by 1988.
For the aluminum industry to generate the necessary capital and
make sufficient power available to meet this shortfall will
require a major effort and some good fortune.

An abundance of bauxite, however, a continued rational
environmental policy, and good demand don't mean much if we lack
the petroleum supplies necessary to form smelting cell electrodes,
the energy to refine and smelt the bauxite, and the capital
necessary to meet the demand. So we're back to the three major
problems--electrical energy, petroleum products, and capital.
These are problems I think the aluminum industry can solve, so
the rest of this talk is my view on how and when solutions can
be found.

Tight supplies of electrical energy will appear in the late
1980's. Even if the nuclear safety issues were resolved this
afternoon, it would be difficult to construct enough nuclear power
plants to allow for a major domestic increase in aluminum smelting
capacity during that period. Any increase in domestic production,
therefore, will have to be heavily based on reducing the energy
required to make a pound of aluminum. One way is to modernize
existing Hall cells. Better control over smelting cell dynamics
may reduce energy requirements by 1/2 kWh/lb, raise the current
efficiency of the cell to 93%, and result in a significant
increase in production. This should happen in the mid 1980's.

Reduction of the unwanted waste heat generated by the
resistances in the smelting cell now enters the picture. Constant
effort is being made to modify design and improve manufacturing
practice to reduce voltage losses in busbar, anode and cathode
connections, and in the electrodes themselves. Major energy loss,
however, is due to the resistance of the electrolyte between the
anode and cathode where 38% of the total energy is converted to
heat that must be rejected to the atmosphere. Reducing this anode-
cathode space is prevented by shorting of the molten aluminum
cathode to the anode as a result of its constant motion under the
influence of vertical magnetic fields and horizontal currents.
Work is underway to counteract the magnetic fields and replace the
liquid aluminum cathode by some form of aluminum wetted refractory
hard metal (TiB_2 ZrB_2) shape. It is hoped that economic and
physical durability problems of these materials can be solved and
such cathodic surfaces can become a commercial reality by 1986.

A longer range solution may be found in the Alcoa Smelting
Process, which uses aluminum chloride as cell feedstock. The

most modern Hall cells use about 6 kWh/lb of aluminum while the
Alcoa Smelting Process cells use less than 4.5 kWh/lb. The cells
are operating beyond our expectations, and to make it more cost-
effective we are redesigning a part of the chemical plant that
feeds them.

In the meantime, aluminum companies will have to look to
the developing countries for places to install new, highly
efficient smelters. One possibility is equatorial countries with
abundant bauxite and a large potential for power development.
Australia and Brazil are two such countries, and expansion has
begun there.

We must also enlarge our recycling efforts. Since it takes
only 5% as much energy to recycle aluminum as it takes to smelt
it from bauxite, it is obvious that used aluminum can become a
major source of new metal. The major recycling effort at present
is in beverage containers where the scrap can be melted and fed
back into cans. In 1980 Alcoa will have recycled about 29% of
the cans manufactured, and we are predicting 30-33% in 1981.
Our goal is to recycle 80% of the cans manufactured. When we
have accomplished this we will go after aircraft and automotive
scrap, which cannot now be recycled to their present alloys.
Downgrading this to casting alloy has limited use, and a low
energy consuming process to restore this reclaimed metal to
metallurgical (99.82 Al) grade is necessary. We believe commercial
purification technology that uses less than 30% of the present
energy for reduction from ore will be available in 1987.

The petroleum coke and pitch problem is another major
restriction facing the industry. By 1985 demand for these
petroleum products may outstrip supplies, and this means we are
going to have to use what is available with greater efficiency.
Alcoa believes that within the decade carbon electrode
technology will have advanced to the point where we can consis-
tantly produce high quality crack-free and dust-free anodes,
significantly reduce production scrap, make use of lower grade
and cheaper materials, and refine our forming methods. This
should decrease carbon anode consumption to 0.4 lb/lb of aluminum
and increase production through higher cell current efficiency.
We have made great strides recently in reducing the energy used
to bake electrodes. Projects are underway to make further
improvements by reducing baking time by one-third while guaranteeing
1100°C baking temperatures and increasing production thermal
efficiency. We think this can be done within five years.

To alleviate the projected petroleum coke shortage predicted
for the mid 1980's, experiments are being successfully conducted
with material made from solvent refined coal liquids. This process

is called SRC-I and is being developed by a consortium of companies
under the auspices of the Department of Energy. The material will
be available from a demonstration plant in 1984.

The ultimate solution--to replace expensive baked carbon
burned in the cell with cheap coal burned at the power plant
while saving energy--is a non-consumable, electrically conductive
cryolite and aluminum resistant anode to replace the consumable,
carbon anodes. In cooperation with the Department of Energy,
programs are now underway to develop such a material and commercial
success is hoped for by the middle of the decade.

At the beginning of this presentation I stated that a world
demand that exceeds supply by 2×10^6 MT was a probability. A
strong market is an advantage to any industry, but some of the
expanding markets for aluminum are critically important. The
automobile industry is one example. Lighter cars use less fuel
than heavy ones, and replacing steel with aluminum is an excellent
way to lightweight automobiles and conserve fuel. Thus, it is
in the National interest to see that sufficient aluminum is
available.

Even if domestic power is available, closing the gap between
demand and supply is going to require an estimated investment of
the order of 7×10^9, an enormous amount of private capital; but
with good management and the productivity improvements being
developed, it is hoped the market needs can be met by the industry.

All-in-all, this forecast has been optimistic--if a bit
guardedly so. Make no mistake, there are problems in the aluminum
industry in the future; but if the hardheaded realists who helped
develop this outlook can be optimistic, then I feel the future
of the aluminum industry is very bright.

DISCUSSION

Elton J. Cairns, University of California, Berkeley: Would you say
a few words about the carbothermic process for aluminum production
and how it compares to the conventional process with regard to primary energy consumption.

Noel Jarrett: One possible answer to all of the problems stated at
the beginning of my talk could be combining the production of carbon
monoxide, aluminum-silicon alloys, metal and solar grade silicon,
and commercially pure aluminum into a single process. An aluminum-
silicon blast furnace using domestic ore (clay, anorthosite) coke
from domestic coal, and oxygen would eliminate the need for electrical energy and be, overall, more energy efficient than producing
the three major products--aluminum, silicon, and carbon monoxide

separately. It would also be less capital intensive and more environmentally manageable.

In cooperation with the Department of Energy, we embarked on
such a program two years ago. To date we have shown that we can
make aluminum-silicon alloys in a shaft furnace using less than 50%
bauxite as the feed and can separate the aluminum and silicon satis-
factorily if we control impurities in the feed. So far, however,
we have been unable to supply more than 30% of the heat requirement
by burning excess coke with oxygen due to the high carbon monoxide
velocity above that percentage. Consequently, some electrical energy
must be introduced in the last stage. This electrical input about
equals the kWh/lb in the Alcoa Smelting Process. Work is continuing,
and while a total elimination of electrical energy does not seem
possible, it is hoped that a process can be developed that will
conserve energy and be cost effective under special circumstances.

SECONDARY BATTERIES

Alvin J. Salkind
CMDNJ - Rutgers Medical School; Visiting Professor of
Chemical Engineering, Rutgers University
Piscataway, New Jersey

The storage battery industry is very segmented; both in
product design and market area. Each technology and market niche
follows a different innovation-development-and market growth
cycle, and the total market consists of overlapping "S" curves.
The dynamism of the storage battery area, to a large extent,
depends on innovations in other industries, similar to the nature
of primary battery industry discussed by Dr. Robert Powers.
Storage battery growth particularly depends on changes in electronics, energy storage, and vehicular propulsion technologies.

The different segments of the industry and their market
status are shown in Table 1. The Automotive Starting-Lighting-
and Ignition (S.L.I.) battery segment represents approximately
80% of the estimated 4.8 billion lead acid battery market for
1979. These estimated numbers do not include production and
sales in the non-free market areas of the world for whom sales
values are not reported. Note that batteries for electric
vehicles, mainly golf carts, sold at approximately $40 per KWH,
the least expensive form of storage battery. The reasons are
that the EV battery is a low rate device compared to an S.L.I.
battery and has a larger capacity, thicker plates, fewer other
components, and fewer assembly steps. None of these prices
include the trade-in value of the product, which can be 10-15%.

The alkaline battery segment of the business is mainly nickel
cadmium cells (over 80%), with some nickel-zinc, nickel-hydrogen,
nickel-iron, silver-zinc, silver-cadmium and silver-hydrogen.
Traditionally, alkaline battery sales are approximately 10-15%
of the value of lead acid battery sales; and at much higher unit

Table 1

STORAGE BATTERY - MARKET SEGMENT

(1979 Dollar in Millions)

Market Segment	Battery System		
	Lead Acid	Alkaline	Other
Vehicle SLI:	3,900 @ $50/kwh	---	Metal-air Metal-halogen
Industrial:			
Standby & U.P.S.	400 @ $100/kwh	300 @ $200/ kwh	Metal-organic
Traction	400	100	
Consumer & Instr.:			
Small Sealed Cells (Power size)	60	180	
Electric Size	---	10	
Energy Storage:		$(N_I C_D$ in USSR)	
Solar	15	---	
L.L.	---	---	
Military & Space:	30 (incl. Sub)	50	
Vehicular Propulsion:			
Electric/ including Golf	46 @ 40	---	
Hybrid	1	1	
Total:	4,800	640	150

($\sim$ $5.6 billion)

prices. It is interesting to note that in the USSR there is a
much higher ratio of nickel-cadmium batteries to lead acid batteries, and probably a ratio of 20% would be a good estimate.

Other secondary systems, such as metal-air, zinc-bromine
and zinc-chlorine, do not have any substantial market at present.

The total free market sales of storage batteries by manufacturers was approximately $5.6 billion in 1979. The retail
value of the product would be about $2\frac{1}{2}$ times this value.

Some problem and change areas for the secondary battery
industry are listed in Table 2. A decade or so ago, there were
more than 200 storage battery manufacturing facilities in the
U.S. These small plants producing 1000-2000 batteries per day
are being replaced by modern new automated factories with production capabilities of 8000-10,000 batteries per day. This trend
is not only for labor saving, but for handling pollution problems.
It encourages changes in process techniques to continuous methods.
There is one facility in Kyoto, Japan, of the Japanese Storage
Battery Company, which would be equal to 10% of the U.S. needs.
The efficiencies of high speed production more than offsets the
extra shipping costs from the reduction in plant sites. However,
there is an optimum distance for shipping a heavy, low value
product, like a S.L.I. battery, compared to the additional costs
of extra plants. The changes in process are also enabling some
changes in product design to be accomplished. In addition, more
research is needed in pollution and recovery systems.

The new growth and opportunity areas, listed in Table 3, are
not in the present large market S.L.I. and industrial segments,
but in electrical or electronic size consumer batteries; energy
storage systems for circuit boards, solar storage, and load
leveling; and in vehicular propulsion, both all electric and
hybrid. I distinguish between the latter two designs because the
construction and performance criteria are quite different. At
this time, I do not believe that sufficient research is being
carried out in the high power density, high temperature-dissipation
battery needed for hybrid applications. It is very possible that
many of the new chemical systems being proposed for EV's will
require a power battery for acceleration and for quickly absorbing
the energy from regenerative braking. These systems will be discussed by Dr. Elton Cairns.

There is a need for rechargeable energy sources in many
proposed medical devices, such as the artificial heart. These
require much higher power and energy levels than such traditional
bioengineering devices as the heart pacer, which are well suited
for primary cells.

Table 2

MAJOR INDUSTRIAL PROBLEMS AND PROCESS NEEDS

PROBLEMS

• <u>Plant Size and Lack of Automation</u>

     1970 > 200 Plants in U.S.

     1990 ∿ 25 Plants in U.S.

     (JSB Plant equal to
     10% of U.S. needs)

• <u>Pollution and Recovery Systems</u>

• Lead Smelting – Needed New
                  Technology

• Lead in Air

• Cadmium – New Electrode Fabrication
            Technology

          Plated and Plastic Electrodes

Table 3

OPPORTUNITIES AND GROWTH AREAS

* Small Power Sources for Circuit Roads

* Power Tools and Small Rechargeables

* Emergency Lighting; Toys

* Energy Storage for Intermittent Power

* Utility Load Leveling

* EV – Electric

* EV – Hybrid

* Medical Devices

Table 4

RECENT APPLICATIONS OF ELECTROCHEMICAL ENGINEERING

* Grid Design

* Diffusion Limitation Recognition

* Expanded Metal Grids

* Continuous Processes

* Plastic Electrodes

* Electroplated Electrodes

* Thin Foil-Layer Grids

Recent applications of the science of electrochemical engineering in the design and fabrication of storage batteries and components are listed in Table 4. The recognition and correction of diffusion limitations in batteries is a most important step. The variation and variability of performance, with higher power demands, is extremely deleterious in high voltage multiple cell applications. The 20% variation from unit-to-unit described in Dr. Karl Kordesch's talk becomes a major cost and energy density factor in such applications as electric vehicles. In traditional production of batteries, the components have been treated as bulk chemicals. Recognition of the surface phenomena controlling factors is necessary to reduce unit variation.

Areas in which R&D are needed are listed in Table 5. The relationship between the activity and cycle life of active materials and the substrate layer is an extremely important one. This is being studied by Dr. McBreen and others for zinc, but it should also be studied for the lead oxide electrode, as well as others. Expanders are materials which aid in maintaining the crystallite size of active materials (negative electrodes) and the electrode porosity. They become more critical as higher power requirements are imposed on electrodes and are crucial to obtaining long cycle life. Most phenomena responsible for the reduction in life of the storage batteries occur during charge. However, the discharge part of the cycling process has received the most attention. Now that electronics are cheap enough that almost every cell could be instrumented, charge control devices are a good method for increasing cycle life.

Both lead acid and nickel cadmium batteries go through multiple charge-discharge "formation" cycles as part of the fabrication process. These are designed to bring the active material to the proper crystal defect structure, surface area, and porosity. This time, labor and energy consuming procedure should be thoroughly examined as a major step in cost reduction and product improvement.

Lastly, some interesting concepts and conjectures for the future are listed below:

1) A gridless lead acid battery. Since the lead is a corrodable substructure, a stable conductive thin film oxide would be more permanent. It could not corrode further and the balance between positive and negative capacity would not change during cycling.

2) Mechanically replaceable electrodes. Studies are now underway in which one electrode, such as the aluminum negative in an aluminum air cell, is mechanically replaced after use. The discharged material can be reformed in an operation external to the battery. Devices are being developed as remote power generators, as well as for electric vehicles.

Table 5

RESEARCH AND DEVELOPMENT NEEDS

* Study of Electrochemical Active Material
    Utilization & Reversibility as a Function
    of:

        Substrate   (McBreen et al. Zinc)
        Doping
        Layer Thickness
        Temperature

* Stabilization & Reversibility of Nickel "4"

* Reversibility of Manganese Dioxide

* Membranes & Separators

* Expanders

* Manufacturing Processes

* Charge Indicators & Charge Control Devices

* Elimination and/or Modification of Formation

* Lead Acid - New Maintenance Free Alloys Capable
    of Deep Discharge Without Building Up
    Passivation Layers

* Intercalate Compounds as Cathodes

3) <u>Parasitic current control in bipolar batteries.</u> One of
the major drawbacks in bipolar batteries (and in bipolar electrochemical cells as well) is the parasitic currents that develop
from having a common electrolyte. A recent patent by Dr. P. Grimes,
which applies an external voltage to the intercell flow paths,
alleviates this problem. The other major problem, in the conductive, inert barrier, still requires improvement.

4) <u>Reversible polymers.</u> Stable polymers capable of multiple
electron changes are now available. Many of them are water sensitive but would be useful in non-aqueous batteries. Years ago, the
azodicarbonamide family of compounds was studied in LeClanché
electrolyte as a lighweight, low-cost cell in dry cell configurations. New methods seem to be emerging for this type of cell.

5) <u>Thin foil techniques.</u> The use of a thin nickel foil substrate for nickel hydroxide active material is under study (INCO).
In this concept, called controlled-micro geometry (CMG) electrodes,
the foils contain minute holes in a definite pattern. The foils
are stacked so that the holes line up and diffusion readily takes
place. This technique appears to be interesting for other systems
as well.

I think the storage battery industry has a bright future, in
spite of the many problems and sparsity of people to solve them.

DISCUSSION

<u>Dr. Elton Cairns, University of California, Berkeley:</u> I am
interested in the new polymeric materials you mentioned for use in
cathodes. Have any estimates been made of the theoretical specific
energies for these materials?

<u>Professor Salkind:</u> Some of the new compounds are proprietary and
calculations have not been revealed. In general, they are poor
conductors, and are now of interest because many new electronic
devices have very low current drains. One new material which has
been recently publicized is polyacetylene.

<u>Dr. R.W. Glazebrook, Shell Research, Ltd., Chester, England:</u>
What is the future for intercalation compounds as reversible
cathodes?

<u>Professor Salkind:</u> They offer interesting opportunities for
cathodes and are being vigorously studied at several research
centers. They were discussed to some extent by Dr. Bruce Hanney
in his talk on vanadium and titanium disulfide.

Although many of the systems seem to be capable of hundreds of cycles, none is in commercial production at present. It appears that the problems in lithium cells are lithium dendrites and other anode problems after several hundred cycles or earlier. There are some problems with electrode swelling, but the major problem, in my view, is that, so far, a mechanism for handling overcharge and/ or overdischarge has not been advanced. With some organic electrolytes, the decomposition products can be toxic. However, the intercalates should be listed as important research areas.

Professor Karl Kordesch, Technical University of Graz, Austria: Is not the 'utilization of active material' the key question in improving cells? An increase from 40 to 80 or 90% would certainly help. Rolled cells, thin plates, etc. should be emphasized.

Professor Salkind: I agree with you completely. Historically, batteries used hand or simple machine fabrication techniques, and components could be made thick but not thin.

Mr. David Nikles, Dept. of Chemistry, Case Western Reserve University: Under what conditions do you see valence four nickel, and how is it stabilized? My own experience in the electrochemistry of some nickel valence 2- chelates is that if the metal is oxidized to N_i^{3+} it immediately oxides something else, usually the ligand.

Professor Salkind: There are good discussions of the crystal structure, stability, and valency of nickel in alkaline media in the reports of H. Bode (in Electrochemical Acta), and the article by Paul Milner and U.B. Thomas (in Volume 5 of Advances in Electrochemistry and Electrochemical Engineering, edited by Tobias and Delahay). There are two brief papers which discuss (gamma) nickel-hydroxide in Progress in Batteries and Solar Cells, Vol. 3 (1980), edited by Kozawa and Kordesch. This is available from Jec Press, Inc., Box 42041, Cleveland, Ohio 44142.

Professor Yeager: The net charge on charging the nickel oxide electrode corresponds to ∿ 3.6 for the valency state at the end of charge, taking into account O_2 cogeneration. The only way to account for this is to postulate a higher valency state of nickel than 3.

Professor Salkind: The assumption has been that nickel 4 is a non-crystalline or weakly crystalline state that does not give good x-ray diffraction patterns. Nickel 3 is also unusual, in that it forms a continuous lattice with nickel 2, and partially charged electrodes give intermediate x-ray patterns, rather than dual patterns of charged and discharged states.

FUTURE TRENDS IN ADVANCED BATTERIES FOR AUTOMOBILE PROPULSION

Elton J. Cairns

University of California, and
Lawrence Berkeley Laboratory
Berkeley, California   94720

I.   INTRODUCTION

In order to establish a basis for perspective with regard to
the discussion below, it is useful to examine the incentive for
advanced battery development in terms of the possible market for
vehicle propulsion batteries. The automobile market in the U.S.
alone is about $100 billion per year, corresponding to about 10
million vehicles per year. If only 10 percent of the market were
to be captured by electric vehicles, and 20% of the cost of the
electric vehicle is the battery, then this would represent a $2
billion per year battery market. This is to be compared to an
estimated $1.5-2 billion per year market for SLI batteries, both
as replacements, and in new vehicles. Thus, only a 10% market
penetration by EV's would more than double the rechargeable
battery market--a rather large incentive!

The impediments to the development of such an attractive
electrochemical energy conversion business can be grouped into
three important categories: performance (W-h/kg, W/kg), durability (cycle life, lifetime), and cost ($/kWh). If the electro-
chemists and electrochemical engineers can solve the problems in
these three categories, then there exists an extremely attractive
opportunity for the development of a multibillion dollar per year
market, starting in the 1980's. General Motors repeatedly has
announced its intention to offer for sale an urban electric auto
(like a Chevette) in 1985, indicating that there could indeed be
an electric vehicle battery market developing. This intention can
materialize only if at least marginally acceptable batteries are
available.

377

The initial performance goals for a marketable electric vehicle (again, according to GM) include a range of about 150 km, and a speed of 80 km/h. This translates into a specific energy of at least 70 Wh/kg, and a peak specific power of 100-150 W/kg. The battery lifetime should be at least 3 years, or 300 deep (80+% depth of discharge) cycles, and the cost should be no greater than $100/kWh.* A number of cells under development have shown the capability of meeting the performance goals, but none has met the performance, durability, and cost goals simultaneously. This is our challenge!

II. BATTERIES FOR THE NEXT TWO DECADES

Reflection on the rate of development and implementation of new electrochemical energy conversion technologies reveals that it requires at least two decades for new systems to reach commercial significance. As an example, the fuel cell was under intensive development for space application throughout the 1960's, but still has not reached commercial markets. Some advanced batteries have been under development for over a decade, and are not very close to commercialization now. Based upon these observations, it is likely that the electric vehicle batteries that may be used in the next two decades are already being investigated.

The advanced batteries that have the best chance of finding application in electric automobiles in the next twenty years include advanced versions of the Pb/PbO_2 cell, with a specific energy above 40 Wh/kg. This cell will probably make use of a circulating electrolyte to promote H_2SO_4 transport into the electrode, and advanced current collector designs to maximize the specific power. Good progress in these areas has been reported already.

The $Fe/KOH/NiOOH$ cell has received renewed development effort in the last decade, with significant gains in specific energy, now approaching 50 Wh/kg, and under test in several electric vehicles. The issues of low efficiency and high cost remain to be resolved, but this battery may be interesting for industrial and mining vehicles, and perhaps some commercial vehicles.

The $Zn/KOH/NiOOH$ cell has been a popular candidate for electric automobiles because of its relatively high specific energy (60-75 Wh/kg), but it suffers from short cycle life of the zinc electrode (100-200 deep cycles), and the relatively high cost of the NiOOH electrode. Because of the high performance of this cell, efforts on the zinc electrode will probably continue.

---

*1981 dollars.

High-temperature systems offer the next significant increase in specific energy, to values above 100 Wh/kg. The high-temperature cell closest to automotive application now is the LiAl/LiCl-KCl/FeS cell, which operates at 450°C. Full-sized cells have demonstrated a specific energy of 100 Wh/kg. Similar (but lower-specific-energy) cells have cycle lives of 300+ deep cycles. The costs now are high, but projections fall below $100/kWh. Even higher performance is available from the $Li_4Si/LiCl-KCl/FeS_2$ cell (450°C), which is in an earlier stage of development: 70-80 Ah single cells of 120-180 Wh/kg, up to 700 deep cycles. A specific energy of 200 Wh/kg seems within reach, and would correspond to a vehicle urban driving range of 400 km, or a shorter range, with a smaller, less costly battery. Other high-temperature possibilities include the Na/S cell (350°C) which uses a sodium-ion conducting ceramic ($Na_2O \cdot 11Al_2O_3$) tube or glass (sodium borate) hollow fiber electrolyte. These cells have irreproducible lives, of up to several hundred cycles, and suffer failure with thermal cycling. The specific energy of Na/S cells falls in the 100-200 Wh/kg range, depending on cell design.

There are other options available, but they seem less promising. These include metal/air cells, which suffer from the low efficiency of the air electrode (large overvoltages for both discharge and recharge reactions, over 0.3 V for each). Fuel cells have the same problem, and all of the flow systems, including metal/halogen systems tend to be complex and bulky.

III. DIRECTIONS FOR THE PERIOD BEYOND THE YEAR 2000

As some of the systems that are known now are implemented, the improvements in specific energy and specific power will probably catalyze our thinking about new opportunities for electrochemical energy conversion, and focus our thinking in a number of areas. It is to be expected that there will be a continuing desire for lighter-weight, more energetic reactants. For use at negative electrodes, hydrogen and lithium will remain attractive; at positive electrodes sulfur, oxygen and perhaps fluorine, and low-equivalent-weight compounds containing them will receive more attention. Electrolytes for such systems will be non-aqueous, with an emphasis on ambient-to-moderate temperatures (up to 200°C). These higher specific energy cells will allow much more design flexibility: lower cost, smaller batteries of moderate driving range, with an option for more range, to 400 or 500 km at added cost. As the electric vehicle field matures, there is to be expected a shift toward fuel-cell powered vehicles, perhaps using H_2 from NH_3, or a similarly convenient source. In principle, the fuel cell vehicle could be a full-performance, general-use vehicle, providing that cost and lifetime goals can be met. Electrocatalysis and materials remain problems.

Many new materials will be required, both active (e.g. reactants, electrocatalysts, electrolytes), and inactive (current collectors, cases, separators). More sophisticated alloys and
compounds will be required to meet all of the conflicting demands.
Some needs include electrolytes (solid and liquid) for use with
pure lithium electrodes (liquid or solid), materials for use with
sulfur as current collectors, seals, containers, special tailored
separators for use with specific electrodes such as lithium,
aluminum, magnesium, zinc. It seems clear that molecular engineering will be pressed into prominence as we tax our materials
capabilities more and more.

Electrochemical cell design will reach much higher levels of
sophistication as cell modeling abilities and computing speed and
capacity increase. Multiparametric optimization of cell designs
will become rather common. Advances in understanding and modeling
cell failure will be accelerated. As cells become more compact,
it will be necessary to place more emphasis on detailed understanding and prediction of heat, mass, and momentum transport in
cells and batteries, in order to avoid catastrophic failures due
to overheating in highly energetic batteries. It will be less
feasible to operate near ambient temperatures, and those capable
of operating at moderately elevated temperatures will be more
prominent.

As batteries become larger and more widely used, maintenance-
free operation becomes essential, a necessity. Many systems will
be sealed, gas-tight, with internal gas recombination.

Molecular engineering techniques should be brought to bear
on the very difficult problems to be solved in the development
of fuel cells. Electrocatalysts of low-cost, plentiful materials
are needed. For the very complex oxygen electrode, it seems probable that a combination of tailored and matched electrocatalysts
will be necessary. The hydrogen electrode also requires less
costly electrocatalysts. If compact, inexpensive hydrogen storage
(or simple, fast preparation from another fuel such as NH_3 or
CH_3OH) is developed, then there will develop a market for fuel-
cell powered vehicles.

The future of electrochemistry and electrochemical engineering
in using our energy resources most effectively is limited only by
our ingenuity!

This work was supported by the U.S. Department of Energy

CONCLUDING REMARKS ON THE SYMPOSIUM

Ernest Yeager

Case Laboratories for Electrochemical Studies
Case Western Reserve University

The speakers with expertise in the various industrial electrochemical areas represented in the Symposium have been generally
quite optimistic concerning new technical innovations as well as
further growth of these industries. For the remainder of the century, applied electrochemistry can be expected to have increasing
importance relative to industry as a whole. The factors responsible
were cited in the opening remarks--the energy problem, the raw
materials problem, the environment problem.

In this Symposium on Electrochemistry in Industry, we have
said little concerning university research and the outlook for
electrochemistry at our universities. Electrochemistry on the
university campus also is in a period of rapid growth and, even
more important, in a period of much innovation. After using almost
exclusively just electrochemical methods for many years, electrochemists are now making extensive use of various complementary in
situ techniques including particularly optical spectroscopic methods, and also ex situ techniques, including quite sophisticated
surface physics electron and ion spectroscopic techniques. There
is much theoretical effort to treat the structure of the electrode-
electrolyte interface, including electrosorption and various electron and atom transfer processes at such interfaces. The science
of electrocatalysis is advancing rapidly both experimentally and
theoretically. In some instances a predictive basis for electrocatalysis is at least partially in place. The computer era and
particularly the microprocessor revolution are having an especially
great impact on electrochemical measurements. Electrochemical
engineering is also rapidly evolving as a formal, well developed
segment of chemical engineering as a whole.

With the rapid growth of electrochemistry as a science and electrochemical engineering as a formal subject after many years of slow growth, the field is attracting some of the brightest graduate and postdoctoral students and no longer just the average student. Furthermore, students are increasingly crossing into electrochemistry from other disciplines besides chemistry, including metallurgy, electrical engineering and physics.

We can end this conference on a quite optimistic note regarding the future of electrochemistry. The problems remaining for electrochemistry to solve, however, constitute a very great challenge. Our nation and the world as a whole would benefit greatly from a quantum jump in the storage batteries available for vehicle propulsion. Fuel cells so far have had only limited success in terrestrial applications; we also need a quantum jump in this area with respect to life, cost and energy efficiency. The majority of the industrial electrolytic processes, particularly aluminum and chloralkali, still suffer from very substantial energy losses which leave much room for improvement. The field of electroorganic chemistry still appears to some of us to be a sleeping giant--with the potential for great impact on synthetic organic chemistry--but still not realized. The corrosion of metals exacts a tremendous cost on our economy (probably over 100 billion dollars per year for this country alone) and this area demands more effective use of our present electrochemical technology as well as the further advancement of our basic knowledge in this area. Corrosion prevention can lead to a lessening of our energy, materials, and environmental problems. Electrochemical sensors are already used extensively in medical applications but other biomedical applications for electrochemistry are still in their infancy--for example, electrodes for transmission of nerve impulses and electrochemical methods for the acceleration of bone healing.

With the increased recognition of the importance of electrochemistry, the rate controlling factor in future growth will be increasingly the availability of scientists and engineers with appropriate interdisciplinary background plus a working knowledge of electrochemistry. It is hoped that industry and government will help the universities of our nation to extend their programs in electrochemistry and to increase the number of students being trained in this area.

INDEX

383